Water Resources and Environmental Depth Reference Manual

for the Civil PE Exam

**Jonathan A. Brant, PhD, PE and
Gerald J. Kauffman, PhD, PE**

The Power to Pass®
www.ppi2pass.com

Professional Publications, Inc. • Belmont, California

Benefit by Registering This Book with PPI

- Get book updates and corrections.
- Hear the latest exam news.
- Obtain exclusive exam tips and strategies.
- Receive special discounts.

Register your book at **ppi2pass.com/register**.

Report Errors and View Corrections for This Book

PPI is grateful to every reader who notifies us of a possible error. Your feedback allows us to improve the quality and accuracy of our products. You can report errata and view corrections at **ppi2pass.com/errata**.

WATER RESOURCES AND ENVIRONMENTAL DEPTH REFERENCE MANUAL FOR THE CIVIL PE EXAM

Current printing of this edition: 5

Printing History

date	edition number	printing number	update
Nov 2014	1	3	Minor corrections.
Oct 2015	1	4	Minor corrections. Minor cover updates.
Oct 2016	1	5	Minor corrections.

Printed in the United States of America.

PPI
1250 Fifth Avenue
Belmont, CA 94002
(650) 593-9119
ppi2pass.com

ISBN: 978-1-59126-095-0

Library of Congress Control Number: 2011929567

F E D C B A

Table of Contents

Preface

The *Water Resources and Environmental Depth Reference Manual* is intended to provide comprehensive coverage of the civil PE water resources and environmental depth exam specifications as presented by the National Council of Examiners for Engineering and Surveying (NCEES). It should be used in conjunction with the *Civil Engineering Reference Manual* (CERM), which covers the wide range of topics on the civil PE breadth exam.

Water resources and environmental engineering is different from many other disciplines of civil engineering. It requires an understanding of naturally occurring processes that respond to unpredictable forces of nature. As nature is unpredictable, calculations are usually based on conservative estimations and frequently rely on the professional judgment of the engineer. Yet, because professional judgment must often be exercised, reported design values will vary from one engineer to the next. The PE exam cannot allow for such variability. Therefore, we've written this book so that the methodologies you'll need to solve problems on the exam are the methodologies presented in each chapter and used in the examples and practice problems.

This book is meant to be a resource for your exam preparation. Therefore, we've done our best to ensure that we've presented the material in this book clearly and accurately. However, if you find a mistake, please let us know. PPI has an errata page on its website, at **ppi2pass.com/errata**, where you can submit suspected errors and view errors already submitted. Valid submitted errors will be posted and incorporated into future printings of this book.

Jonathan A. Brant, PhD, PE
Gerald J. Kauffman, PhD, PE

Acknowledgments

Thanks to Michael Bateman, PE; Steve C. Chiesa, PhD, PE; Tim O. Moore II, PhD, PE; and Thomas W. Schreffler, QEP, PE, for performing the technical review of this book, and to Ken Li, Patrick Albrecht, and Todd Fisher for checking the calculations. Thanks also goes to the PPI Product Development and Implementation staff, including Sarah Hubbard, director of product development and implementation; Cathy Schrott, production services manager; Megan Synnestvedt and Jenny Lindeburg King, editorial project managers; Tyler Hayes, Chelsea Logan, Scott Marley, Magnolia Molcan, and Julia White, copy editors; Kate Hayes, production associate; Tom Bergstrom, technical illustrator; and Amy Schwertman, cover designer. Thank you to Michael R. Lindeburg, PE, for the gracious use of material from the *Civil Engineering Reference Manual*.

Gerald would like to thank the students in the CIEG 440, Water Resources Engineering course, at the University of Delaware, Department of Civil and Environmental Engineering. Their enthusiasm and yearning for learning provided the inspiration for assembling this book. Gerald also wishes to express gratitude to his wife DeEtte, who gave her unconditional support during the writing of this book.

Jon would like to express sincerest gratitude to all the people who have supported his contributions to this book. Without their assistance and personal encouragement, this work would not have been possible. Jon would like to thank his loving wife Beth and daughter Olivia. Without their unyielding support, nothing would be possible. He would also like to thank Dr. Wane Schneiter and Dr. Amy Childress for serving as sources of inspiration throughout his career. Finally, Jon thanks both of his parents for all their untold sacrifices, which have provided him with all of the opportunities he has been blessed with in life.

Jonathan A. Brant, PhD, PE
Gerald J. Kauffman, PhD, PE

References

CODES AND REFERENCES USED ON THE EXAM

The water resources and environmental depth section of the civil PE exam is not based on specific codes or references. However, the minimum recommended library for the exam consists of the *Civil Engineering Reference Manual* (CERM) and the *Water Resources and Environmental Depth Reference Manual* (for the breadth and depth sections of the exam, respectively).

In addition to CERM and this book, it is recommended you bring *Urban Hydrology for Small Watersheds* (*TR-55*) and the *Precipitation-Frequency Atlas of the United States* with you to the exam. Though this book presents many worksheets from *TR-55*, *TR-55* contains additional data on rainfall and soils that you may need for the exam but that is too voluminous to include in this book. (For uniformity, *TR-55* worksheets reproduced in this book have been modified to reflect this book's nomenclature.) The *Precipitation-Frequency Atlas of the United States* is published by the National Oceanic and Atmospheric Administration (NOAA). In particular, you should bring with you to the exam NOAA Atlas 2 and Atlas 14, which contain precipitation depth maps for 6 hr and 24 hr storms. Links to the *TR-55*, the NOAA atlases, and other resources are available online at **ppi2pass.com/CEwebrefs**.

REFERENCES USED IN THIS BOOK

The following references were used to prepare this book. You may also find them useful references to bring with you to the exam.

Hydrologic Unit Maps. WPS 2294. Paul R. Seaber, F. Paul Kapinos, and George L. Knapp. U.S. Geological Survey.

Open Channel Hydraulics. V. T. Chow. The Blackburn Press.

Precipitation-Frequency Atlas of the United States. NOAA Atlas 14, Vol. 1 and Vol. 2. National Oceanic and Atmospheric Administration (NOAA).[1]

Precipitation-Frequency Atlas of the Western United States. NOAA Atlas 2. National Oceanic and Atmospheric Administration (NOAA).[1]

Recommended Standards for Wastewater Facilities. Ten States Standards. Great Lakes—Upper Mississippi River Board.[1]

Recommended Standards for Water Works. Ten States Standards. Great Lakes—Upper Mississippi River Board.[1]

Technical Manual for Stream Encroachment in New Jersey. New Jersey Dept. of Environmental Protection.

Urban Hydrology for Small Watersheds. TR-55. Natural Resources Conservation Service.[1]

Water Resources Engineering. Ray K. Linsley, Joseph B. Franzini, David L. Freyberg, and George Tchobanoglous. McGraw-Hill.

Water Surface Profiles. Vol. 6. U.S. Army Corps of Engineers.

Water Supply and Pollution Control. John W. Clark, Warren Viessman, Jr., and Mark J. Hammer. Harper & Row.

[1]A link to a downloadable version is provided at **ppi2pass.com/CEwebrefs**.

Introduction

ABOUT THIS BOOK

The *Water Resources and Environmental Depth Reference Manual* covers the water resources and environmental depth section of the civil PE exam administered by the National Council of Examiners for Engineering and Surveying (NCEES). This section of the exam is intended to assess your knowledge of design procedures and field practice.

This book is written with the exam in mind. Major topics, equations, and example problems are presented, and practice problems are given at the end of each chapter. Common resources, such as *TR-55* and the NOAA atlases, are used in examples and practice problems to increase your familiarity with these resources before you need them on the exam.

This book's eleven chapters are organized into three topics covering the following exam specifications.

- Hydraulics—Closed Conduit

    Energy and/or continuity equation (e.g., Bernoulli), pressure conduit (e.g., single pipe, force mains), closed pipe flow equations (Hazen-Williams, Darcy-Weisbach), friction and/or minor losses, pipe network analysis (e.g., pipeline design, branch networks, loop networks), pump application and analysis, cavitation, transient analysis (e.g., water hammer), closed conduit flow measurement, momentum equation (e.g., thrust blocks, pipeline restraints)

- Hydraulics—Open Channel

    Open-channel flow (e.g., Manning's equation), culvert design, spillway capacity, energy dissipation (e.g., hydraulic jump, velocity control), stormwater collection (stormwater inlets, gutter flow, street flow, storm sewer pipes), floodplain/floodway, subcritical and supercritical flow, open channel flow measurement, gradually varied flow

- Hydrology

    Storm characterization (rainfall measurement and distribution), storm frequency, hydrograph application and development, synthetic hydrographs, rainfall intensity-duration-frequency (IDF) curves, time of concentration, runoff analysis (rational method, NRCS method), gauging stations (runoff frequency analysis, flow calculations), depletions (e.g., transpiration, evaporation, infiltration), sedimentation, erosion, detention/retention ponds

- Groundwater and Well Fields

    Aquifers (e.g., characterization), groundwater flow (Darcy's Law, seepage analysis), well analysis (steady flow only), groundwater control (drainage, construction dewatering), water quality analysis, groundwater contamination

- Wastewater Treatment

    Wastewater flow rates (e.g., municipal, industrial, commercial), unit operations and processes, primary treatment (e.g., bar screens, clarification), secondary clarification, chemical treatment, collection systems (e.g., lift stations, sewer network, infiltration, inflow), National Pollutant Discharge Elimination System (NPDES) permitting, effluent limits, biological treatment, physical treatment, solids handling (e.g., thickening, drying processes), digesters, disinfection, nitrification and/or denitrification, operations (e.g., odor control, corrosion control, compliance), advanced treatment (e.g., nutrient removal, filtration, wetlands), beneficial reuse (e.g., liquids, biosolids, gas)

- Water Quality

    Stream degradation (e.g., thermal, base flow, TDS, TSS, BOD, COD), oxygen dynamics (e.g., oxygenation, deoxygenation, oxygen sag curve), risk assessment and management, toxicity, biological contaminants (e.g., algae, mussels), chemical contaminants (e.g., organics, heavy metals), bioaccumulation, eutrophication, indicator organisms and testing, sampling and monitoring (e.g., QA/QC, laboratory procedures)

- Water Treatment

    Demands, hydraulic loading, storage (raw and treated water), sedimentation, taste and odor control, rapid mixing, coagulation and flocculation, filtration, disinfection, softening, advanced treatment (e.g., membranes, activated carbon, desalination), distribution systems

ABOUT THIS BOOK'S UNIT CONVERSIONS

Unit conversion is one of the most error-prone components of the water resources and environmental depth section of the civil PE exam, so we've included all unit conversions throughout the solutions. The situations in which conversions are needed are numerous. For example, the United States Geological Survey stream gauge network provides records of stream flow in cubic feet per second, while a water distribution network engineer requires flow in million gallons per day and gallons per minute. Precipitation is measured by the United States National Weather Service in inches, yet engineers typically estimate stormwater runoff in units of cubic feet per second. Therefore, you should pay special attention to which units are specified in a problem and for the final answer. Make sure you convert appropriately. It would be a shame to get a problem wrong because you converted to cubic feet per minute instead of cubic feet per second.

As long as U.S. government agencies continue to provide water resources data in customary U.S. units, it is likely that most hydrology and hydraulics problems on the PE exam will be in customary U.S. units. However, NCEES specifies that SI units will also appear on the exam. Therefore, correct unit conversion is essential to mastering the water resources and environmental depth exam. Common unit conversions are provided in App. 1.B and App. 1.C as an aid.

HOW TO USE THIS BOOK

This book provides a comprehensive, targeted review of the material on the water resources and environmental depth section of the civil PE exam, and is designed to be used in conjunction with the *Civil Engineering Reference Manual* as your primary breadth exam review resource. Start by reviewing the exam topics (listed in this introduction) and familiarizing yourself with the content and format of this book by looking at the table of contents, the index, and scanning the chapters. Each chapter begins with a nomenclature list of the chapter's major variables and their units and ends with practice problems related to the presented concepts. Significant terms and concepts have been indexed to provide a method of easily finding information and data. Common acronyms and their definitions are listed in App. 1.A for quick reference. Unit conversions, national drinking water standards, and selected Ten States Standards are also given in the appendices.

The chapters are grouped into three topics. While any concept can be reviewed and referenced individually, successive chapters within each topic build on concepts previously presented. Decide on a study schedule, assess your strengths and weaknesses, and determine how much time to spend reviewing each chapter. Read the chapter, solving the example problems and reviewing the presented solutions as you go. Then solve the end-of-chapter practice problems: Restrain yourself from looking at the solutions until you've tried solving each problem on your own. The practice problems are designed to give you experience applying relevant equations, data, and theories to a given problem. Compare your solving approach against that provided in the solution. With practice, you will be able to quickly decide which data and equations are applicable to the problem at hand.

Topic I: Water Resources

Chapter

1. Hydrology
2. Closed Conduit Hydraulics
3. Open Channel Hydraulics
4. Groundwater Engineering
5. Water Treatment

1 Hydrology

Nomenclature

a	storm constant	in/hr	cm/h
A	area	ac	ha
A_d	watershed area	ac	ha
A_g	drainage area to stream gauge	mi^2	km^2
A_s	drainage area to point with unknown peak flow	mi^2	km^2
b	storm constant	min	min
c	storm constant	–	–
c	rational runoff coefficient	–	–
C	mean pollutant concentration	mg/L	mg/L
CN	runoff curve number	–	–
d	depth	in	cm
E	evaporation	in	cm
ET	evapotranspiration	in	cm
$f(n_t, N)$	probability of a storm or flood occurring n_t times within N yr	–	–
F_p	pond and swamp adjustment factor	–	–
h	height	ft	m
i	rainfall intensity	in/hr	cm/h
I	infiltration	in	cm
I_a	initial abstraction	in	cm
L	annual pollutant load	lbm/yr	kg/yr
L	length of flow	ft	m
n	Manning roughness coefficient	–	–
n_t	number of times a storm or flood occurs in time frame	–	–
N	number of years in time frame	yr	yr
p	probability of a storm occurring once in time frame	–	–
P	precipitation (rainfall)	in	cm

q	discharge	ft^3/sec	m^3/s
q_p	peak discharge	ft^3/sec	m^3/s
q_u	unit peak discharge	ft^3/ sec-mi^2-in	m^3/ s·km^2·cm
Q	flow rate	ft^3/sec	m^3/s
Q_g	peak flow at stream gauge	ft^3/sec	m^3/s
Q_p	peak runoff	ft^3/sec	m^3/s
Q_s	peak flow to site	ft^3/sec	m^3/s
R	hydraulic radius	ft	m
R	runoff	in	cm
R_s	runoff coefficient	–	–
S	slope	ft/ft	m/m
S	potential maximum retention	in	cm
ΔS	change in moisture storage	in	cm
t	time	sec	s
t_c	time of concentration	hr	h
t_p	time to peak runoff	hr	h
t_r	return or recurrence interval of a storm	yr	yr
T	transpiration	in	cm
v	velocity	ft/sec	m/s
V	volume	ft^3	m^3

Subscripts

ave	average
ch	channel flow
cum	cumulative
incr	incremental
sc	shallow concentrated flow
sf	sheet flow

1. INTRODUCTION[1]

Hydrology is the study of water as it is dispersed and circulated between the earth and the atmosphere. The *hydrologic cycle*, or *water budget*, tracks how the precipitation, P, that falls on the earth collects in a waterway (runoff, R), permeates into the ground (infiltration, I), is stored in the ground (i.e., is the change in moisture storage, ΔS), and then is dispersed back into the atmosphere through evaporation (E) or plant transpiration (T). The sum of evaporation and transpiration are often represented as one term, *evapotranspiration* (ET). A water resources engineer is most interested in the runoff and infiltration terms of the hydrologic cycle. Estimates of runoff are needed to design storm sewers, culverts, and stormwater basins. Infiltration is necessary to design

---

[1]Abbreviations used in this book are given in App. 1.A. Flow rate, velocity, and volumetric unit conversions are found in App. 1.B and App. 1.C.

Figure 1.1 *The Hydrologic Cycle*

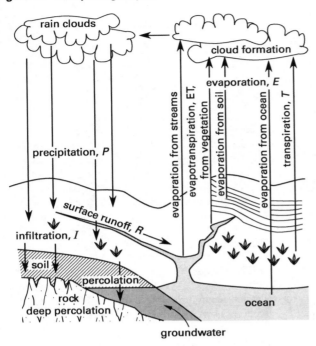

Adapted from NRCS. *Hydrology National Engineering Handbook.* Part 630. September 1997. USDA.

groundwater facilities, such as septic systems and recharge basins.

The science of hydrology is defined by the *water budget equation*, Eq. 1.1, of the hydrologic cycle (see Fig. 1.1). The water budget is typically expressed as a depth.

$$P = R + I + \text{ET} - \Delta S \qquad 1.1$$

2. WATERSHED MANAGEMENT

RAINFALL ≠ RUNOFF (EXCEPT 100% IMPERVIOUS)

ΔS = in-out

The *watershed* is the fundamental hydrologic unit for managing water resources. It is defined as the geographical area that drains into a waterway, such as a lake, stream, river, or ocean.

Watersheds come in all sizes, ranging from the catchment basin of a city block to the largest watershed on earth, the Amazon Basin. They can be categorized by the hierarchy described in Table 1.1. The terms *basin* and watershed are used interchangeably. Depending on the project area of interest, watersheds can be nested. For instance, Fig. 1.2(a) shows a 0.5 mi^2 urban drainage catchment in Newark, Delaware, nested within the 7 mi^2 Cool Run sub-watershed, which in turn is nested within the 98 mi^2 White Clay Creek watershed. Figure 1.2(b) shows that the White Clay Creek watershed is one of the four major streams in the 565 mi^2 Christina River sub-basin. Figure 1.2(c) shows that the Christina River sub-basin is part of the 13,000 mi^2 Delaware River basin.

Table 1.1 *Watershed Hierarchy*

unit	area (mi^2)	example, area (mi^2)
catchment	0.5–1.0	urban drainage, 0.5
sub-watershed	1.0–10	Cool Run, 7
watershed	10–100	White Clay Creek, 98
sub-basin	100–1000	Christina sub-basin, 565
basin	> 1000	Delaware River, 13,000

(Multiply mi^2 by 2.59 to obtain km^2.)

Figure 1.2 *Nested Watersheds in the Delaware River Basin*

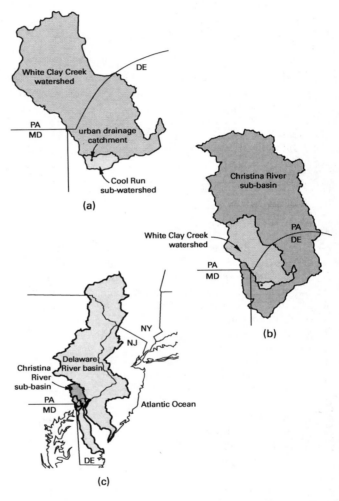

Adapted from maps created by University of Delaware, Water Resources Agency, 2007.

The United States Geological Survey (USGS) has delineated 21 major watersheds (river basins) in the United States (see Fig. 1.3). Watershed boundaries rarely coincide with political boundaries. Because the physical watersheds do not follow state boundaries, engineers must know both the scientific and the socio-political aspects of water resources management. States in the same watershed may have different stormwater and floodplain regulations. The interstate and interdepartmental management of the U.S. river and stream systems

Figure 1.3 *Major Watersheds (River Basins) in the United States*

Reprinted from *Hydrologic Unit Maps*. Water Supply Paper 2294. 1994. USGS.

Figure 1.4 *Wilson Run Watershed Boundary*

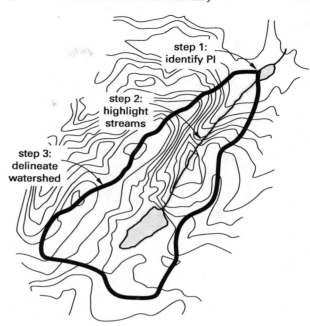

Adapted from *Wilmington North Quadrange Topographic Map*. 1993. USGS.

adds to the challenge and complexity of watershed and water resource engineering.

Local watersheds can be delineated on a topographic map according to the following three steps (see Fig. 1.4).

step 1: Identify the *point of interest* (PI) at the outlet of the waterway in question.

step 2: Highlight contributing streams and waterways on the topographic map.

step 3: Starting at the PI, delineate the watershed boundary on the topographic map. The watershed boundary crosses perpendicular to the contour lines. Look for closed contour lines at the top of the ridgelines, which usually indicate the watershed boundary.

Once the boundaries of the watershed are delineated, the drainage area, A_s, the time of concentration, t_c, and the runoff coefficient, R_s, can be calculated. The curve number, CN, for computing runoff (see Sec. 1.11) can also be calculated. Chapter 2 explains the drainage area, the time of concentration, and the runoff coefficient to delineate floodplain areas, calculate design flow, and design storm sewers and roadway culverts.

Example 1.1

In Colorado, Joe Wright Creek (6.9 mi^2) flows into the North Fork of the Cache La Poudre River (539 mi^2), which in turn flows into the South Platter River (14,627 mi^2). Classify each waterway in accordance with the accepted watershed hierarchy.

Solution

Referring to Table 1.1, Joe Wright Creek is a sub-watershed, the North Fork of the Cache La Poudre River is a sub-basin, and the South Platter River is a basin.

3. PRECIPITATION

Parameters of precipitation are used to calculate rainfall intensity, i, for runoff analysis (see Sec. 1.11). The United States National Weather Service (USNWS) compiles precipitation statistics for the United States in the following categories.

- total depth (inches)
- duration (days or hours)
- intensity (inches per hour)
- return interval (e.g., 10 yr storm, 50 yr storm)
- rainfall distribution types

Figure 1.5 depicts the mean annual precipitation in the United States from 1961 to 1990. Annual precipitation is highest (more than 60 in/yr) along the Gulf Coast and in the southeastern United States, where it is fueled by tropical moisture, and in the Northwest, where it is powered by the moist Pacific marine currents. Proceeding inland, the climate becomes drier, leading to the semiarid Great Plains (10–20 in/yr), and drier still in the desert Southwest (4–10 in/yr). In mountainous areas, precipitation is greater due to the lifting, or *orographic*, effect that causes rising air to cool and moisture to condense. Table 1.2 summarizes average annual precipitation for major cities in the United States from 1961 to 1990.

Water Resources

Figure 1.5 *Mean Annual Precipitation in the Continental United States (inches per year)*

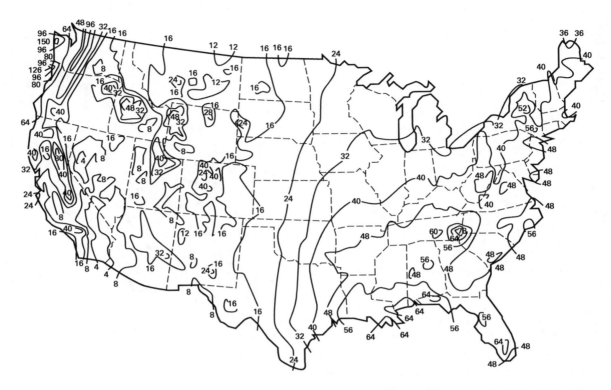

Adapted from National Oceanic and Atmospheric Administration. National Climatic Data Center. *Climate Atlas of the United States Version 2.0.* Period of Record 1961–1990, 2007. USDC.

Table 1.2 *Average Annual Precipitation in Major U.S. Cities*

city	average annual precipitation (in)
Atlanta	50.77
Boston	41.51
Chicago	35.82
Cleveland	36.63
Denver	15.40
Houston	46.07
Kansas City	37.62
Los Angeles	14.77
Miami	55.91
Minneapolis	28.32
New Orleans	61.88
New York	47.25
Philadelphia	41.41
Phoenix	7.66
Salt Lake City	16.18
San Francisco	19.71
Seattle	38.60
St. Louis	37.51
Washington, D.C.	38.63

(Multiply in by 2.54 to obtain cm.)

Reprinted from National Climatic Data Center. Period of Record 1961–1990. USDC.

Example 1.2

According to the map shown, what is the approximate mean annual precipitation for Chicago, Illinois?

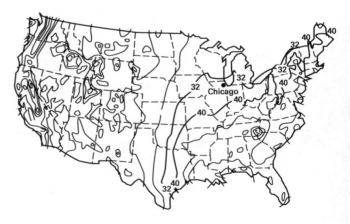

Solution

Referring to the given map, the mean annual precipitation for southern Wisconsin is 32 in and is 40 in for southern Illinois. Chicago is located approximately 2/5 of the distance between the 32 in contour line and the 40 in contour line. Interpolating between the two precipitation contour lines, the mean annual precipitation for Chicago is

$$32 \text{ in} + \left(\tfrac{2}{5}\right)(40 \text{ in} - 32 \text{ in}) = 35.2 \text{ in}$$

Table 1.2 gives a more exact value of 35.82 in.

4. STORM FREQUENCY

When designing water resource structures, such as storm sewers and culverts, it is necessary to account for the *storm frequency* (also known as *recurrence interval* or *return interval*). Engineers factor in storm or flood frequency (i.e., a 100 yr storm[2] or 25 yr flood) to estimate the probability that hydraulic structures can safely convey design flows.

The frequency or return interval, t_r, of a storm is the inverse of the probability, p, and is found by Eq. 1.2.

$$t_r = \frac{1}{p} \qquad \textit{1.2}$$

A 100 yr storm has a probability of $1/t_r$ or $1/100$ per year, which is a 0.01 or a 1% chance of occurring in any given single year. Table 1.3 relates storm return interval to probability.

Table 1.3 *Storm Return Interval and Probability of Occurring Once in Any Given Year*

return interval, t_r (yr)	probability, p
2	0.50
5	0.20
10	0.10
25	0.04
50	0.02
100	0.01
500	0.002

To minimize risk to life and property, engineers employ the *return interval criteria* summarized in Table 1.4 to conservatively design water resources engineering structures. For instance, to design a storm sewer, an engineer would calculate the 10 yr design flow using the rational method or the NRCS curve number (CN) method, as described in Sec. 1.11.

---

[2]The *100 yr storm* is often incorrectly defined as a storm likely to occur once every 100 yr. While it is unlikely for a 100 yr storm to occur twice within a short period, it can occur. Therefore, it is more appropriate to describe a 100 yr storm by its probability, which is a 1% chance of occurring within any given year.

Table 1.4 *Water Resources Design Return Interval Criteria*

return interval, t_r (yr)	design criteria
10	storm sewer design
50	small culvert and bridge design (low traffic)
100	large culvert and bridge design (high traffic)
	floodplain delineation
	stormwater detention basins
500	dam safety

Equation 1.3 determines the probability of a storm or flood occurring n_t times in N years.

$$f(n_t, N) = \frac{N! p^{n_t} (1 - p)^{N - n_t}}{n_t!(N - n_t)!} \qquad \textit{1.3}$$

Example 1.3

What is the probability that a 100 yr storm will occur twice in 4 years?

Solution

Using Eq. 1.2,

$$p = \frac{1}{t_r} = \frac{1}{100 \text{ yr}}$$
$$= 0.01$$

Using Eq. 1.3,

$$f(n_t, N) = \frac{N! p^{n_t} (1 - p)^{N - n_t}}{n_t!(N - n_t)!}$$

$$f(2, 4) = \frac{(4!)(0.01)^2 (1 - 0.01)^{4-2}}{(2!)(4 - 2)!}$$

$$= \frac{(4)(3)(2)(1)(0.0001)(0.9801)}{(2)(1)(2)(1)}$$

$$= 0.000588 \quad (0.06\%)$$

There is a 0.06% probability, or a 1 in 1700 chance, that a 100 yr storm would occur twice in 4 years.

5. CHARACTERISTICS OF PRECIPITATION

Total Precipitation Depths

Total precipitation depths for 24 hr duration storms in the central and eastern United States can be determined from design storm charts contained in *TR-55: Urban Hydrology for Small Watersheds* (*TR-55*). Figure 1.6 displays the 100 yr, 24 hr duration rainfall depth contour lines. The *TR-55* includes rainfall depth maps for 24 hr storms with return intervals of 2 yr, 5 yr, 10 yr, 25 yr, 50 yr, and 100 yr.

Figure 1.6 *Typical 100 yr, 24 hr Precipitation Depth Curves*

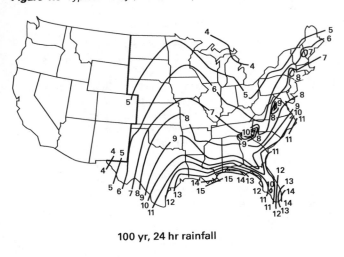

100 yr, 24 hr rainfall

Reprinted from *TR-55: Urban Hydrology for Small Watersheds.* NRCS. 1986. USDA.

Figure 1.7 *Rainfall Distribution in the United States*

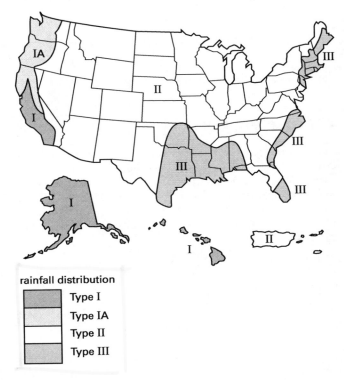

rainfall distribution

Type I
Type IA
Type II
Type III

Reprinted from *TR-55: Urban Hydrology for Small Watersheds.* NRCS. 1986. USDA.

Example 1.4

Using Fig. 1.6, what is the 100 yr, 24 hr precipitation depth for Philadelphia?

Solution

Interpolate between southeastern Pennsylvania, where the rainfall depth is 7 in, and southern Delaware, where the rainfall depth is 8 in. The 100 yr, 24 hr precipitation depth for Philadelphia, Pennsylvania is 7.5 in.

Rainfall Distribution Types

The *TR-55* classifies rainfall distribution types in the United States according to the map shown in Fig. 1.7. The NRCS has identified four typical rainfall distributions in the United States. *Type I* and *Type IA* represent Pacific states such as Alaska, California, Hawaii, Oregon, and Washington, which have the least intense rainfall, wet winters, and dry summers. Most of the United States has a *Type II* rainfall distribution, which means that over 90% of the rainfall is distributed within the middle 12 hr of a 24 hr storm. The runoff hydrograph (see Sec. 1.6) that results from a Type II rainfall distribution will usually start after the tenth hour of a 24 hr duration storm and recede by hour 24. *Type III* represents the Gulf of Mexico and the Atlantic coastal areas where tropical storms bring large amounts of rainfall in 24 hr.

Hyetographs

Precipitation patterns are characterized using storm *hyetographs*, which summarize the time interval, t, the rainfall depth, d, and the intensity of precipitation

events, i. Table 1.5 records incremental and cumulative precipitation depth and intensity data for a typical storm. Incremental and cumulative rainfall intensity are calculated by dividing the depth by the time interval, as shown in columns 4 and 5.

Table 1.5 *Precipitation Depth and Intensity Data for a Typical Storm*

time interval, t (hr)	incr. depth, d_{incr} (in)	cum. depth, d_{cum} (in)	incr. intensity, i_{incr} (in/hr)	cum. intensity, i_{cum} (in/hr)
0	0.0	0.0	0.0	0.0
1	0.0	0.0	0.0	0.0
2	0.1	0.1	0.1	0.1
3	0.2	0.3	0.2	0.15
4	0.3	0.6	0.3	0.2
5	0.2	0.8	0.2	0.2
6	0.1	0.9	0.1	0.18

(Multiply in by 2.54 to obtain cm.)
(Multiply in/hr by 2.54 to obtain cm/h.)

Figure 1.8 plots a storm hyetograph from the data in Table 1.5. The hyetograph relates incremental precipitation depth on the vertical axis to time on the horizontal axis.

HYETOGRAPH ≠ HYDROGRAPH

Figure 1.8 *Typical Storm Hyetograph*

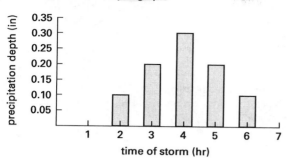

Figure 1.9 *Theissen Precipitation Diagram*

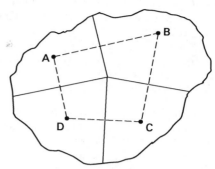

Example 1.5

From the storm hyetograph depicted in Fig. 1.8, what is the maximum rainfall intensity? In what time interval does it occur?

Solution

The maximum rainfall intensity is 0.3 in/hr, occurring in hour 4 of the storm.

Theissen Network

$$\frac{\Sigma PA}{\Sigma A}$$

Over large areas or watersheds, precipitation can vary depending on elevation, proximity to large waterways, and other factors. If multiple precipitation gauges are available, the precipitation at each station can be weighted in proportion to the area each station represents by using the *Theissen network* (see Fig. 1.9). The Theissen network is constructed according to the following steps.

step 1: Plot the location of each precipitation gauge, or station. Identify each gauge as A, B, C, and so on.

step 2: Connect each station to the two nearest stations with straight lines.

step 3: Delineate perpendicular bisecting lines halfway between adjacent gauges.

step 4: Define a series of polygons with the perpendicular bisecting lines. (For instance, in Fig. 1.9, there are Theissen polygons A, B, C, and D.)

step 5: Calculate the area of each Theissen polygon, then sum their areas.

step 6: Multiply the Theissen area by the precipitation depth at each station to determine the product (the volume of the precipitation).

step 7: Sum the products (precipitation volumes).

step 8: Determine the average precipitation depth per unit area in the watershed by dividing the sum of the products (precipitation volumes) by the sum of the Theissen areas.

Example 1.6

For the watershed depicted in Fig. 1.9, the precipitation is 4.0 in at gauge A, 5.0 in at gauge B, 3.5 in at gauge C, and 4.5 in at gauge D. The area of Theissen polygon A is 10 mi², B is 18 mi², C is 15 mi², and D is 11 mi². Calculate the average precipitation, P, per square mile of the watershed.

Solution

Construct the following table. The weighted precipitation is the Theissen area (column 2) multiplied by the precipitation (column 3).

station	Theissen area (mi²)	precipitation, P (in)	weighted precipitation (mi²-in)
A	10	4.0	40.0
B	18	5.0	90.0
C	15	3.5	52.5
D	11	4.5	49.5
sum	54	17.0	232

The average precipitation is the sum of the weighted precipitation values in column 4 divided by the sum of the Theissen areas in column 2.

$$P = \frac{232 \text{ mi}^2\text{-in}}{54 \text{ mi}^2}$$
$$= 4.3 \text{ in}$$

6. HYDROGRAPHS

Hydrographs depict the runoff, or flow, from a storm as a function of time. Water resources engineers use hydrographs when the volume and rate of runoff are necessary for design. For example, the volume of runoff is needed to estimate the required storage volume and inflow and outflow rates at a stormwater detention pond. The volume of runoff can be calculated as the area under the hydrograph curve.

Hydrographs are composed of two flow components: (1) *base flow*, which is the movement of groundwater to a stream or waterway, and (2) *overland flow*, which is the surface runoff to a waterway during and after a precipitation event. Runoff hydrographs can be constructed directly from real-time USGS stream gauge data or from synthetic unit hydrographs, which are described in the following subsections.

Commonly, base flow is separated from the overland flow component using the *straight-line method*. Base flow originates from groundwater and is used to define the minimal flow left in streams to protect habitats. Overland flow is the component of the hydrograph that is used to design hydraulic structures, such as culverts and weirs. A horizontal line is drawn starting from the point where the hydrograph slope starts to rise and ending at the point where the falling hydrograph slope starts to level off, as depicted in Fig. 1.10. The portion of the hydrograph below the dashed line is the base flow component, and the portion above the line is the overland flow component.

Figure 1.10 Hydrograph Depicting Base and Overland Flow

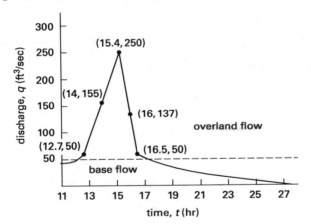

Once overland flow is separated from base flow in the hydrograph, the volume of overland runoff can be calculated. The hydrograph gives the discharge, q, in unit volume per second, and the time, t, in hours. Based on a triangular overland flow distribution, the volume of the overland runoff, V, can be estimated using the values of discharge and time from the hydrograph in Eq. 1.4.

$$V = \frac{\Delta t \Delta q}{2} \qquad \text{(AREA OF TRIANGLE)} \qquad 1.4$$

Unit Hydrograph

A *unit hydrograph* is used to design stormwater detention ponds. It is developed to account for 1 in (2.54 cm) of precipitation falling evenly over the entire area of a watershed.

The average volume, V, of overland runoff for a unit hydrograph is the product of the watershed area, A_d, and the average depth, $P_{\text{ave,excess}}$, of the excess precipitation, as shown in Eq. 1.5.

$$V = A_d P_{\text{ave,excess}} \qquad 1.5$$

The unit hydrograph of a storm can be defined by dividing the flow values by the precipitation depth of the storm.

Example 1.7

A watershed area is 0.5 mi². Define the unit hydrograph for the hydrograph coordinates in Fig. 1.10, then define the watershed's hydrograph for a 2 in storm.

Solution

Use Eq. 1.4.

$$V = \frac{\Delta t \Delta q}{2}$$

$$= \frac{\left(\begin{array}{c} (16.5 \text{ hr} - 12.7 \text{ hr})\left(250 \, \frac{\text{ft}^3}{\text{sec}} - 50 \, \frac{\text{ft}^3}{\text{sec}}\right) \\ \times \left(60 \, \frac{\text{min}}{\text{hr}}\right)\left(60 \, \frac{\text{sec}}{\text{min}}\right) \end{array} \right)}{2}$$

$$= 1{,}368{,}000 \text{ ft}^3$$

Use Eq. 1.5.

$$V = A_d P_{\text{ave,excess}}$$

$$P_{\text{ave,excess}} = \frac{V}{A_d} = \frac{(1{,}368{,}000 \text{ ft}^3)\left(12 \, \frac{\text{in}}{\text{ft}}\right)}{(0.5 \text{ mi}^2)\left(640 \, \frac{\text{ac}}{\text{mi}^2}\right)\left(43{,}560 \, \frac{\text{ft}^2}{\text{ac}}\right)}$$

$$= 1.18 \text{ in}$$

Referring to the following table, column 3 defines the unit hydrograph for a 0.5 mi² watershed area. The unit hydrograph is calculated by dividing each hydrograph runoff point in column 2 by the average precipitation, $P_{\text{ave,excess}}$, which is 1.18 in.

Column 4 defines the hydrograph coordinates for a 2.0 in storm in the watershed, which are calculated by multiplying the unit hydrograph runoff calculated in column 3 by the storm precipitation, 2.0 in.

time (hr)	1.18 in hydrograph runoff (ft³/sec)	unit hydrograph runoff (ft³/sec-in)	2.0 in hydrograph runoff (ft³/sec)
12.7	50	50/1.18 = 42.4	(42.4)(2.0) = 84.7
14	155	155/1.18 = 131.4	(131.4)(2.0) = 262.7
15.4	250	250/1.18 = 211.9	(211.9)(2.0) = 423.7
16	137	137/1.18 = 116.1	(116.2)(2.0) = 232.2
16.5	50	50/1.18 = 42.4	(42.4)(2.0) = 84.7

Synthetic Unit Triangular Hydrograph

In ungauged watersheds, *synthetic hydrographs* are used to design reservoirs and detention ponds and can be estimated using the *NRCS synthetic unit triangular hydrograph method* (see Fig. 1.11). The coordinates of the synthetic hydrograph are defined by the peak runoff, Q_p, time to peak runoff, t_p, and time of concentration, t_c.

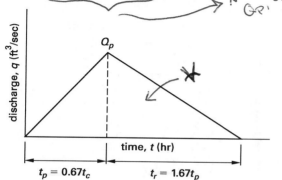

Figure 1.11 *Synthetic Unit Triangular Hydrograph*

NEED t_c & Q_p

The *peak runoff* is the maximum rate of runoff flow. (See Sec. 1.11 for how to calculate peak runoff.) The time of concentration is the maximum time for runoff to travel from the farthest and uppermost point in a watershed downstream to the outlet point of the watershed. The storm duration is equal to the time of concentration for small watersheds (1 mi^2). The ascending leg of the hydrograph is the *time to peak runoff*, t_p, as measured from the beginning of overland runoff hydrograph as defined in Fig. 1.11 and Eq. 1.6.

$$t_p = 0.67t_c \qquad 1.6$$

ONLY FOR SYNTHETIC UNIT TRIANGULAR HYDROGRAPH

The receding leg of the hydrograph is measured after the time of peak runoff using Eq. 1.7.

$$t_r = 1.67t_p \qquad 1.7$$

Example 1.8

Define a synthetic unit triangular hydrograph where the peak runoff is estimated as $100 \text{ ft}^3/\text{sec}$ and the time of concentration is 1.5 hr.

Solution

Plot the peak runoff at $100 \text{ ft}^3/\text{sec}$. Use Eq. 1.6 to define the time for the ascending limb of the hydrograph as t_p, the time to peak runoff.

$$t_p = 0.67t_c$$
$$= (0.67)(1.5 \text{ hr})$$
$$= 1.01 \text{ hr}$$

Using Eq. 1.7, define the time of the receding limb of the hydrograph.

$$t_r = 1.67t_p$$
$$= (1.67)(1.01 \text{ hr})$$
$$= 1.69 \text{ hr}$$

Plot the synthetic unit triangular hydrograph as shown.

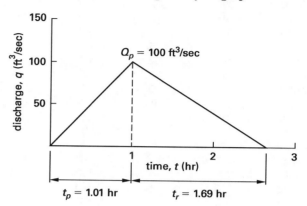

7. RESERVOIR VOLUME (RIPPL DIAGRAM)

The minimum required volume of a water tank or reservoir can be estimated by creating a *Rippl diagram* according to the following steps, and as illustrated in Ex. 1.9.

step 1: Tabulate days of accumulation (column one) and the cumulative volume (column two).

step 2: Plot cumulative volume (y-axis) versus time (x-axis).

step 3: Determine the half-maximum accumulation and mark the point on the curve. The half-maximum accumulation is the intersection of cumulative volume and time at the point where time is one-half of the total duration.

step 4: Draw a straight line through the origin and the half-maximum point and continue it through the top of the plotted cumulative volume curve. This is the average volume line.

step 5: Draw two lines parallel to the average volume line. The first line, tangent 1, runs through times earlier than the cumulative volume curve, except at the point that it touches the curve. Tangent 2 runs through times later than the cumulative volume curve, except at the point that it touches the curve.

step 6: Starting at the point where the lowest parallel line intersects the x-axis, draw a vertical line upward to intersect with the highest parallel line.

step 7: Starting at the intersection of the vertical line with the highest parallel line, draw a horizontal line to intersect the y-axis. The *minimum required volume* is given by that point.

Example 1.9

Estimate the minimum required volume of a reservoir with a cumulative volume relationship as defined in the following table.

time (days)	cumulative volume (MG)	average volume (MG)	tangent 1 (MG)	tangent 2 (MG)
0	0	0	–	30
1	6	15	–	45
2	12	30	0	60
3	18	45	15	75
4	28	60	30	90
5	46	75	45	105
6	68	90	60	120
7	98	105	75	135
8	130	120	90	150
9	155	135	105	165
10	175	150	120	180
11	192	165	135	195
12	205	180	150	210
13	215	195	165	225
14	222	210	180	240
15	225	225	195	255

Solution

The data for step 1 is tabulated in the problem statement and plotted as a Rippl diagram, which is labeled with the associated step numbers. The minimum required volume is found in step 7 to be 60 MG.

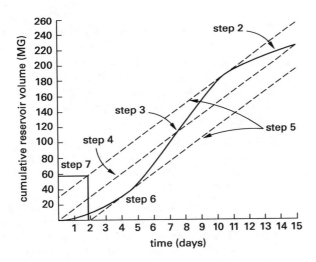

8. RAINFALL INTENSITY-DURATION-FREQUENCY RELATIONS

Intensity-Duration-Frequency Curves

Rainfall intensity, i, and storm duration (given by time of concentration, t_c) are related by *intensity-duration-frequency* (IDF) curves published by the USNWS, the United States Department of Agriculture (USDA), and

local or state governments for specific geographic regions of the United States. Most state and local governments as well as the Departments of Transportation publish IDF curves in stormwater manuals. Figure 1.12 provides a typical IDF curve published by the USDA for New Castle County, Delaware.

Figure 1.12 *Rainfall Intensity-Duration-Frequency Curve for New Castle County, Delaware*

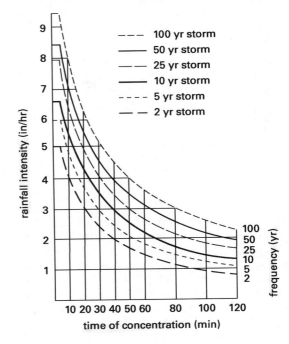

Adapted from *Surface Water Management Code.* 2007. USDA and New Castle County.

Example 1.10

The time of concentration for a watershed is 60 min. Using Fig. 1.12, estimate the rainfall intensity for a 10 yr storm.

Solution

For a time of concentration of 60 min, the 10 yr storm intensity is 2.2 in/hr.

Intensity-Duration-Frequency Constants

In locations for which rainfall intensity-duration-frequency curves are not available, the engineer may use Eq. 1.8 and constant values (see Table 1.6) to estimate rainfall intensity for a 10 yr storm event. The 10 yr intensity, i_{10}, is used to design storm sewers in urban areas.

$$i_{10} = \frac{a}{(t+b)^c} \qquad 1.8$$

The intensity-duration-frequency constants a, b, and c for several cities are listed in Table 1.6. These constants

Table 1.6 *Intensity-Duration-Frequency Constants for Cities in the United States for a 10 yr Storm*

city	a (in/hr)	b (min)	c
Atlanta	64.1	8.16	0.76
Chicago	60.9	9.56	0.81
Cleveland	47.6	8.86	0.79
Denver	50.8	10.5	0.84
Houston	98.3	9.30	0.80
Los Angeles	10.9	1.15	0.51
Miami	79.9	7.24	0.73
New York	51.4	7.85	0.75
Santa Fe	32.2	8.54	0.76
St. Louis	61.0	8.96	0.78

(Multiply in by 2.54 to obtain cm.)

Adapted from G. W. Wenzel. American Geophysical Union. *Rainfall for Urban Stormwater Design, Urban Stormwater Hydrology.* 1982. Water Resources Monograph Board.

Table 1.7 *Typical Monthly Evaporation Data*

month	Wilmington, DE (in)	Bakersfield, CA (in)	Blue River, OK (in)
Jan	0.4	1.4	1.7
Feb	0.5	2.2	2.2
Mar	1.2	4.1	3.0
Apr	1.8	5.9	4.0
May	3.0	8.3	5.0
Jun	4.0	9.5	5.6
Jul	4.8	9.9	5.7
Aug	5.4	8.8	5.2
Sep	4.8	6.6	4.4
Oct	3.2	4.5	3.3
Nov	2.0	2.2	2.4
Dec	0.6	1.3	1.8

(Multiply in by 2.54 to obtain cm.)

Adapted from National Oceanic and Atmospheric Administration. *NOAA Technical Report NWS 33* and *NOAA Technical Report NWS 34.* 1982.

relate rainfall depth for a 10 yr storm with the time of concentration. To work with these constants, the duration of the storm under consideration, time *t,* is given in minutes (e.g., 60 min for a 1 hr duration storm). The resulting intensity is in units of inches per hour.

Example 1.11

Using Table 1.6, calculate the rainfall intensity for a 10 yr, 1 hr storm in Denver.

Solution

Use Eq. 1.8 and values from Table 1.6.

$$i_{10} = \frac{a}{(t+b)^c}$$

$$= \frac{50.8 \ \dfrac{\text{in}}{\text{hr}}}{(60 \text{ min} + 10.5 \text{ min})^{0.84}}$$

$$= 1.42 \text{ in/hr}$$

9. EVAPOTRANSPIRATION

Transpiration, evaporation, and infiltration are components of precipitation loss in the hydrologic cycle. In transpiration, water is absorbed by plants and released as vapor into the atmosphere through the leaves. In data, transpiration is usually combined with evaporation and is collectively known as *evapotranspiration,* ET.

ET is defined as all water vapor emitted to the atmosphere from plants and other sources and can be estimated from pan evaporation data. Evaporation increases with warmth and plant foliage during the spring, summer, and early fall months, and decreases during the cooler months after the leaves have fallen off the trees and plants. Table 1.7 summarizes typical monthly evaporation data for selected regions of the United States. Evaporation peaks during the warm summer months and declines during the cooler winter months.

Figure 1.13 depicts the average annual lake evaporation in the United States, as compiled by the USDC. In arid parts of the United States, such as Arizona, evaporation can exceed 40 in, or three times the annual precipitation, substantially depleting lakes and reservoirs in the Western states. In the desert Southwest, engineers must plan for evaporation when designing water budgets for reservoirs. Irrigation is usually needed to sustain crops and lawns in areas west of the 100th meridian (from North Dakota through Nebraska to Texas), where evaporation exceeds precipitation by more than a two-to-one margin. Arid and semiarid cities where annual evaporation substantially exceeds precipitation are Phoenix, Los Angeles, San Francisco, Salt Lake City, and Denver (see Table 1.8).

Figure 1.13 *Mean Annual Lake Evaporation in the Continental United States*

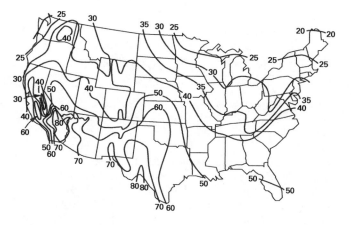

Reprinted from *Climate Atlas of the United States.* Period of Record 1946–1955, 1968. USDC.

Table 1.8 Comparison of Annual Precipitation and Evaporation for Major Cities in the United States

city	annual precipitation (in)	annual evaporation (in)
Atlanta	50.8	43
Boston	41.5	28
Chicago	35.8	30
Cleveland	36.6	30
Denver	15.4	40
Houston	46.1	53
Kansas City	37.6	45
Los Angeles	14.8	40
Miami	55.9	50
Minneapolis	28.3	30
New Orleans	61.9	50
New York	47.2	31
Philadelphia	41.4	34
Phoenix	7.7	60
Salt Lake City	16.2	40
San Francisco	19.7	40
Seattle	38.6	25
St. Louis	37.5	36
Washington, D.C.	38.6	35

(Multiply in by 2.54 to obtain cm.)

Adapted from National Climatic Data Center. Period of Record 1961–1990. USDC.

10. INFILTRATION

Infiltration is the component of precipitation loss that recharges the groundwater table. The NRCS divides soils into four major *hydrologic soil groups* (HSG) of A, B, C, and D, depending on their runoff potential and infiltration rate, as shown in Table 1.9.

Table 1.9 Hydrologic Soil Groups

HSG group	soil type	infiltration rate (in/hr)
A	sands and loams that have low runoff potential	high (> 0.3)
B	silt loams	moderate (0.15–0.3)
C	clay loams	low (0.05–0.15)
D	silts and clays	very low (0–0.05)

(Multiply in by 2.54 to obtain cm.)

Reprinted from *TR-55: Urban Hydrology for Small Watersheds*. NRCS. 1986. USDA.

Hydrologic soil groups for watersheds can be obtained from the NRCS *TR-55: Urban Hydrology for Small Watersheds*, which is available at **www.ppi2pass.com/CEwebrefs**. Soil surveys also provide hydrologic soil groups as shown in Table 1.10, which is an excerpted table of hydrologic soil groups for Stockbridge (HSG C) and Georgia (HSG C) soils from the Bennington County, Vermont soil survey. Soil survey maps are examined to identify the soil name. Soil names are then compared to a soil survey table or to the index of soils in App. A of the *TR-55* to obtain HSG A, B, C, or D.

For example, using Table 1.10, the hydrologic soil group for soil map symbol 64B (Stockbridge) and soil map symbol 66B (Georgia) is C.

11. RUNOFF ANALYSIS

In the hydrologic cycle, precipitation results in runoff, infiltration, and/or evapotranspiration. *Runoff* is the fraction of water from precipitation, snow melt, or irrigation that empties into uncontrolled surface streams, rivers, drains, or sewers. Water resources engineers define runoff to size hydraulic structures, such as storm-water ponds, storm sewers, and culverts. Environmental engineers are interested in controlling the contaminants carried in runoff.

The term "runoff" is sometimes used interchangeably with the terms "flow" and "discharge." Runoff rate is measured in cubic feet per second, million gallons per day, or gallons per minute. The unit of cubic feet per second is commonly used to measure runoff entering or moving through waterways, and to size pipes and storm sewers. The units of million gallons per day and gallons per minute are commonly used in water distribution design using closed conduit flow or flow under pressure.

Runoff may be measured directly from a stream gauge or estimated using rainfall-runoff equations, such as the *rational method* or the *TR-55* method. Runoff may be classified as direct runoff or base runoff according to the time of its appearance after rainfall or melting snow. Runoff can also be categorized by its source as surface runoff, storm interflow, groundwater runoff, urban runoff, or agricultural runoff. The amount and composition of the runoff is determined by meteorological conditions and the physical properties of the affected land surface.

Water resources engineers commonly perform *runoff analysis* using the following two preferred methods.

1. *Rational method*—for small watersheds with areas less than 200 ac.

2. *NRCS curve number method*—for watershed areas larger than 0.3 mi^2 ($\approx$ 200 ac) and less than 10 mi^2. For design problems pertaining to watersheds greater than 10 mi^2, the watershed is subdivided into sub-watersheds of less than 10 mi^2, and the curve number method is applied.

The Rational Method for Calculating Runoff

The *rational method* was developed during the 1850s. It calculates the peak runoff, Q_p, using Eq. 1.9.

$$Q_p = ciA_d \quad\quad 1.9$$

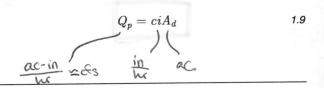

Table 1.10 Hydrologic Soil Groups for Typical Soils in Bennington County, Vermont

map symbol and soil name	hydrologic group	month	water table upper limit	water table lower limit	kind
64B: Stockbridge	C	Jan–Dec	>6.0	>6.0	–
64C: Stockbridge	C	Jan–Dec	>6.0	>6.0	–
64D: Stockbridge	C	Jan–Dec	>6.0	>6.0	–
65C: Stockbridge	C	Jan–Dec	>6.0	>6.0	–
65D: Stockbridge	C	Jan–Dec	>6.0	>6.0	–
66A: Georgia	C	Jan–May	1.2–2.5	1.3–2.7	perched
		Jun–Oct	>6.0	>6.0	–
		Nov–Dec	1.2–2.5	1.3–2.7	perched
66B: Georgia	C	Jan–May	1.2–2.5	1.3–2.7	perched
		Jun–Oct	>6.0	>6.0	–
		Nov–Dec	1.2–2.5	1.3–2.7	perched
66C: Georgia	C	Jan–May	1.2–2.5	1.3–2.7	perched
		Jun–Oct	>6.0	>6.0	–
		Nov–Dec	1.2–2.5	1.3–2.7	perched

Adapted from *Soil Survey of Bennington County, Vermont.* NRCS. 2006. USDA.

The *rational runoff coefficient* is c. Rainfall intensity, i, is given in inches per hour, and the watershed area, A_d, is given in acres.

The units for Q are acre-inch per hour, which is approximately equal to cubic feet per second. Consequently, the units for Q are commonly given as cubic feet per second. The rational method, Eq. 1.9, is one of the formulas in engineering where the units do not cancel out.

Runoff coefficients estimate the percentage or ratio of runoff to total precipitation from a particular storm precipitation event. Approximate runoff coefficient, c, values for different land use conditions, which can be obtained from charts in water resources textbooks, are also given in Table 1.11.

Figure 1.14 is a flow chart showing the required information, choices, and steps in the rational method.

The *watershed area*, A_d, is measured from topographic maps, in terms of acres. *Rainfall intensity*, i, is derived from IDF curves. The storm duration is assumed to be equal to the time of concentration for the small watersheds to which the rational method applies. The *time of concentration*, the maximum time for runoff to travel from the farthest and uppermost point in a watershed to the outlet, has contributions from sheet flow, t_{sf}, shallow concentrated flow, t_{sc}, and channel flow, t_{ch} (see Fig. 1.15).

Time of concentration can be relatively estimated using many equations found in water resources engineering textbooks. A preferred and precise method for deriving the time of concentration is Eq. 1.10, derived from the *TR-55* Worksheet 3 shown in Fig. 1.16.

$$t_c = t_{sf} + t_{sc} + t_{ch} \qquad 1.10$$

t_c is the *total time of concentration* (in hours) for a particular watershed. On the worksheet, the components

Table 1.11 Runoff Coefficients for the Rational Method

land use	runoff coefficient, c
business: downtown areas	0.70–0.95
neighborhood areas	0.50–0.70
residential	
single-family areas	0.30–0.50
multi units, detached	0.40–0.60
multi units, attached	0.60–0.75
suburban	0.25–0.40
residential (> 1.2 ac lots)	0.30–0.45
apartment dwelling areas	0.50–0.70
industrial	
light areas	0.50–0.80
heavy areas	0.60–0.90
parks, cemeteries	0.10–0.25
playgrounds	0.20–0.40
railroad yard areas	0.20–0.40
unimproved areas	0.10–0.30
streets	
asphalt	0.70–0.95
concrete	0.80–0.95
drives and walks	0.75–0.85
roofs	0.75–0.85

Reprinted from *Urban Drainage Manual.* 2009. Federal Highway Administration.

of t_c are found under sheet flow, t_{sf}; shallow concentrated flow, t_{sc}; and channel flow, t_{ch}.[3]

The travel time (in hours) for the sheet flow component is found from Eq. 1.11.

$$t_{sf} = \frac{(0.007 \text{ hr})(nL)^{0.8}}{P_2^{0.5} S^{0.4}} \qquad 1.11$$

---

[3]The worksheet refers to each of these components as T_t. However, this book uses subscripts to distinguish the component travel times.

Figure 1.14 Rational Method Flow Chart

```
┌─────────────────────────────────────────┐
│ given a runoff coefficient, c, for a     │
│ given land use, or c_total for a         │
│ composite area; given an estimate of     │
│ the area, A_d, in acres; and given the   │
│ design frequency of storm                │
└─────────────────────────────────────────┘
                    │
            ◇ Is the watershed
              area less than
              0.3 mi² (≈200 ac)? ◇
           │yes              │no
           ▼                 ▼
┌──────────────────┐   ◇ Is there      ◇
│ determine time   │yes  sufficient data
│ of concentration,│◄─── to reliably
│ t_c, using TR-55 │    compute time of
│ Worksheet 3      │    concentration?
└──────────────────┘        │no
           │                ▼
           │        ┌──────────────────┐
           │        │ time of          │
           │        │ concentration,   │
           │        │ t_c, equals      │
           │        │ storm duration   │
           │        └──────────────────┘
           ▼                │
┌──────────────────────────────────────┐
│ determine storm intensity, i, rainfall│
│ intensity-duration-frequency curves   │
└──────────────────────────────────────┘
                    │
                    ▼
┌──────────────────────────────────────┐
│ calculate peak flow                   │
│ runoff Q_p = ciA_d                    │
└──────────────────────────────────────┘
```

Figure 1.15 Time of Concentration in a Watershed

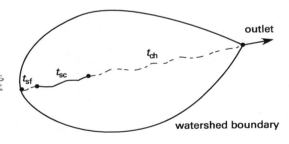

The *Manning roughness coefficient, n,* is found from Table 1.12 (*TR-55* Table 3-1).

The length of the sheet flow, L, is given in feet, with a 300 ft maximum. The 2 yr, 24 hr precipitation depth, P_2, given in inches, is found from Fig. 1.17 (*TR-55* Fig. B-3). The slope of the sheet flow, S, is given as foot per foot.

The travel time (in hours) for the shallow concentrated flow component is calculated from Eq. 1.12.

$$t_{sc} = \frac{L}{v} \qquad 1.12$$

L is the flow length of the shallow concentrated flow (in feet). The velocity, v (in feet per second), is taken from

Fig. 1.18 (*TR-55* Fig. 3-1), depending on the slope and whether the surface is paved or unpaved.

The travel time for the channel flow component is similar to the shallow concentrated flow component and is found from Eq. 1.13.

$$t_{ch} = \frac{L}{v} \qquad 1.13$$

The velocity, v, in the channel is derived from the *Manning equation* in Eq. 1.14. Values of the Manning roughness coefficient, n, are given in Table 1.12.

$$v = \left(\frac{1.00}{n}\right) R^{2/3} \sqrt{S} \qquad \text{[SI]} \quad 1.14(a)$$

$$v = \left(\frac{1.49}{n}\right) R^{2/3} \sqrt{S} \qquad \text{[U.S.]} \quad 1.14(b)$$

Many watersheds are occupied by more than one land use or land cover condition. Such varied watersheds require the calculation of a *composite runoff coefficient* for the rational method. The composite c_{total} value may be calculated by prorating the coefficient by the amount of land cover and then dividing by the total watershed area. For example, in a watershed with two different areas of land use, the composite c_{total} value is

$$c_{total} = \frac{\sum c_i A_i}{\sum A_i} = \frac{c_1 A_1 + c_2 A_2}{A_1 + A_2} \qquad 1.15$$

Once the time of concentration is calculated, select the rainfall intensity, i, from an IDF curve, such as the one in Fig. 1.12.

Example 1.12

Using Table 1.11, calculate the composite runoff coefficient for a 100 ac watershed occupied by 50 ac of single family residential, 25 ac of light industrial, and 25 ac of park land uses.

Solution

The c values are entered in the following table. To be conservative, select the highest c values in the range of values given in the table.

land use	c	area (ac)
single family residential	0.5	50
light industrial	0.8	25
park land	0.25	25

Use Eq. 1.15.

$$
\begin{aligned}
c_{total} &= \frac{\sum c_i A_i}{\sum A_i} \\
&= \frac{(0.5)(50 \text{ ac}) + (0.8)(25 \text{ ac}) + (0.25)(25 \text{ ac})}{50 \text{ ac} + 25 \text{ ac} + 25 \text{ ac}} \\
&= 0.51
\end{aligned}
$$

Figure 1.16 *Computing Time of Concentration**

| Project | | By | | Date |
| Location | | Checked | | Date |

Check one: ☐ Present ☐ Developed

Check one: ☐ t_c ☐ T_t through subarea

Notes: Space for as many as two segments per flow type can be used for each worksheet. Include a map, schematic, or description of flow segments.

Sheet flow (Applicable to t_c only)

Segment ID

1. Surface description (Table 1.12) ....

2. Manning's roughness coefficient, n (Table 1.12) ...

3. Flow length, L (total $L \leq 300$ ft) .... ft

4. Two year, 24 hour rainfall, P_2 .... in

5. Land slope, S.. ft/ft

EQ. 1.11 6. $t_{sf} = \dfrac{(0.007 \text{ hr})(nL)^{0.8}}{P_2^{0.5}S^{0.4}}$ Compute t_{sf} .... hr ☐ + ☐ = ☐

Shallow concentrated flow

Segment ID

7. Surface description (paved or unpaved) ....

8. Flow length, L.. ft

9. Watercourse slope, S.. ft/ft

10. Average velocity, v (Fig. 1.18) .... ft/sec

EQ. 1.12 11. $t_{sc} = \dfrac{L}{v}$ Compute t_{sc} .... hr ☐ + ☐ = ☐

Channel flow

Segment ID

12. Cross-sectional flow area, A_{ft^2} .... ft^2

13. Wetted perimeter, WP .... ft

14. Hydraulic radius, $R = \dfrac{A_{ft^2}}{WP}$, compute R .... ft

15. Channel slope, S.. ft/ft

16. Manning's roughness coefficient, n .... ft/ft

17. $v = \left(\dfrac{1.49}{n}\right) R^{2/3} \sqrt{S}$ Compute v ......ft/sec

18. Flow length, L.. ft

EQ. 1.13 19. $t_{ch} = \dfrac{L}{v}$ Compute t_{ch}.......... hr ☐ + ☐ = ☐

20. Watershed or subarea, t_c (add t_{sf}, t_{sc}, and t_{ch} from steps 6, 11, and 19) .... hr

*TR-55 Worksheet 3 gives each time component as T_t. This book uses subscripts to differentiate between the component travel times. Reprinted from *TR-55: Urban Hydrology for Small Watersheds*. NRCS. Worksheet 3. 1986. USDA.

Table 1.12 Manning Roughness Coefficients for Sheet Flow

surface description	n^a
smooth surfaces (concrete, gravel, asphalt, or bare soil)	0.011
fallow (no residue)	0.05
cultivated soils:	
residue cover $\leq 20\%$	0.06
residue cover $> 20\%$	0.17
grass:	
short grass prairie	0.15
dense grasses[b]	0.24
bermuda grass	0.41
range (natural)	0.13
woods[c]:	
light underbrush	0.40
dense underbrush	0.80

[a]The n values are a composite of information compiled by Engman (1986).
[b]Includes species such as weeping lovegrass, bluegrass, buffalo grass, blue grama grass, and native grass mixtures.
[c]When selecting n, consider cover to a height of about 0.1 ft. This is the only part of the plant cover that will obstruct sheet flow.

Reprinted from *TR-55: Urban Hydrology for Small Watersheds.* NRCS. Table 3-1. 1986. USDA.

Figure 1.17 2 yr, 24 hr Rainfall in the Continental United States

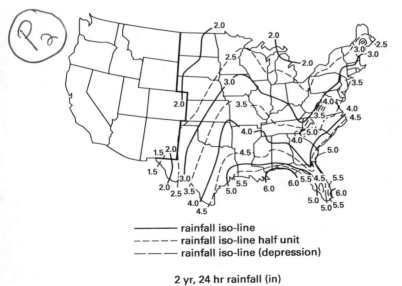

2 yr, 24 hr rainfall (in)

Reprinted from *TR-55: Urban Hydrology for Small Watersheds.* NRCS. Fig. B-3. 1986. USDA.

Figure 1.18 Average Velocities for Calculating Shallow Concentrated Flow

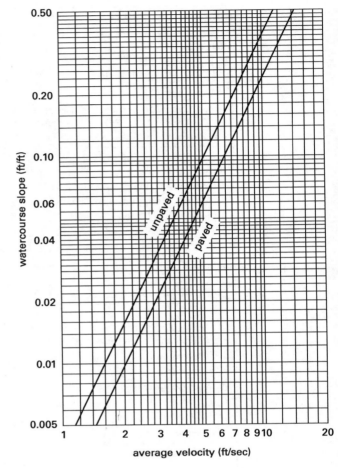

Reprinted from *TR-55: Urban Hydrology for Small Watersheds.* NRCS. Fig. 3-1. 1986. USDA.

Example 1.13

Using the rational method and the rainfall IDF curve given in Fig. 1.12, estimate the 100 yr runoff from a 100 ac watershed covered by single-family residential land use. The dimensions for time of concentration are

> sheet flow
>   length: 300 ft
>   surface description: dense grass
>   Manning roughness coefficient: 0.24
>   2 yr, 24 hr rainfall: 3.2 in
>   slope: 0.07 ft/ft
> shallow concentrated flow
>   surface: unpaved
>   flow length: 1000 ft
>   slope: 0.1 ft/ft
> channel flow
>   velocity: 8 ft/sec
>   length: 3100 ft

Solution

The first step is to find the intensity for calculating runoff from a 100 yr storm. To find intensity, i, calculate the time of concentration using Fig. 1.16 (*TR-55* Worksheet 3). The travel time for the sheet flow component (no more than 300 ft long) is given by Eq. 1.11.

$$t_{sf} = \frac{(0.007 \text{ hr})(nL)^{0.8}}{P_2{}^{0.5} S^{0.4}}$$

$$= \frac{(0.007 \text{ hr})\Big((0.24)(300 \text{ ft})\Big)^{0.8}}{(3.2 \text{ in})^{0.5}\Big(0.07 \frac{\text{ft}}{\text{ft}}\Big)^{0.4}}$$

$$= 0.347 \text{ hr}$$

From Fig. 1.18, based on a slope of 0.1 ft/ft and an unpaved surface, the velocity, v, for the sheet flow component is 5 ft/sec. The travel time for the shallow concentrated flow component is given by Eq. 1.12.

$$t_{sc} = \frac{L}{v} = \frac{1000 \text{ ft}}{\Big(3600 \frac{\text{sec}}{\text{hr}}\Big)\Big(5 \frac{\text{ft}}{\text{sec}}\Big)}$$

$$= 0.0556 \text{ hr}$$

The travel time for the channel flow component is given by Eq. 1.13.

$$t_{ch} = \frac{L}{v} = \frac{3100 \text{ ft}}{\Big(3600 \frac{\text{sec}}{\text{hr}}\Big)\Big(8 \frac{\text{ft}}{\text{sec}}\Big)}$$

$$= 0.108 \text{ hr}$$

The time of concentration is calculated using Eq. 1.10.

$$t_c = t_{sf} + t_{sc} + t_{ch}$$

$$= (0.347 \text{ hr} + 0.0556 \text{ hr} + 0.108 \text{ hr})\Big(60 \frac{\text{min}}{\text{hr}}\Big)$$

$$= 31 \text{ min}$$

From the IDF curve in Fig. 1.12, for a time of concentration, t_c, of 31 min, the intensity, i, for a 100 yr storm is 5 in/hr.

Estimate the coefficient of runoff for single family land use using Table 1.11.

$$c = 0.4$$

From Eq. 1.9,

$$Q_{100 \text{ yr}} = ciA_d$$

$$= (0.4)\Big(5 \frac{\text{in}}{\text{hr}}\Big)(100 \text{ ac})$$

$$= 200 \text{ ac-in/hr} \quad (\approx 200 \text{ ft}^3/\text{sec})$$

② The NRCS Curve Number Method

The *NRCS curve number method* (formerly known as the SCS method) accounts for precipitation losses, such as transpiration, evaporation, and infiltration, as *initial abstraction*, I_a. The amount of runoff depends on the watershed area and slope, soil permeability, and the amount of impervious cover. The *curve number* is derived from the impervious cover of a particular land use or land cover condition. Curve numbers increase with increasing runoff. The *runoff curve number*, CN, is obtained for particular land use and hydrologic soil group from Table 1.13 through Table 1.15 (*TR-55* Tables 2-2a, b, and c).

The potential maximum retention after runoff begins, S, is obtained from the curve number using Eq. 1.16.

$$S = \frac{1000 \text{ in}}{\text{CN}} - 10 \text{ in} \qquad \textit{1.16}$$

The initial abstraction quantity, I_a, is found from Eq. 1.17.

$$I_a = 0.2S \qquad \textit{1.17}$$

SEE TABLE 1.16 p.1-23

Example 1.14

Using Table 1.15, define the initial abstraction losses from transpiration, evaporation, and infiltration for a watershed with wooded land and HSG B soils.

Solution

The curve number, CN, is 55 for wooded land and HSG B soils. Find the potential maximum retention, S, using Eq. 1.16.

$$S = \frac{1000 \text{ in}}{\text{CN}} - 10 \text{ in}$$

$$= \frac{1000 \text{ in}}{55} - 10 \text{ in}$$

$$= 8.18 \text{ in}$$

Using Eq. 1.17, the initial abstraction losses are

$$I_a = 0.2S$$

$$= (0.2)(8.18 \text{ in})$$

$$= 1.64 \text{ in}$$

Table 1.13 Runoff Curve Numbers for Urban Areas[a]

cover description		curve numbers for hydrologic soil group			
cover type and hydrologic condition	average point impervious area[b]	A	B	C	D
fully developed urban areas (vegetation established)					
open space (lawns, parks, golf courses, cemeteries, etc.):[c]					
poor condition (grass cover < 50%)		68	79	86	89
fair condition (grass cover 50% to 75%)		49	69	79	84
good condition (grass cover > 75%)		39	61	74	80
impervious areas:					
paved parking lots, roofs, driveways, etc. (excluding right-of-way)		98	98	98	98
streets and roads: paved; curbs and storm sewers (excluding right-of-way)		98	98	98	98
paved; open ditches (including right-of-way)		83	89	92	93
gravel (including right-of-way)		76	85	89	91
dirt (including right-of-way)		72	82	87	89
western desert urban areas:					
natural desert landscaping (pervious areas only)[d]		63	77	85	88
artificial desert landscaping (impervious weed barrier, desert shrub with 1 in to 2 in sand or gravel mulch and basin borders)		96	96	96	96
urban districts:					
commercial and business	85	89	92	94	95
industrial	72	81	88	91	93
residential districts by average lot size:					
1/8 ac or less (town houses)	65	77	85	90	92
1/4 ac	38	61	75	83	87
1/3 ac	30	57	72	81	86
1/2 ac	25	54	70	80	85
1 ac	20	51	68	79	84
2 ac	12	46	65	77	82
developing urban areas					
newly graded areas (pervious areas only, no vegetation)[e]		77	86	91	94
idle lands (CNs are determined using cover types)					

[a]Average runoff condition, and $I_a = 0.2S$.
[b]The average percent impervious area shown was used to develop the composite CNs. Other assumptions are as follows: impervious areas are directly connected to the drainage system, impervious areas have a CN of 98, and pervious areas are considered equivalent to open space in good hydrologic condition. CNs for other combinations of conditions may be calculated using *TR-55* Fig. 2-3 or Fig. 2-4.
[c]CNs shown are equivalent to those of pasture. Composite CNs may be calculated for other combinations of open space cover type.
[d]Composite CNs for natural desert landscaping should be calculated using *TR-55* Fig. 2-3 or Fig. 2-4 based on the impervious area percentage (CN = 98) and the pervious area CN. The pervious area CNs are assumed equivalent to desert shrub in poor hydrologic condition.
[e]Composite CNs to use for the design of temporary measures during grading and construction should be computed using *TR-55* Fig. 2-3 or Fig. 2-4 based on the degree of development (impervious area percentage) and the CNs for the newly graded pervious areas.

Reprinted from *TR-55: Urban Hydrology for Small Watersheds*. NRCS. Table 2-2a. 1986. USDA.

Table 1.14 Runoff Curve Numbers for Cultivated Agricultural Lands[a]

cover description			curve numbers for hydrologic soil group			
cover type	treatment[b]	hydrologic condition[c]	A	B	C	D
fallow	bare soil	–	77	86	91	94
	crop residue cover (CR)	poor	76	85	90	93
		good	74	83	88	90
row crops	straight row (SR)	poor	72	81	88	91
		good	67	78	85	89
	SR + CR	poor	71	80	87	90
		good	64	75	82	85
	contoured (C)	poor	70	79	84	88
		good	65	75	82	86
	C + CR	poor	69	78	83	87
		good	64	74	81	85
	contoured & terraced (C&T)	poor	66	74	80	82
		good	62	71	78	81
	C&T + CR	poor	65	73	79	81
		good	61	70	77	80
small grain	SR	poor	65	76	84	88
		good	63	75	83	87
	SR + CR	poor	64	75	83	86
		good	60	72	80	84
	C	poor	63	74	82	85
		good	61	73	81	84
	C + CR	poor	62	73	81	84
		good	60	72	80	83
	C&T	poor	61	72	79	82
		good	59	70	78	81
	C&T + CR	poor	60	71	78	81
		good	58	69	77	80
close-seeded or broadcast legumes or rotation meadow	SR	poor	66	77	85	89
		good	58	72	81	85
	C	poor	64	75	83	85
		good	55	69	78	83
	C&T	poor	63	73	80	83
		good	51	67	76	80

[a]Average runoff condition, and $I_a = 0.2S$.
[b]Crop reside cover applies only if residue is on at least 5% of the surface throughout the year.
[c]Hydraulic condition is based on a combination of factors that affect infiltration and runoff, including (a) density and canopy of vegetative areas, (b) amount of year-round cover, (c) amount of grass or close-seeded legumes, (d) percentage of residue cover on the land surface (good 20%), and (e) degree of surface roughness.

Poor: Factors impair infiltration and tend to increase runoff.
Good: Factors encourage average and better than average infiltration and tend to decrease runoff.

Reprinted from *TR-55: Urban Hydrology for Small Watersheds*. NRCS. Table 2-2b. 1986. USDA.

Water Resources

Table 1.15 *Runoff Curve Numbers for Other Agricultural Lands*[a]

cover description		curve numbers for hydrologic soil group			
cover type	hydrologic condition	A	B	C	D
pasture, grassland, or	poor	68	79	86	89
range—continuous	fair	49	69	79	84
forage for grazing[b]	good	39	61	74	80
meadow—continuous grass, protected from grazing and generally mowed for hay	–	30	58	71	78
brush—brush-weed-grass	poor	48	67	77	83
mixture with brush	fair	35	56	70	77
the major element[c]	good	30[d]	48	65	73
woods—grass	poor	57	73	82	86
combination (orchard	fair	43	65	76	82
or tree farm)[e]	good	32	58	72	79
woods[f]	poor	45	66	77	83
	fair	36	60	73	79
	good	30[d]	55	70	77
farmsteads—buildings, lanes, driveways, and surrounding lots	–	59	74	82	86

[a]Average runoff condition, and $I_a = 0.2S$.
[b]Poor: 50% ground cover or heavily grazed with no mulch.
Fair: 50% to 75% ground cover and not heavily grazed.
Good: 75% ground cover and lightly or only occasionally grazed.
[c]Poor: 50% ground cover.
Fair: 50% to 75% ground cover.
Good: 75% ground cover.
[d]Actual curve number is less than 30; use CN = 30 for runoff calculations.
[e]CNs shown were calculated for areas with 50% woods and 50% grass (pasture) cover. Other combinations of conditions may be calculated from the CNs for woods and pasture.
[f]Poor: Forest litter, small trees, and brush are destroyed by heavy grazing or regular burning.
Fair: Woods are grazed but not burned, and some forest litter covers the soil.
Good: Woods are protected from grazing, and litter and brush adequately cover the soil.

Reprinted from *TR-55: Urban Hydrology for Small Watersheds.* NRCS. Table 2-2c. 1986. USDA.

NRCS *TR-55* Applications of the USDA-NRCS Runoff Curve Number Method

The NRCS runoff curve number (CN) method for calculating runoff for watershed areas larger than 0.3 mi^2 ($\approx$ 200 ac) and less than 10 mi^2 is described in the *TR-55*. The *TR-55* provides the following three important worksheets to calculate runoff using the NRCS method.

1. Figure 1.16 (*TR-55* Worksheet 3) calculates the time of concentration, t_c.

2. Figure 1.19 (*TR-55* Worksheet 2) calculates the curve number, CN, for various soil groups and land cover conditions. Column 1 lists the soil name and hydrologic soil group. Column 2 lists the cover description of the land use or land cover in the watershed. Column 3 lists the curve number for the particular watershed depending on the land cover. The curve number is found in Table 1.13 through Table 1.15. Column 4 lists the area of each land use. Column 5 lists the product of the CN times the area. Figure 1.19 is then used to estimate the composite curve number for the particular watershed.

3. Figure 1.20 (*TR-55* Worksheet 4) calculates the graphical peak discharge, q_p.

The following parameters are needed to calculate the peak discharge using Fig. 1.20.

- Drainage area, A_{mi^2}, is the watershed area of the project.

- Runoff curve number, CN, is calculated from Fig. 1.19 (*TR-55* Worksheet 2).

- Time of concentration, t_c (in hours), is calculated from Fig. 1.16 (*TR-55* Worksheet 3).

- Rainfall distribution (Type I, II, or III) is determined using the map of the United States depicted in Fig. 1.7.

Water Resources

Figure 1.19 *Calculating Curve Number and Runoff*

Project		By		Date
Location		Checked		Date

Check one: ☐ Present   ☐ Developed

1. Runoff curve number

soil name and hydrologic group Table 1.9	cover description (cover type, treatment, and hydrologic condition; percent impervious; unconnected/connected impervious area ratio)	CN[1] Table 1.13 / Table 1.14 / Table 1.15			area ☐ ac ☐ mi² ☐ %	product of CN × area

[1]Use only one CN source per line.

Totals

$\text{CN (weight)} = \dfrac{\text{total product}}{\text{total area}} = \underline{\hspace{2cm}} = \underline{\hspace{2cm}}$ Use CN

2. Runoff

	storm #1	storm #2	storm #3
frequency.. yr			
rainfall, P (24 hr)..in			
runoff, Q_{in} ....in			

(Use *P* and CN with Fig. 1.20 or Eq. 1.11 and Eq. 1.17)

Reprinted from *TR-55: Urban Hydrology for Small Watersheds*. NRCS. Worksheet 2. 1986. USDA.

- Pond and swamp areas are calculated as a percentage of the drainage area, A_{mi^2}.

- Frequency is the design storm return interval for which the calculation is being made. The frequency can range from 2 yr to 500 yr, depending on the item being designed, as shown in Table 1.4.

- Rainfall, P, is determined from a precipitation map, such as Fig. 1.6 for a 24 hr storm.

- Initial abstraction, I_a, is calculated from Table 1.16 for a particular CN.

- I_a/P is the initial abstraction divided by the 24 hr rainfall.

- Unit peak discharge, q_u (in cubic feet per second-square mile-inch), is determined from Fig. 1.21.

- Runoff, Q_{in} (in inches), is read from Table 1.17 or Fig. 1.22. Runoff can also be calculated by Eq. 1.18.

$$Q_{\text{in}} = \frac{(P - 0.2S)^2}{P + 0.8S} \qquad 1.18$$

P. 1-24

- Find the potential maximum retention after runoff begins, S, using Eq. 1.16.

- The pond and swamp adjustment factor, F_p, is 1.0 if there are no ponds or swamps in the watershed.

- The peak discharge, 100 yr runoff, q_p (in cubic feet per second), is found from Eq. 1.19.

$$q_p = q_u A_{\text{mi}^2} Q_{\text{in}} F_p \qquad 1.19$$

TR-55: table 4-2 F_p

Water Resources

Figure 1.20 Calculating Peak Discharge

Project	By		Date
Location	Checked		Date

Check one: ☐ Present   ☐ Developed

1. data

   drainage area .... A_{mi^2} = _____ mi² (acres/640)

   runoff curve number .... CN = _____ (from Fig. 1.19)

   time of concentration .... t_c = _____ hr (from Fig. 1.16)

   rainfall distribution .... = _____ (Fig. 1.7)

   pond and swamp areas spread
   throughout watershed........................ = _____ percent of A_{mi^2} (___acres or mi² covered)

	storm #1	storm #2	storm #3
2. frequency.. yr			
3. rainfall, P (24 hr).................................in			
4. initial abstraction, I_a ....in (Use CN with Table 1.16)			
5. compute I_a/P..			
6. unit peak discharge, q_u ....ft³/sec-mi²-in (Use t_c and I_a/P with Fig. 1.21)			
7. runoff, Q_{in}....................................in (From Worksheet 2) Fig. 1.22			
8. pond and swamp adjustment factor, F_p .... (Factor is 1.0 for 0% pond and swamp area.)			
9. peak discharge, q_p..........................ft³/sec (Where $q_p = q_u A_{mi^2} Q_{in} F_p$.)			

Reprinted from *TR-55: Urban Hydrology for Small Watersheds.* NRCS. Worksheet 4. 1986. USDA.

Example 1.15

A 1000 ac watershed in Wilmington, Delaware contains 200 ac single family residential (1/4 ac) lots, 300 ac of row crops in good condition, and 500 ac of wooded area in good condition. According to the county NRCS soil mapping, all soils in the watershed are Downer soils, HSG B. The time of concentration is 0.5 hr. The watershed has a Type II rainfall distribution and is not covered by wetlands. Use the NRCS CN method and Fig. 1.19 to calculate the 100 yr runoff for the watershed.

Solution

Using Fig. 1.19 and Table 1.13, calculate the composite CN.

soil name/ HSG	land cover	CN	area, A (ac)	product (CN × A)
Downer/B	SF residential (1/4 ac)	75	200	15,000
Downer/B	row crop, good condition	78	300	23,400
Downer/B	wooded, good condition	55	500	27,500
	total		1000	65,900

The composite CN is

$$CN = \frac{65{,}900 \text{ ac}}{1000 \text{ ac}} = 65.9 \quad (66)$$

Using Fig. 1.20, calculate the NRCS peak runoff.

step 1: data

drainage area, A_{mi^2}

$$A_{\mathrm{mi}^2} = \frac{1000 \text{ ac}}{640 \ \dfrac{\text{ac}}{\text{mi}^2}} = 1.56 \text{ mi}^2$$

runoff curve number, CN = 66

time of concentration, $t_c = 0.5$ hr

rainfall distribution = Type II

pond and swamp areas = 0% of A_{mi^2}

step 2: frequency = 100 yr

step 3: 24 hr rainfall, $P = 7.5$ in (interpolated from Fig. 1.6 for Wilmington, Delaware)

step 4: initial abstraction, $I_a = 1.030$ in (Table 1.16 for CN 66)

step 5: calculate I_a/P

$$\frac{I_a}{P} = \frac{1.030 \text{ in}}{7.5 \text{ in}} = 0.14$$

step 6: unit peak discharge, $q_u = 500$ ft^3/sec-mi^2-in (see Fig. 1.21, for a t_c of 0.5 hr and an I_a/P of 0.14)

Table 1.16 I_a Values for Runoff Curve Numbers

curve number	I_a (in)	curve number	I_a (in)
40	3.000	70	0.857
41	2.878	71	0.817
42	2.762	72	0.778
43	2.651	73	0.740
44	2.545	74	0.703
45	2.444	75	0.667
46	2.348	76	0.632
47	2.255	77	0.597
48	2.167	78	0.564
49	2.082	79	0.532
50	2.000	80	0.500
51	1.922	81	0.469
52	1.846	82	0.439
53	1.774	83	0.410
54	1.704	84	0.381
55	1.636	85	0.353
56	1.571	86	0.326
57	1.509	87	0.299
58	1.448	88	0.273
59	1.390	89	0.247
60	1.333	90	0.222
61	1.279	91	0.198
62	1.226	92	0.174
63	1.175	93	0.151
64	1.125	94	0.128
65	1.077	95	0.105
66	1.030	96	0.083
67	0.985	97	0.062
68	0.941	98	0.041
69	0.899		

(Multiply in by 2.54 to obtain cm.)

Reprinted from *TR-55: Urban Hydrology for Small Watersheds.* NRCS. 1986. USDA.

Figure 1.21 Unit Peak Discharge

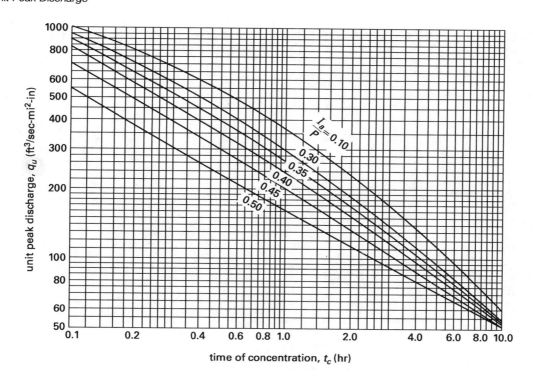

Reprinted from *TR-55: Urban Hydrology for Small Watersheds.* NRCS. Exhibit 4-II. 1986. USDA.

Table 1.17 *Runoff Depth for Selected Curve Numbers*

rainfall (in)	curve number												
	40	45	50	55	60	65	70	75	80	85	90	95	98
	runoff depth (in)												
1.0	0.00	0.00	0.00	0.00	0.00	0.00	0.00	0.03	0.08	0.17	0.32	0.56	0.79
1.2	0.00	0.00	0.00	0.00	0.00	0.00	0.03	0.07	0.15	0.27	0.46	0.74	0.99
1.4	0.00	0.00	0.00	0.00	0.00	0.02	0.06	0.13	0.24	0.39	0.61	0.92	1.18
1.6	0.00	0.00	0.00	0.00	0.01	0.05	0.11	0.20	0.34	0.52	0.76	1.11	1.38
1.8	0.00	0.00	0.00	0.00	0.03	0.09	0.17	0.29	0.44	0.65	0.93	1.29	1.58
2.0	0.00	0.00	0.00	0.02	0.06	0.14	0.24	0.38	0.56	0.80	1.09	1.48	1.77
2.5	0.00	0.00	0.02	0.08	0.17	0.30	0.46	0.65	0.89	1.18	1.53	1.96	2.27
3.0	0.00	0.02	0.09	0.19	0.33	0.51	0.71	0.96	1.25	1.59	1.98	2.45	2.77
3.5	0.02	0.08	0.20	0.35	0.53	0.75	1.01	1.30	1.64	2.02	2.45	2.94	3.27
4.0	0.06	0.18	0.33	0.53	0.76	1.03	1.33	1.67	2.04	2.46	2.92	3.43	3.77
4.5	0.14	0.30	0.50	0.74	1.02	1.33	1.67	2.05	2.46	2.91	3.40	3.92	4.26
5.0	0.24	0.44	0.69	0.98	1.30	1.65	2.04	2.45	2.89	3.37	3.88	4.42	4.76
6.0	0.50	0.80	1.14	1.52	1.92	2.35	2.81	3.28	3.78	4.30	4.85	5.41	5.76
7.0	0.84	1.24	1.68	2.12	2.60	3.10	3.62	4.15	4.69	5.25	5.82	6.41	6.76
8.0	1.25	1.74	2.25	2.78	3.33	3.89	4.46	5.04	5.63	6.21	6.81	7.40	7.76
9.0	1.71	2.29	2.88	3.49	4.10	4.72	5.33	5.95	6.57	7.18	7.79	8.40	8.76
10.0	2.23	2.89	3.56	4.23	4.90	5.56	6.22	6.88	7.52	8.16	8.78	9.40	9.76
11.0	2.78	3.52	4.26	5.00	5.72	6.43	7.13	7.81	8.48	9.13	9.77	10.39	10.76
12.0	3.38	4.19	5.00	5.79	6.56	7.32	8.05	8.76	9.45	10.11	10.76	11.39	11.76
13.0	4.00	4.89	5.76	6.61	7.42	8.21	8.98	9.71	10.42	11.10	11.76	12.39	12.76
14.0	4.65	5.62	6.55	7.44	8.30	9.12	9.91	10.67	11.39	12.08	12.75	13.39	13.76
15.0	5.33	6.36	7.35	8.29	9.19	10.04	10.85	11.63	12.37	13.07	13.74	14.39	14.76

(Multiply in by 2.54 to obtain cm.)

*Interpolate the values shown to obtain runoff depths for CNs or rainfall amounts not shown.

Reprinted from *TR-55: Urban Hydrology for Small Watersheds*. NRCS. Table 2-1. 1986. USDA.

step 7: runoff, $Q_{in} = 3.6$ in (see Fig. 1.22 for a P of 7.5 in and a CN of 66)

step 8: pond and swamp adjustment factor, $F_p = 1.0$

step 9: peak discharge, q_p

$$q_p = q_u A_{mi^2} Q_{in} F_p$$
$$= \left(500 \; \frac{ft^3}{sec\text{-}mi^2\text{-}in}\right)(1.56 \text{ mi}^2)(3.6 \text{ in})(1.0)$$
$$= 2808 \text{ ft}^3/sec$$

Figure 1.22 *Solution of Runoff Equation*

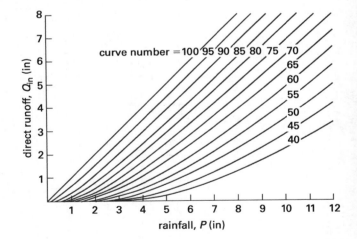

Reprinted from *TR-55: Urban Hydrology for Small Watersheds*. NRCS. Figure 2-1. 1986. USDA.

12. GAUGING STATIONS

Along streams for which the USGS has established stream gauges, discharges can be estimated at various locations upstream and downstream using the ratio of discharge to drainage area shown in Eq. 1.20.

$$\frac{Q_s}{A_s} = \frac{Q_g}{A_g}$$

$$Q_s = \frac{Q_g A_s}{A_g}$$

1.20

A_g: stream gauge area for
A_s: area along stream
Q_s: flow along stream

Example 1.16

The 100 yr peak discharge at the USGS stream gauge along the Red River is 200 ft³/sec. The stream at that location has a drainage area of 150 mi². Determine the 100 yr peak discharge at a site upstream where the drainage area is 100 mi².

Solution

Use Eq. 1.20.

$$Q_s = \frac{Q_g A_s}{A_g} = \frac{\left(200 \ \frac{\text{ft}^3}{\text{sec}}\right)(100 \ \text{mi}^2)}{150 \ \text{mi}^2}$$

$$= 133 \ \text{ft}^3/\text{sec}$$

13. SEDIMENTATION

Sedimentation diminishes the flood-carrying capacity of streams and waterways. Sedimentation can also pollute streams and damage the stream bottom habitat by increasing turbidity.

The *total suspended solids* (TSS) loads produced by a watershed can be estimated by the simple method developed by T. R. Schueler in 1987. The annual pollutant load, L (in units mass per year), is calculated from the drainage area, A (in acres), the mean annual precipitation, P (in inches), the runoff coefficient, R_s, from Table 1.18, and the mean total suspended solids (TSS) pollutant concentration, C (in milligrams per liter). The conversion factor 0.226, in units of liter-pound per acre-inch-milligram, converts to units of pounds per year.

$$L = 0.226 A P R_s C \qquad 1.21$$

SCHUELER METHOD:
USED TO ESTIMATE SEDIMENT
LOADING FOR DIFFERENT
TYPES OF LAND USE

Table 1.18 Total Suspended Solids Load Variables

land use	mean TSS concentration, C (mg/L)	runoff coefficient, R_s
single family residential (1/2 ac per dwelling unit, du, and greater)	140	0.30
high density/multi-family residential	180	0.65
office	175	0.60
industrial	251	0.72
transportation/utility	350	0.90
commercial	168	0.85
institutional	128	0.55
open space/parks	20	0.20
wooded/forested	20	0.20
agriculture	300	0.30

Adapted from T. R. Schueler. *Controlling Urban Runoff: A Practical Manual for Planning and Designing Urban BMPs.* Publication No. 87703. 1987. Metropolitan Washington Council of Governments.

Example 1.17

Calculate the TSS load for a 198 ac watershed in Boston with the following land covers: 99 ac of forest, 16 ac of residential with 1/2 ac lots, and 83 ac of agricultural hay.

Solution

Identify the values of R_s and C from Table 1.18. From Table 1.2, the annual precipitation for Boston is 41.5 in/yr. Calculate the sediment loads for each of the land uses according to the simplified method, Eq. 1.21.

$$L = 0.226 A P R_s C$$

$$L_{\text{forest}} = \left(0.226 \ \frac{\text{L-lbm}}{\text{ac-in-mg}}\right)(99 \ \text{ac})\left(41.5 \ \frac{\text{in}}{\text{yr}}\right)$$
$$\times (0.20)\left(20 \ \frac{\text{mg}}{\text{L}}\right)$$
$$= 3714 \ \text{lbm/yr}$$

$$L_{\text{residential}} = \left(0.226 \ \frac{\text{L-lbm}}{\text{ac-in-mg}}\right)(16 \ \text{ac})\left(41.5 \ \frac{\text{in}}{\text{yr}}\right)$$
$$\times (0.30)\left(140 \ \frac{\text{mg}}{\text{L}}\right)$$
$$= 6303 \ \text{lbm/yr}$$

$$L_{\text{agriculture}} = \left(0.226 \ \frac{\text{L-lbm}}{\text{ac-in-mg}}\right)(83 \ \text{ac})\left(41.5 \ \frac{\text{in}}{\text{yr}}\right)$$
$$\times (0.30)\left(300 \ \frac{\text{mg}}{\text{L}}\right)$$
$$= 70{,}061 \ \text{lbm/yr}$$

The TSS load is the sum of the forest, residential, and agricultural loads.

$$\text{TSS load} = L_{\text{forest}} + L_{\text{residential}} + L_{\text{agriculture}}$$

$$= 3714 \ \frac{\text{lbm}}{\text{yr}} + 6303 \ \frac{\text{lbm}}{\text{yr}} + 70{,}061 \ \frac{\text{lbm}}{\text{yr}}$$

$$= 80{,}078 \ \text{lbm/yr}$$

The unit TSS load is the TSS load per acre.

$$\text{unit TSS load} = \frac{80{,}078 \ \dfrac{\text{lbm}}{\text{yr}}}{198 \ \text{ac}}$$

$$= 404 \ \text{lbm/ac-yr}$$

14. EROSION

Soil is eroded by the movement of water and wind over the surface of the ground. *Erosion* starts with sheet flow over the surface from a watershed area of as little as 10 ac. When the watershed area starts to exceed about 20 ac, the runoff has enough tractive force to carve out rills and rivulets, causing shallow concentrated flow. When the watershed area starts to exceed 40 ac, the flow is large enough to carve out incised stream channels. Over 90% of the annual soil erosion in a watershed occurs with large storms exceeding 1 in of precipitation.

Depending on the watershed, soil erosion rates can range between 200 lbm/ac-yr and 2000 lbm/ac-yr. Soil transport in streams varies with the square of the channel velocity. Soil erosion potential is based on four factors.

- soil type
- land use/land cover
- topography/slope
- climate

Most counties and states in the United States have adopted soil erosion and sediment control regulations in addition to the standards developed by the USDA NRCS and the local soil conservation districts. Construction sites and agricultural areas tend to produce the highest soil erosion rates. Typical soil erosion controls at construction sites include hay bales, silt fences, grass filter swales, sediment control ponds, and vegetative ground cover. Soil erosion and sediment loads can be estimated for watersheds using the simplified method, Eq. 1.21.

Soil Type

Soil types differ in the following factors that affect erodibility.

- soil texture (fine, medium, coarse)
- organic material
- structure
- permeability

Well-drained sandy or gravel soils with high permeability and interlocked particle texture are less prone to erosion. Fine sand and silts are the most erodible soils. Soils become less erodible with increasing clay content.

Land Use/Land Cover

Forests and grasslands have low erosion potential compared to unstabilized agricultural lands, developing suburban/urban land, and wooded areas. Erosion occurs when soil is left bare. Therefore, the most effective soil erosion control measure is vegetative ground cover, such as trees, shrubs, plants, and grass turf.

Topography/Slope

Erosion is highest in large, steeply sloped watersheds. As the length of the stream increases, the flow increases, which increases the erosion potential. Soil erosion increases with watershed slope and area.

Climate

The frequency, intensity, and duration of rainfall are factors in determining the rate of runoff, and hence, erosion. Frequent, high intensity, long duration storms which exceed at least 1 in of precipitation produce large amounts of soil erosion. When the ground is frozen or snow covered, erosion is temporally abated. However, when the ground thaws or the snow melts, soils are again particularly prone to erosion. Erosion at construction sites or farms is particularly severe during the spring thaw if the soils are not stabilized with vegetative cover.

PRACTICE PROBLEMS

1. In the mid-Atlantic region of the United States, a meteorological station records the following readings in one year: the annual precipitation is 45 in, infiltration is 10 in, evapotranspiration is 24 in, and change in moisture storage is 0 in. Calculate annual runoff in acre-feet if the watershed area is 1 mi^2.

2. The hydrograph shown is associated with a watershed area of 1.0 mi^2. Define the unit hydrograph for the hydrograph, then define the hydrograph for the watershed for a 3.0 in storm.

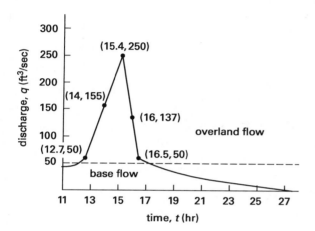

3. Define a synthetic unit triangular hydrograph in which the peak runoff is estimated as 150 ft^3/sec and the time of concentration is 2.0 hr.

4. Use the rational method to estimate the 10 yr runoff from a 200 ac watershed developed for neighborhood business land use. The dimensions for the time of concentration are the following.

> sheet flow
>> length: 300 ft
>> surface description: dense grass
>> 2 yr, 24 hr rainfall: 3.2 in
>> slope: 0.02 ft/ft
> shallow concentrated flow
>> surface: unpaved
>> flow length: 500 ft
>> slope: 0.1 ft/ft
> channel flow
>> velocity: 3 ft/sec
>> length: 2000 ft

5. A 500 ac watershed in Delaware comprises 100 ac single family residential (1/3 ac lots), 100 ac of row crops in good condition, and 300 ac of wooded area in good condition. The time of concentration is 1.0 hr. Assume a Type II rainfall distribution and that none of the watershed is covered by wetlands. According to

the county NRCS soil mapping, all soils in the watershed are Downer soils (HSG B). Calculate the 100 yr runoff using the NRCS CN method.

6. The 100 yr peak discharge at the USGS stream gauge along the White River is 10,000 ft^3/sec, and the stream gauge has a drainage area of 900 mi^2. Determine the 100 yr peak discharge at a site along the White River with a drainage area of 1000 mi^2.

7. Calculate the total suspended sediment load for a 1000 ac watershed with the following land covers: forest (500 ac), residential 1/2 ac lots (200 ac), and agricultural hay (300 ac). The normal annual precipitation is 50 in/yr.

Problem 8 and Prob. 9 refer to the following illustration of a 500 ac Wilmington, Delaware watershed.

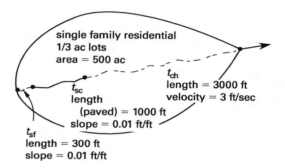

8. Given a Manning coefficient of 0.03, and single family residential (1/3 ac lots) land use, underlain by HSG B soils, calculate the 100 yr peak discharge for the watershed using the rational method.

9. Assuming there are no ponds or swamps in the watershed, and assuming that the watershed has a total time to concentration of 0.56 hr, calculate the 100 yr peak discharge for Wilmington, Delaware, using the TR55 method.

10. If the annual precipitation is 40 in, the runoff is 10 in, and evapotranspiration is 22 in, what is the infiltration?

(A) 8 in

(B) 30 in

(C) 50 in

(D) 70 in

11. What is the 100 yr, 24 hr precipitation depth for Atlanta, Georgia?

(A) 6 in

(B) 7 in

(C) 8 in

(D) 10 in

12. What is the maximum incremental intensity of a storm based on the following precipitation data?

hour of storm	incremental precipitation depth (in)
0.0	0.0
0.2	0.1
0.4	0.2
0.6	0.5
0.8	0.2
1.0	0.1

(A) 0.50 in/hr

(B) 1.0 in/hr

(C) 1.5 in/hr

(D) 2.5 in/hr

13. What is the percentage probability of a 50 yr storm occurring once in any given year?

(A) 2%

(B) 20%

(C) 50%

(D) 200%

14. What is the 100 yr flow from a 200 ac watershed in a downtown area where the time of concentration is 0.5 hr?

(A) 80 ft^3/sec

(B) 600 ft^3/sec

(C) 800 ft^3/sec

(D) 1000 ft^3/sec

15. In Delaware, what is the time of concentration for a 300 ft long sheet flow component over dense grass with a slope of 0.07 ft/ft?

(A) 0.2 hr

(B) 0.3 hr

(C) 0.5 hr

(D) 0.7 hr

16. What is the runoff curve number for a 1000 ac watershed that is 50% single family residential (1/2 ac lots) and 50% open space (grass cover exceeds 75%), assuming HSG C soils?

(A) 74

(B) 75

(C) 76

(D) 77

17. A 10 yr discharge of 550 ft^3/sec is measured at the stream gauge that drains 52 mi^2. What is the 10 yr peak discharge along a stream from a watershed area of 88 mi^2?

(A) 8.3 ft^3/sec

(B) 330 ft^3/sec

(C) 650 ft^3/sec

(D) 930 ft^3/sec

SOLUTIONS

1. Rearrange the water budget equation, Eq. 1.1, and solve for runoff, R.

$$R = P - I - \text{ET} + \Delta S$$

$$= \left(\frac{45 \text{ in} - 10 \text{ in} - 24 \text{ in} + 0 \text{ in}}{12 \frac{\text{in}}{\text{ft}}} \right)$$

$$\times (1 \text{ mi}^2) \left(640 \frac{\text{ac}}{\text{mi}^2} \right)$$

$$= 587 \text{ ac-ft}$$

2. Find the average precipitation for the watershed. Use Eq. 1.4.

$$V = \frac{\Delta t \Delta q}{2}$$

$$= \frac{(16.5 \text{ hr} - 12.7 \text{ hr}) \left(250 \frac{\text{ft}^3}{\text{sec}} - 50 \frac{\text{ft}^3}{\text{sec}} \right)}{\times \left(60 \frac{\text{min}}{\text{hr}} \right) \left(60 \frac{\text{sec}}{\text{min}} \right)}{2}$$

$$= 1{,}368{,}000 \text{ ft}^3$$

Use Eq. 1.5.

$$P_{\text{ave,excess}} = \frac{V}{A_d}$$

$$= \frac{(1{,}368{,}000 \text{ ft}^3) \left(12 \frac{\text{in}}{\text{ft}} \right)}{(1.0 \text{ mi}^2) \left(640 \frac{\text{ac}}{\text{mi}^2} \right) \left(43{,}560 \frac{\text{ft}^2}{\text{ac}} \right)}$$

$$= 0.6 \text{ in}$$

The coordinates for the 1.0 in unit hydrograph are calculated by dividing each runoff point (column 2) by 0.6 in, as shown in column 3 of the following table.

The hydrograph for a 3.0 in storm in the watershed is calculated by multiplying the unit hydrograph runoff (column 3) by 3.0 in, shown in column 4 of the following table.

time (hr)	0.6 in hydrograph runoff (ft³/sec)	unit hydrograph runoff (ft³/sec-in)	3.0 in hydrograph runoff (ft³/sec)
12.7	50	$50/0.6 = 83.33$	$(83.33)(3.0) = 250$
14	155	$155/0.6 = 258.3$	$(258.3)(3.0) = 775$
15.4	250	$250/0.6 = 416.7$	$(416.7)(3.0) = 1250$
16	137	$137/0.6 = 228.3$	$(228.3)(3.0) = 685$
16.5	50	$50/0.6 = 83.33$	$(83.33)(3.0) = 250$

3. The peak runoff is plotted at 150 ft³/sec.

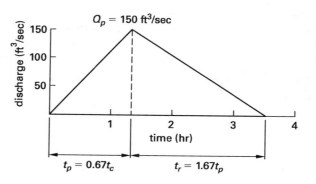

Using Eq. 1.6, the ascending limb of the hydrograph is found from the time to peak.

$$t_p = 0.67 t_c = (0.67)(2.0 \text{ hr})$$

$$= 1.34 \text{ hr}$$

Use Eq. 1.7 to find the receding limb of the hydrograph.

$$t_r = 1.67 t_p = (1.67)(1.34 \text{ hr})$$

$$= 2.2 \text{ hr}$$

4. To compute the time of concentration, use Fig. 1.16. According to Eq. 1.10,

$$t_c = t_{\text{sf}} + t_{\text{sc}} + t_{\text{ch}}$$

The travel time for the sheet flow component, no more than 300 ft long, is given by Eq. 1.11, using the Manning coefficient from Table 1.12.

$$t_{\text{sf}} = \frac{(0.007 \text{ hr})(nL)^{0.8}}{P_2{}^{0.5} S^{0.4}}$$

$$= \frac{(0.007 \text{ hr}) \left((0.24)(300 \text{ ft}) \right)^{0.8}}{(3.2 \text{ in})^{0.5} \left(0.02 \frac{\text{ft}}{\text{ft}} \right)^{0.4}}$$

$$= 0.572 \text{ hr}$$

The travel time for the shallow concentrated flow component is given by Eq. 1.12. Based on a slope of 0.1 ft/ft, an unpaved surface, and Fig. 1.18, velocity, v, is 5 ft/sec.

$$t_{sc} = \frac{L}{v}$$
$$= \frac{500 \text{ ft}}{\left(3600 \frac{\text{sec}}{\text{hr}}\right)\left(5 \frac{\text{ft}}{\text{sec}}\right)}$$
$$= 0.0275 \text{ hr}$$

The travel time for the channel flow component is given by Eq. 1.13.

$$t_{ch} = \frac{L}{v}$$
$$= \frac{2000 \text{ ft}}{\left(3600 \frac{\text{sec}}{\text{hr}}\right)\left(3 \frac{\text{ft}}{\text{sec}}\right)}$$
$$= 0.185 \text{ hr}$$

The time of concentration is given by Eq. 1.10.

$$t_c = t_{sf} + t_{sc} + t_{ch}$$
$$= (0.573 \text{ hr} + 0.0278 \text{ hr} + 0.185 \text{ hr})\left(60 \frac{\text{min}}{\text{hr}}\right)$$
$$= 47.09 \text{ min} \quad (47 \text{ min})$$

From the IDF curve in Fig. 1.12, for a t_c of 47 min, the intensity, i, for a 10 yr storm is 2.6 in/hr. The area, A_d, is given as 200 ac. From Table 1.11, c is 0.6 for neighborhood land use. From Eq. 1.9,

$$Q_{10 \text{ yr}} = ciA_d$$
$$= \frac{(0.6)\left(2.6 \frac{\text{in}}{\text{hr}}\right)(200 \text{ ac})\left(43{,}560 \frac{\text{ft}^2}{\text{ac}}\right)}{\left(12 \frac{\text{in}}{\text{ft}}\right)\left(3600 \frac{\text{sec}}{\text{hr}}\right)}$$
$$= 314.6 \text{ ft}^3/\text{sec} \quad (315 \text{ ft}^3/\text{sec})$$

5. Using Fig. 1.19, calculate the composite CN. The individual CNs are found in Table 1.13 through Table 1.15.

soil name/HSG	land cover	CN	area, A (ac)	product, $CN \times A$ (ac)
Downer/B	SF residential, 1/3 ac	72	100	7200
Downer/B	row crop, good condition	78	100	7800
Downer/B	wooded, good condition	55	300	16,500
		total	500	31,500

The composite curve number is

$$CN = \frac{31{,}500 \text{ ac}}{500 \text{ ac}} = 63$$

Using Fig. 1.20, calculate the NRCS peak runoff.

step 1: data

drainage area, A_{mi^2}

$$A_{\text{mi}^2} = \frac{500 \text{ ac}}{640 \frac{\text{ac}}{\text{mi}^2}}$$
$$= 0.781 \text{ mi}^2$$

runoff curve number, CN = 63 (composite CN)

time of concentration, $t_c = 1.0$ hr (given)

rainfall distribution = Type II (given)

pond and swamp areas = 0% of A_{mi^2}

step 2: frequency = 100 yr

step 3: 24 hr rainfall, $P = 7.5$ in (see Fig. 1.6)

step 4: initial abstraction, $I_a = 1.175$ in (Table 1.16 for CN 63)

step 5: calculate I_a/P

$$\frac{I_a}{P} = \frac{1.175 \text{ in}}{7.5 \text{ in}} = 0.16$$

step 6: unit peak discharge, $q_u = 340$ ft³/sec-mi²-in (see Fig. 1.21 for a t_c of 1.0 hr and an I_a/P of 0.16)

step 7: runoff, $Q_{\text{in}} = 3.4$ in (see Fig. 1.22, for a P of 7.5 in and CN 63)

step 8: pond and swamp adjustment factor, $F_p = 1.0$

step 9: peak discharge, q_p

$$q_p = q_u A_{\text{mi}^2} Q_{\text{in}} F_p$$
$$= \left(340 \frac{\text{ft}^3}{\text{sec-mi}^2\text{-in}}\right)(0.781 \text{ mi}^2)(3.4 \text{ in})(1.0)$$
$$= 903 \text{ ft}^3/\text{sec}$$

6. The 100 yr flow at the site is given by Eq. 1.20.

$$Q_s = \frac{Q_g A_s}{A_g}$$
$$= \frac{\left(10{,}000 \frac{\text{ft}^3}{\text{sec}}\right)(1000 \text{ mi}^2)}{900 \text{ mi}^2}$$
$$= 11{,}111 \text{ ft}^3/\text{sec}$$

7. Calculate the sediment loads for each of the land uses according to the simplified method. Substitute the given precipitation and acreages, as well as values of the runoff coefficient, R_s, and the mean TSS concentration, C, from Table 1.18, into Eq. 1.21.

$$L = 0.226APR_sC$$

$$L_{\text{forest}} = \left(0.226 \; \frac{\text{L-lbm}}{\text{ac-in-mg}}\right)(500 \text{ ac})\left(50 \; \frac{\text{in}}{\text{yr}}\right)$$
$$\times (0.20)\left(20 \; \frac{\text{mg}}{\text{L}}\right)$$
$$= 22{,}600 \text{ lbm/yr}$$

$$L_{\text{residential}} = \left(0.226 \; \frac{\text{L-lbm}}{\text{ac-in-mg}}\right)(200 \text{ ac})\left(50 \; \frac{\text{in}}{\text{yr}}\right)$$
$$\times (0.30)\left(140 \; \frac{\text{mg}}{\text{L}}\right)$$
$$= 94{,}920 \text{ lbm/yr}$$

$$L_{\text{agriculture}} = \left(0.226 \; \frac{\text{L-lbm}}{\text{ac-in-mg}}\right)(300 \text{ ac})\left(50 \; \frac{\text{in}}{\text{yr}}\right)$$
$$\times (0.30)\left(300 \; \frac{\text{mg}}{\text{L}}\right)$$
$$= 305{,}100 \text{ lbm/yr}$$

The TSS load is

$$L_{\text{forest}} + L_{\text{residential}} + L_{\text{agriculture}}$$
$$= 22{,}600 \; \frac{\text{lbm}}{\text{yr}} + 94{,}920 \; \frac{\text{lbm}}{\text{yr}} + 305{,}100 \; \frac{\text{lbm}}{\text{yr}}$$
$$= 422{,}620 \text{ lbm/yr}$$

The unit TSS load is the TSS load per acre.

$$\frac{422{,}620 \; \frac{\text{lbm}}{\text{yr}}}{1000 \text{ ac}} = 423 \text{ lbm/ac-yr}$$

8. To find the intensity, calculate the time of concentration. Taking the time of concentration parameters from the illustration,

    sheet flow
        slope: 0.01 ft/ft
        Manning coefficient: 0.03 (given)
        length: 300 ft
        precipitation: 3.5 in (Wilmington,
        Delaware, 2 yr, 24 hr rainfall from
        Fig. 1.17)
    shallow concentrated flow
        slope: 0.01 ft/ft (paved)
        length: 1000 ft
    channel flow
        velocity: 3 ft/sec (given)
        length: 3000 ft

Using Fig. 1.16 and Eq. 1.11, determine time of concentration, t_c.

$$t_{\text{sf}} = \frac{(0.007 \text{ hr})(nL)^{0.8}}{P_2{}^{0.5}S^{0.4}}$$
$$= \frac{(0.007 \text{ hr})\big((0.03)(300 \text{ ft})\big)^{0.8}}{(3.5 \text{ in})^{0.5}\left(0.01 \; \frac{\text{ft}}{\text{ft}}\right)^{0.4}}$$
$$= 0.137 \text{ hr}$$

For a slope, S, of 0.01 ft/ft on a paved surface, velocity, v, is 2 ft/sec (see Fig. 1.18). From Eq. 1.12,

$$t_{\text{sc}} = \frac{L}{\text{v}} = \frac{1000 \text{ ft}}{\left(3600 \; \frac{\text{sec}}{\text{hr}}\right)\left(2 \; \frac{\text{ft}}{\text{sec}}\right)}$$
$$= 0.139 \text{ hr}$$

From Eq. 1.13,

$$t_{\text{ch}} = \frac{L}{\text{v}} = \frac{3000 \text{ ft}}{\left(3600 \; \frac{\text{sec}}{\text{hr}}\right)\left(3 \; \frac{\text{ft}}{\text{sec}}\right)}$$
$$= 0.278 \text{ hr}$$

According to Eq. 1.10,

$$t_c = t_{\text{sf}} + t_{\text{sc}} + t_{\text{ch}}$$
$$= (0.137 \text{ hr} + 0.139 \text{ hr} + 0.278 \text{ hr})\left(60 \; \frac{\text{min}}{\text{hr}}\right)$$
$$= 33 \text{ min}$$

Determine rainfall intensity, i, using Fig. 1.12. For a t_c of 33 min and a 100 yr storm, $i = 4.9$ in/hr.

To determine the runoff coefficient, c, refer to Table 1.11 for single family residential. Select a value for c of 0.5.

Determine 100 yr runoff using Eq. 1.9.

$$Q_{100\,\text{yr}} = ciA_d$$
$$= \frac{(0.5)\left(4.9 \; \frac{\text{in}}{\text{hr}}\right)(500 \text{ ac})\left(43{,}560 \; \frac{\text{ft}^2}{\text{ac}}\right)}{\left(12 \; \frac{\text{in}}{\text{ft}}\right)\left(3600 \; \frac{\text{sec}}{\text{hr}}\right)}$$
$$= 1235 \text{ ft}^3/\text{sec}$$

9. Using Fig. 1.19,

step 1: data

drainage area, A_{mi^2}

$$A_{mi^2} = \frac{500 \text{ ac}}{640 \frac{\text{ac}}{\text{mi}^2}} = 0.781 \text{ mi}^2$$

CN = 72 (for HSG B and 1/3 ac residential land use, Table 1.13 through Table 1.15)

$t_c = 0.56 \text{ hr}$

rainfall distribution = Type II (see Fig. 1.7)

pond and swamp areas = 0% of A_{mi^2}

step 2: frequency = 100 yr

step 3: 100 yr, 24 hr rainfall, $P = 7.5$ in (See Fig. 1.6 for Wilmington, Delaware)

step 4: $I_a = 0.778$ in for a CN of 72 (Table 1.16)

step 5: I_a/P

$$\frac{I_a}{P} = \frac{0.778 \text{ in}}{7.5 \text{ in}} = 0.10$$

step 6: unit peak discharge, $q_u = 500 \text{ ft}^3/\text{sec-mi}^2\text{-in}$ (see Fig. 1.21, for a t_c of 0.56 hr and an I_a/P of 0.10)

step 7: runoff, $Q_{in} = 4.3$ in (Table 1.17, interpolating for a P of 7.5 in and a CN of 72)

step 8: pond/swamp adjustment factor, $F_p = 1.0$

step 9: 100 yr peak discharge, q_p

$$q_p = q_u A_{mi^2} Q_{in} F_p$$

$$= \left(500 \ \frac{\text{ft}^3}{\text{sec-mi}^2\text{-in}}\right)(0.781 \text{ mi}^2)(4.3 \text{ in})(1.0)$$

$$= 1679 \text{ ft}^3/\text{sec}$$

10. Rearrange the water budget equation, Eq. 1.1, and solve for infiltration, I.

$$I = P - R - ET + \Delta S$$
$$= 40 \text{ in} - 10 \text{ in} - 22 \text{ in} + 0 \text{ in}$$
$$= 8 \text{ in}$$

The answer is (A).

11. According to Fig. 1.6, the 100 yr, 24 hr precipitation depth for Atlanta is 8 in.

The answer is (C).

12. The incremental intensity, i, is the increment in precipitation depth divided by the time interval. Using the data in the table, the maximum occurs between hours 0.4 and 0.6.

$$i = \frac{0.5 \text{ in}}{0.6 \text{ hr} - 0.4 \text{ hr}}$$
$$= 2.5 \text{ in/hr}$$

The answer is (D).

13. The probability is related to the number of years in the recurrence interval by Eq. 1.2. Rearranging,

$$p = \frac{1}{t_r} = \frac{1}{50 \text{ yr}}$$
$$= 0.02 \quad (2\%)$$

The answer is (A).

14. Determine the rational runoff coefficient, c, by referring to Table 1.11 for downtown area land use.

$$c = 0.8$$

Determine rainfall intensity, i, for a 100 yr storm, from Fig. 1.12.

$$t_c = 0.5 \text{ hr} \quad (30 \text{ min})$$
$$i = 5.2 \text{ in/hr}$$

Use Eq. 1.9.

$$Q_{100\,yr} = ciA_d$$

$$= \frac{(0.8)\left(5.2 \ \frac{\text{in}}{\text{hr}}\right)(200 \text{ ac})\left(43,560 \ \frac{\text{ft}^2}{\text{ac}}\right)}{\left(12 \ \frac{\text{in}}{\text{ft}}\right)\left(3600 \ \frac{\text{sec}}{\text{hr}}\right)}$$

$$= 832 \text{ ft}^3/\text{sec} \quad (800 \text{ ft}^3/\text{sec})$$

The answer is (C).

15. Use Eq. 1.11 to find the time of concentration for sheet flow. Table 1.12 gives a Manning roughness coefficient of 0.24 for dense grass, and Fig. 1.17 gives a 2 yr, 24 hr precipitation depth of 3.5 in for Delaware.

$$t_{sf} = \frac{(0.007 \text{ hr})(nL)^{0.8}}{P_2{}^{0.5} S^{0.4}}$$

$$= \frac{(0.007 \text{ hr})\Big((0.24)(300 \text{ ft})\Big)^{0.8}}{(3.5 \text{ in})^{0.5}\Big(0.07 \ \frac{\text{ft}}{\text{ft}}\Big)^{0.4}}$$

$$= 0.33 \text{ hr} \quad (0.3 \text{ hr})$$

The answer is (B).

16. Using Fig. 1.19 and selecting the CNs from Table 1.13, calculate the composite CN.

HSG	land cover	CN	area, A (ac)	CN $\times$ A
C	SF residential, 1/2 ac	80	500	40,000
C	open space, good condition	74	500	37,000
	total		1000 ac	77,000

The composite CN is

$$CN = \frac{77,000}{1000 \text{ ac}} = 77$$

The answer is (D).

17. Using the ratio of drainage areas, Eq. 1.20, the 10 yr flow at the site is

$$Q_s = \frac{Q_g A_s}{A_g}$$

$$= \frac{\Big(550 \ \frac{\text{ft}^3}{\text{sec}}\Big)(88 \text{ mi}^2)}{52 \text{ mi}^2}$$

$$= 931 \text{ ft}^3/\text{sec} \quad (930 \text{ ft}^3/\text{sec})$$

The answer is (D).

2 Closed Conduit Hydraulics

Nomenclature

A	area	ft^2	m^2
A_b	bearing area of thrust block	ft^2	m^2
C	Hazen-Williams roughness coefficient	–	–
C	pressure wave velocity, speed of sound in water, 4720 (1440)	ft/sec	m/s
C_f	flow coefficient	–	–
C_p	pressure wave velocity after closed gate	ft/sec	m/s
D	diameter	ft	m
e	roughness of interior pipe surface	ft	m
E	modulus of elasticity	lbf/in^2	Pa
f	friction factor	–	–
F	force	lbf	N
g	gravitational acceleration	ft/sec^2	m/s^2
h	height or head	ft	m
K	friction loss coefficient	–	–
L	length	ft	m
$\dot{m}$	mass flow rate	lbm/sec	kg/s
NPSHA	net positive suction head available	ft	m
NPSHR	net positive suction head required	ft	m
p	pressure	lbf/in^2	Pa
p_{atm}	atmospheric pressure	lbf/in^2	Pa
$p_{bearing}$	bearing capacity of soil	lbf/ft^2	Pa
P	pump power	ft-lbf/sec	kg·m/s
q	flow correction factor	gal/min	m^3/min
Q	flow rate	ft^3/sec	m^3/s
r	geometric radius of pipe	ft	m
Re	Reynolds number	–	–
S	slope	ft/ft	m/m
t	time	sec	s
t_h	pipe thickness	in	mm
TDH	total dynamic head	ft	m
v	velocity	ft/sec	m/s
v_{1x}	velocity at point 1 in x direction	ft/sec	m/s
V	volume	ft^3	m^3
z	elevation	ft	m

Symbols

γ	specific weight	lbf/ft^3	N/m^3
η	pump efficiency	–	–
ρ	water density	lbm/ft^3	kg/m^3
ν	kinematic viscosity	ft^2/sec	m^2/s

Subscripts

F	friction loss
L	loss
m	momentum or manometer fluid
n	brake input
p	pipe, pressure, or pump
r	reaction
s	supplied
vap	vapor

1. INTRODUCTION

Hydraulics is the study of the flow of water through natural and engineered systems. *Closed conduit flow* is flow under pressure. In water resources engineering, this flow usually involves pipes or storage tanks in a drinking water distribution network, a reservoir pump storage pipeline, or a pump station in a wastewater treatment plant. Fundamental equations of closed conduit flow include the *conservation of energy (Bernoulli) equation*, the *continuity equation*, the *Darcy-Weisbach equation*, and the *Hazen-Williams equation*.

2. WATER PRESSURE

The *absolute pressure* of water at rest is the sum of the *gage pressure* (hydrostatic pressure) and atmospheric pressure, as shown in Eq. 2.1. γ is the specific weight of water and p_{atm} is atmospheric pressure. The specific weight of water at 50°F is 62.4 lbf/ft^3 (999.7 kg/m^3 at 10°C). Standard atmospheric pressure is 14.7 lbf/in^2

$(1.013 \times 10^5$ Pa). The physical properties of water at atmospheric pressure are given in App. 2.A and App. 2.B.

$$p = \gamma h + p_{\text{atm}} \qquad 2.1$$

Example 2.1

The water level in the open water supply tank shown is at 100 ft above mean sea level (msl), and the base of the tank is at 50 ft msl. Calculate the absolute water pressure at the base of the tank.

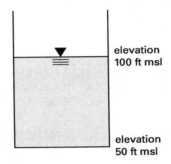

Solution

The height of the water column is

$$h = 100 \text{ ft} - 50 \text{ ft} = 50 \text{ ft}$$

Atmospheric pressure is not (essentially) dependent on elevation. The absolute pressure at the base of the tank is found using Eq. 2.1.

$$
\begin{aligned}
p &= \gamma h + p_{\text{atm}} \\
&= \frac{\left(62.4 \ \dfrac{\text{lbf}}{\text{ft}^3}\right)(50 \text{ ft})}{\left(12 \ \dfrac{\text{in}}{\text{ft}}\right)^2} + 14.7 \ \frac{\text{lbf}}{\text{in}^2} \\
&= 36.4 \text{ lbf/in}^2
\end{aligned}
$$

3. ENERGY EQUATION

Two of the most essential equations in hydraulics and water resources engineering are the energy and continuity equations. These equations are based on the principles of conservation of energy and mass. That is, energy is conserved in a closed pipe system, and what flows into a pipe section must flow out (see Fig. 2.1).

The *conservation of energy equation* (also called the *Bernoulli equation*), Eq. 2.2, for flow between two points along a pipe is

$$\frac{p_1}{\gamma} + \frac{v_1^2}{2g} + z_1 = \frac{p_2}{\gamma} + \frac{v_2^2}{2g} + z_2 + h_L \qquad 2.2$$

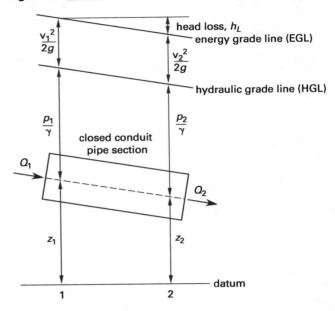

Figure 2.1 *Closed Conduit Flow Energy Equation Parameters*

The terms $v_1^2/2g$ and $v_2^2/2g$ are called the *velocity heads*. The *head loss*, h_L, includes pipe friction losses calculated by the Darcy-Weisbach equation or the Hazen-Williams equation (see Sec. 2.6 and Sec. 2.7), plus minor head losses from turbulence at bends, tees, valves, and gates in a pipeline system (see Sec. 2.10).

4. CONTINUITY EQUATION

The *continuity equation* (also called the *conservation of mass equation*), Eq. 2.3, for two or more points along a channel or pipeline is written in terms of flows, Q_1, or, because flow is area times velocity, in terms of Av.

$$
\begin{aligned}
Q_1 = Q_2 &= Q_n \\
&= A_1 v_1 = A_2 v_2 = A_n v_n
\end{aligned} \qquad 2.3
$$

Example 2.2

A large pipe (1) splits into two smaller pipe branches (2 and 3), as depicted in the illustration. The flow through pipe 1 is 5 ft^3/sec, and the diameter of pipe 1 is 24 in. The flow through pipe 2 is 2 ft^3/sec, and the diameter of pipe 2 is 12 in. The diameter of pipe 3 is 18 in. The flow fills the circular pipes. What are the velocity in pipe 1 and the flow through pipe 3?

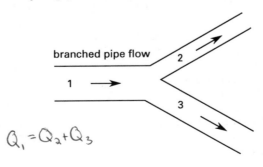

Solution

Because the water fills the circular pipe, the area, A, of the flow is πr^2. Use Eq. 2.3.

$$Q_1 = A_1 v_1$$

$$v_1 = \frac{Q_1}{A_1}$$

$$= \frac{5 \frac{ft^3}{sec}}{\pi \left(\frac{24 \text{ in}}{(2)\left(12 \frac{\text{in}}{ft}\right)} \right)^2}$$

$$= 1.6 \text{ ft/sec}$$

The flow through pipe 3 is

$$Q_3 = Q_1 - Q_2$$

$$= 5 \frac{ft^3}{sec} - 2 \frac{ft^3}{sec}$$

$$= 3 \text{ ft}^3/\text{sec}$$

$\dot{m} = Q\rho$
↑mass flow rate

Example 2.3

A 12 in diameter concrete pipe at section 1 expands to a 24 in pipe at section 2. The velocity through pipe 1 is 10 ft/sec. What is the velocity through pipe 2?

Solution

The area of pipe 1 is

$$A_1 = \pi r^2 = \pi \left(\frac{D}{2}\right)^2$$

$$= \pi \left(\frac{12 \text{ in}}{(2)\left(12 \frac{\text{in}}{ft}\right)} \right)^2$$

$$= 0.785 \text{ ft}^2$$

Using Eq. 2.3, the flow through pipe 1 is

$$Q = A_1 v_1$$

$$= (0.785 \text{ ft}^2)\left(10 \frac{ft}{sec}\right)$$

$$= 7.85 \text{ ft}^3/\text{sec}$$

The velocity through pipe 2 is

$$Q = A_2 v_2$$

$$v_2 = \frac{Q}{A_2} = \frac{Q}{\pi \left(\frac{D}{2}\right)^2}$$

$$= \frac{7.85 \frac{ft^3}{sec}}{\pi \left(\frac{24 \text{ in}}{(2)\left(12 \frac{\text{in}}{ft}\right)} \right)^2}$$

$$= 2.5 \text{ ft/sec}$$

5. MOMENTUM EQUATION

Pipelines are impacted by static forces and dynamic thrust forces, particularly where flow changes direction in a pipe. *Static forces* include the weight of the pipeline, the weight of the fluid, and the weight of the soil surrounding the pipe. *Dynamic thrust forces* occur at the pipeline elbows, tees, bends, valves, and dead ends where flow in the pipe changes direction. Wherever the flow changes direction in a pipe, the resulting forces must be resisted by concrete thrust blocks or other types of pipeline restraints that provide suitable soil bearing capacity. The force associated with a change in the direction of flow is given by the *conservation of momentum equation*, Eq. 2.4.

DIRECTIONAL VECTOR (x & y)

resultant

$F_r = \sqrt{F_x^2 + F_y^2}$

$$F = \dot{m}(v_2 - v_1) \qquad \text{[SI]} \quad \textbf{2.4(a)}$$

$$F = \frac{\dot{m}(v_2 - v_1)}{g_c} \qquad \text{[U.S.]} \quad \textbf{2.4(b)}$$

Thrust Blocks and Pipeline Restraints

Thrust block design is based on the three conservation equations of hydraulics: the conservation of energy (see Eq. 2.2), the conservation of mass (see Eq. 2.3), and the conservation of momentum (see Eq. 2.4).

The forces acting on the pipe where flow changes direction can be calculated using Eq. 2.5 through Eq. 2.7. The *pressure force* is

$$F_p = pA \qquad \qquad \textbf{2.5}$$

The *momentum force* is

$$F_m = \rho Q(v_2 - v_1) \qquad \text{[SI]} \quad \textbf{2.6(a)}$$

$$F_m = \frac{\gamma Q(v_2 - v_1)}{g} \qquad \text{[U.S.]} \quad \textbf{2.6(b)}$$

The *reaction force* is

$$F_r = F_m - F_p \qquad \qquad \textbf{2.7}$$

• $(\cos\theta - 1)$ for F_x
• $(\sin\theta)$ for F_y
θ = pipe bend ∠

The *bearing area*, A_b, of a pipe thrust block must be equal to the reaction force, F_r, divided by the bearing capacity of the soil, $p_{bearing}$, as shown in Eq. 2.8.

$$A_b = \frac{F_r}{p_{bearing}} \qquad 2.8$$

Many water companies specify a minimum soil bearing capacity for thrust block design. A common industry-wide standard uses a minimum soil bearing capacity of 1500 lbf/ft^2 (7323 kg/m^2).

Example 2.4

In the ductile iron pipeline tee junction shown, the pressure entering pipe 1 is 40 lbf/in^2, and pipe head losses and minor losses in the tee junction are zero. Soil pressure is limited to 1500 lbf/ft^2. Estimate the forces and the size of a thrust block system needed for the pipeline tee junction.

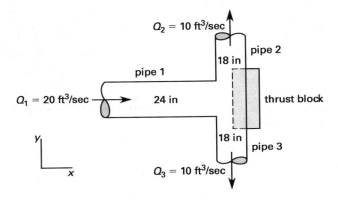

Solution

The velocity in pipe 1 is

$$v_1 = \frac{Q_1}{A_1} = \frac{Q_1}{\pi\left(\frac{D}{2}\right)^2} = \frac{20 \, \frac{\text{ft}^3}{\text{sec}}}{\pi\left(\frac{24 \text{ in}}{(2)\left(12 \, \frac{\text{in}}{\text{ft}}\right)}\right)^2}$$

$$= 6.37 \text{ ft/sec}$$

The velocity in pipe 2 and pipe 3 is

$$v_2 = v_3 = \frac{Q_2}{A_2} = \frac{Q_2}{\pi\left(\frac{D}{2}\right)^2} = \frac{10 \, \frac{\text{ft}^3}{\text{sec}}}{\pi\left(\frac{18 \text{ in}}{(2)\left(12 \, \frac{\text{in}}{\text{ft}}\right)}\right)^2}$$

$$= 5.66 \text{ ft/sec}$$

Calculate the pressure in pipes 2 and 3 using Eq. 2.2.

$$\frac{p_1}{\gamma} + \frac{v_1^2}{2g} + z_1 = \frac{p_2}{\gamma} + \frac{v_2^2}{2g} + z_2 + h_L$$

$$z_1 = z_2$$

$$h_L = 0 \text{ ft}$$

$$p_1 = 40 \text{ lbf/in}^2$$

Solve Eq. 2.2 for p_2.

$$p_2 = p_1 + \frac{\gamma(v_1^2 - v_2^2)}{2g}$$

$$= 40 \, \frac{\text{lbf}}{\text{in}^2}$$

$$+ \frac{\left(62.4 \, \frac{\text{lbf}}{\text{ft}^3}\right)\left(\left(6.37 \, \frac{\text{ft}}{\text{sec}}\right)^2 - \left(5.66 \, \frac{\text{ft}}{\text{sec}}\right)^2\right)}{(2)\left(32.2 \, \frac{\text{ft}}{\text{sec}^2}\right)\left(12 \, \frac{\text{in}}{\text{ft}}\right)^2}$$

$$= 40.06 \text{ lbf/in}^2$$

The flows and diameters of pipes 2 and 3 are equal.

$$p_2 = p_3 = 40.06 \text{ lbf/in}^2$$

Calculate the pressure force in the x direction. The forces in the x direction through pipes 2 and 3 are zero.

$$F_{px} = F_{px1} + F_{px2} + F_{px3}$$

$$= p_{x1}A_1 + 0 + 0$$

$$= p_{x1}\left(\pi\left(\frac{D}{2}\right)^2\right)$$

$$= \left(40 \, \frac{\text{lbf}}{\text{in}^2}\right)\pi\left(\frac{24 \text{ in}}{2}\right)^2$$

$$= 18{,}096 \text{ lbf}$$

Calculate the pressure forces in the y direction. The force in the y direction in pipe 1 is zero.

$$F_{py} = F_{py1} + F_{py2} + F_{py3}$$

$$= 0 + (-p_{y2}A_2) + p_{y3}A_3$$

$$= 0 + \left(-40.06 \, \frac{\text{lbf}}{\text{in}^2}\right)\left(12 \, \frac{\text{in}}{\text{ft}}\right)^2 (1.78 \text{ ft}^2)$$

$$+ \left(40.06 \, \frac{\text{lbf}}{\text{in}^2}\right)\left(12 \, \frac{\text{in}}{\text{ft}}\right)^2 (1.78 \text{ ft}^2)$$

$$= 0$$

Calculate the momentum force on the pipe tee junction.

$$F_{mx} = \frac{\gamma Q}{g}(v_{x1} - v_{x2})$$

$$= \left(\frac{\left(62.4 \ \frac{\text{lbf}}{\text{ft}^3}\right)\left(20 \ \frac{\text{ft}^3}{\text{sec}}\right)}{32.2 \ \frac{\text{ft}}{\text{sec}^2}}\right)\left(6.4 \ \frac{\text{ft}}{\text{sec}} - 0 \ \frac{\text{ft}}{\text{sec}}\right)$$

$$= 248 \ \text{lbf}$$

Calculate the momentum force in the y direction.

$$F_{my} = \frac{\gamma Q}{g}(v_2 - v_3)$$

$$= \left(\frac{\left(62.4 \ \frac{\text{lbf}}{\text{ft}^3}\right)\left(10 \ \frac{\text{ft}^3}{\text{sec}}\right)}{32.2 \ \frac{\text{ft}}{\text{sec}^2}}\right)\left(5.7 \ \frac{\text{ft}}{\text{sec}} - 5.7 \ \frac{\text{ft}}{\text{sec}}\right)$$

$$= 0$$

Calculate the reaction force in the x direction.

$$F_{rx} = F_{mx} - F_{px}$$

$$= 248 \ \text{lbf} - 18{,}096 \ \text{lbf}$$

$$= -17{,}848 \ \text{lbf}$$

Calculate the reaction force in the y direction.

$$F_{ry} = F_{my} - F_{py}$$

$$= 0 \ \text{lbf} - 0 \ \text{lbf}$$

$$= 0$$

Calculate the bearing area of the pipe thrust block using Eq. 2.8.

$$A_b = \frac{F_r}{p_{\text{bearing}}} = \frac{17{,}848 \ \text{lbf}}{1500 \ \frac{\text{lbf}}{\text{ft}^2}}$$

$$= 11.90 \ \text{ft}^2$$

Specify a concrete thrust block with an area greater than A_b.

$$(3 \ \text{ft})(4 \ \text{ft}) = 12 \ \text{ft}^2 > 11.90 \ \text{ft}^2$$

6. THE DARCY-WEISBACH EQUATION

The Darcy-Weisbach equation, Eq. 2.9, is used to estimate the head loss caused by pipe friction.

-GASES OR
COMPLEX FLUIDS,
- HIGH PRESSURES
NONSTANDARD
WATER PROPERTIES

$$h_L = \frac{fLv^2}{2gD} \qquad 2.9$$

Values of the friction factor, f, are determined from the *Moody friction factor chart* (see Fig. 2.2) using the relative roughness, e/D, of the pipe and the Reynolds number, Re. The *kinematic viscosity* of water, ν, can be found from App. 2.A or App. 2.B. The Reynolds number is calculated from Eq. 2.10.

$$\text{Re} = \frac{Dv}{\nu} \qquad 2.10$$

The Moody friction factor chart lists the *specific roughness*, e, for several materials (see Fig. 2.2). To use the Moody friction factor chart, the *relative roughness*, e/D, is needed.

Example 2.5

Use the Darcy-Weisbach equation to estimate the head loss through a 5 ft diameter, 2000 ft long concrete pipe that conveys 50 ft³/sec of 50°F water.

Solution

To find the friction factor, f, for the Darcy-Weisbach equation, calculate the Reynolds number using Eq. 2.10. Calculate the velocity from the given values of the flow and area.

$$A = \pi\left(\frac{D}{2}\right)^2$$

$$= \pi\left(\frac{5 \ \text{ft}}{2}\right)^2$$

$$= 19.6 \ \text{ft}^2$$

$$v = \frac{Q}{A} = \frac{50 \ \frac{\text{ft}^3}{\text{sec}}}{19.6 \ \text{ft}^2}$$

$$= 2.55 \ \text{ft/sec}$$

At 50°F, the kinematic viscosity of water is 1.41×10^{-5} ft²/sec. The Reynolds number is

$$\text{Re} = \frac{Dv}{\nu}$$

$$= \frac{(5 \ \text{ft})\left(2.55 \ \frac{\text{ft}}{\text{sec}}\right)}{1.41 \times 10^{-5} \ \frac{\text{ft}^2}{\text{sec}}}$$

$$= 9.04 \times 10^5$$

Calculate the relative roughness of the concrete pipe, taking the roughness of the interior pipe surface, e, to be 0.001 ft.

$$\frac{e}{D} = \frac{0.001 \ \text{ft}}{5 \ \text{ft}}$$

$$= 0.0002$$

From the Moody friction factor chart in Fig. 2.2, for a Reynolds number, Re, of 9.04×10^5 and a relative roughness, e/D, of 0.0002, the friction factor, f, is 0.015.

Figure 2.2 *Moody Friction Factor Chart*

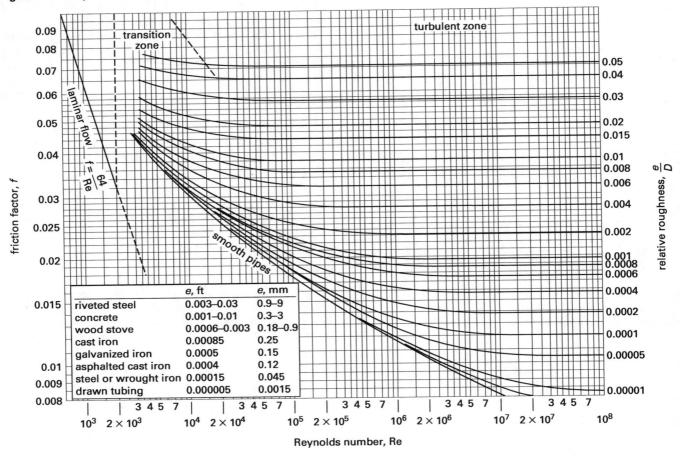

Reprinted with permission from L. F. Moody. "Friction Factor for Pipe Flow," ASME Transactions. Vol. 66. copyright © 1944. Published by the American Society of Mechanical Engineers.

Using the Darcy-Weisbach equation, the head loss is

$$h_L = \frac{fLv^2}{2gD}$$

$$= \frac{(0.015)(2000 \text{ ft})\left(2.55 \ \dfrac{\text{ft}}{\text{sec}}\right)^2}{(2)\left(32.2 \ \dfrac{\text{ft}}{\text{sec}^2}\right)(5 \text{ ft})}$$

$$= 0.606 \text{ ft}$$

7. THE HAZEN-WILLIAMS EQUATION

Flow under pressure through a pipe or closed conduit is expressed by the *Hazen-Williams equation*, Eq. 2.11. One of the principal parameters in this equation is the *Hazen-Williams roughness coefficient, C,* an indication of the friction caused by the interior of a pipe. Smoother pipes, such as new ductile iron pipes, have higher C values than older brick-lined pipes with more interior roughness. Therefore, newer and smoother

pipes can convey more flow than older and rougher pipes. Table 2.1 gives the Hazen-Williams roughness coefficients for various pipe types.

It is necessary to use correct units when using the Hazen-Williams equation.

$$h_{L,\text{ft}} = \frac{10.44 L_{\text{ft}} Q_{\text{gpm}}^{1.85}}{C^{1.85} D_{\text{in}}^{4.87}} \qquad 2.11$$

$$h_{L,\text{ft}} = \frac{L_{\text{ft}} Q_{\text{gpm}}^{1.85}}{17{,}076 \, C^{1.85} D_{\text{ft}}^{4.87}} \qquad 2.12$$

Equation 2.13 is the Hazen-Williams equation in terms of velocity.

$$h_{L,\text{ft}} = \frac{3.022 L_{\text{ft}} v_{\text{ft/sec}}^{1.85}}{C^{1.85} D_{\text{ft}}^{1.17}} \qquad 2.13$$

ONLY TURBULENT

WATER DISTRIBUTION SYSTEMS

Table 2.1 Hazen-Williams Roughness Coefficients

pipe type	C
brick	100
cast iron, 20 yr old	100
riveted steel	110
ductile iron, 20 yr old	110
clay pipe	110
cast iron, 5 yr old	120
welded steel	120
wood plank	120
concrete	130
cast iron, new	130
ductile iron, new	130
plastic	140

Reprinted from L. A. Rossman. EPANET2 Users Manual. September 2000. EPA.

Example 2.6

Calculate the pressure head loss through a 100 ft long, 12 in diameter, new ductile iron pipe, in which the flow rate is 1.44 MGD.

Solution

Convert units of flow to gallons per minute.

$$Q = \frac{\left(1.44 \ \frac{\text{MG}}{\text{day}}\right)\left(1,000,000 \ \frac{\text{gal}}{\text{MG}}\right)}{\left(24 \ \frac{\text{hr}}{\text{day}}\right)\left(60 \ \frac{\text{min}}{\text{hr}}\right)}$$

$$= 1000 \ \text{gal/min}$$

The pressure head loss is calculated by the Hazen-Williams equation, Eq. 2.12.

$$h_L = \frac{L_{\text{ft}} \, Q_{\text{gpm}}^{1.85}}{17,076 \, C^{1.85} D_{\text{ft}}^{4.87}}$$

$$= \frac{(100 \ \text{ft})\left(1000 \ \frac{\text{gal}}{\text{min}}\right)^{1.85}}{(17,076)(130)^{1.85}(1 \ \text{ft})^{4.87}}$$

$$= 0.26 \ \text{ft}$$

8. PARALLEL PIPING

Frequently in a water distribution network or a reservoir pipeline, one pipe splits into two or more pipes, so designers must calculate the flow through pipes in parallel. If the flows from the parallel pipes are later rejoined, the conservation of energy equation implies that the head loss for flow entering and leaving the parallel pipes must be the same in each pipe. Therefore, for two or more pipes in parallel, the energy (or Bernoulli) equation is

combined with the Hazen-Williams equation for head loss (see Eq. 2.14).

$$h_L = h_{L1} = h_{L2} = h_{L3} \qquad 2.14$$

Given the three-pipe network in Fig. 2.3, the continuity equation is rewritten as Eq. 2.15.

$$Q = Q_1 + Q_2 + Q_3 \qquad 2.15$$

Figure 2.3 Piping System in Parallel

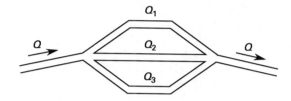

The flows through each pipe can be calculated according to the following iterative process.

1. Calculate the head loss in each pipe using the Hazen-Williams equation.

2. Assume one value of head loss for all the pipes.

3. Solve for Q_1, Q_2, and Q_3 using the continuity equation in terms of $h_L = h_{L1} = h_{L2} = h_{L3}$.

4. If the continuity equation, $Q = \sum v_i A_i$, is not solved, assume a new head loss through each pipe and repeat the first three steps.

Example 2.7

Determine the flow through each pipe in the parallel system shown in Fig. 2.3, assuming the flow is 10,000 gal/min and given the following data.

pipe	L (ft)	D (in)	C
1	120	12	100
2	100	12	100
3	80	12	100

Solution

Solve for head losses through each pipe using the Hazen-Williams equation, Eq. 2.12. For pipe 1,

$$h_{L1} = \frac{L_{\text{ft}} \, Q_{\text{gpm}}^{1.85}}{17,076 \, C^{1.85} D_{\text{ft}}^{4.87}}$$

$$= \frac{(120 \ \text{ft}) Q_1^{1.85}}{(17,076)(100)^{1.85}(1 \ \text{ft})^{4.87}}$$

$$= 0.0000014 \, Q_1^{1.85}$$

For pipe 2,

$$h_{L2} = \frac{L_{\text{ft}} Q_{\text{gpm}}^{1.85}}{17{,}076 \, C^{1.85} D_{\text{ft}}^{4.87}}$$

$$= \frac{(100 \text{ ft}) Q_2^{1.85}}{(17{,}076)(100)^{1.85}(1 \text{ ft})^{4.87}}$$

$$= 0.00000117 Q_2^{1.85}$$

For pipe 3,

$$h_{L3} = \frac{L_{\text{ft}} Q_{\text{gpm}}^{1.85}}{17{,}076 \, C^{1.85} D_{\text{ft}}^{4.87}}$$

$$= \frac{(80 \text{ ft}) Q_3^{1.85}}{(17{,}076)(100)^{1.85}(1 \text{ ft})^{4.87}}$$

$$= 0.000000935 Q_3^{1.85}$$

By trial and error, find a consistent value for the head loss, h_L. Vary h_L and solve for the flow, Q, for pipes 1, 2, and 3. Assume h_L is 1.0 ft, 5.0 ft, then 3.0 ft. Repeat the calculation cycle if $Q \neq Q_1 + Q_2 + Q_3$.

An error of $\pm 5\%$ is tolerable. For the final iteration, assume that the head loss, h_L, is 3.5 ft.

Solving for the flow in pipe 1,

$$h_{L1} = 3.5 \text{ ft} = 0.0000014 Q_1^{1.85}$$

$$Q_1^{1.85} = 2{,}500{,}000$$

$$Q_1 = 2873 \text{ gal/min}$$

Solving for the flow in pipe 2,

$$h_{L2} = 3.5 \text{ ft} = 0.00000117 Q_2^{1.85}$$

$$Q_2^{1.85} = 2{,}991{,}452$$

$$Q_2 = 3166 \text{ gal/min}$$

Solving for the flow in pipe 3,

$$h_{L3} = 3.5 \text{ ft} = 0.000000935 Q_3^{1.85}$$

$$Q_3^{1.85} = 3{,}743{,}315$$

$$Q_3 = 3574 \text{ gal/min}$$

By the continuity equation,

$$Q = Q_1 + Q_2 + Q_3$$

$$= 2873 \, \frac{\text{gal}}{\text{min}} + 3166 \, \frac{\text{gal}}{\text{min}} + 3574 \, \frac{\text{gal}}{\text{min}}$$

$$= 9613 \text{ gal/min} \quad [\approx 10{,}000 \text{ gal/min, OK}]$$

9. BRANCHED PIPE NETWORKS

Figure 2.4 depicts a branched pipe network where three pipes are each connected to three reservoirs. Disregarding velocity head (valid for low velocity flow) and pressure head (valid for unpressurized and open channel flow), the head loss due to friction is calculated from the Bernoulli equation as Eq. 2.16 through Eq. 2.18. The reservoir elevations, z_A, z_C, and z_D, are usually given. Unknown are the elevation at the intersection of the pipes, z_B, and the flow quantity and direction through each of the pipes Q_1, Q_2, and Q_3. Equation 2.12 gives the head loss through each pipe.

For pipe 1,

$$h_{L1} = z_A - z_B \qquad \textit{2.16}$$

For pipe 2,

$$h_{L2} = z_B - z_C \qquad \textit{2.17}$$

For pipe 3,

$$h_{L3} = z_B - z_D \qquad \textit{2.18}$$

Figure 2.4 Branched Pipes to Reservoir's Network

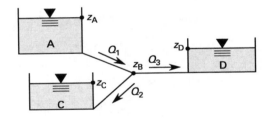

Once the flows in the pipe system reach equilibrium, the continuity equation becomes Eq. 2.19.

$$\sum Q = Q_1 - Q_2 - Q_3 = 0 \qquad \textit{2.19}$$

Through an iterative process, the flows in each pipe can be calculated according to the following steps.

step 1: Assume Q_1 through pipe 1.

step 2: Assume a water surface elevation, z_B, at the intersection of the pipes.

step 3: Solve for Q_1, Q_2, and Q_3 using the continuity equation. If the sum of the flows is not equal to zero, then assume a new z_B and recalculate.

step 4: Repeat steps 1, 2, and 3 until the continuity equation is satisfied.

Example 2.8

For the reservoir system illustrated, the elevations of the reservoir water surfaces are $z_A = 110$ ft, $z_C = 50$ ft, and $z_D = 65$ ft. Given the parameters listed in the following table, estimate the flow in each of the supply pipes.

pipe	L (ft)	D (in)	C
1	100	12	100
2	100	12	100
3	100	12	100

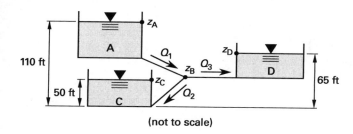

(not to scale)

Solution

Using Eq. 2.16 through Eq. 2.18, substitute the given values for each pipe into the Hazen-Williams equation to find three expressions for z_B in terms of the flow in each pipe.

The head loss in each pipe is the difference in the height at the intersection, z_B, and the end of the pipe, z_A, z_C, or z_D.

For the head loss in pipe 1,

$$z_A - z_B = \frac{L_{ft} Q_{gpm}^{1.85}}{17,076 \, C^{1.85} D_{ft}^{4.87}}$$

$$110 \text{ ft} - z_B = \frac{(100 \text{ ft}) Q_1^{1.85}}{(17,076)(100)^{1.85}(1 \text{ ft})^{4.87}}$$

$$= 0.00000117 Q_1^{1.85}$$

For the head loss in pipe 2,

$$z_B - z_C = \frac{L_{ft} Q_{gpm}^{1.85}}{17,076 \, C^{1.85} D_{ft}^{4.87}}$$

$$z_B - 50 \text{ ft} = \frac{(100 \text{ ft}) Q_2^{1.85}}{(17,076)(100)^{1.85}(1 \text{ ft})^{4.87}}$$

$$= 0.00000117 Q_2^{1.85}$$

For the head loss in pipe 3,

$$z_B - z_D = \frac{L_{ft} Q_{gpm}^{1.85}}{17,076 \, C^{1.85} D_{ft}^{4.87}}$$

$$z_B - 65 \text{ ft} = \frac{(100 \text{ ft}) Q_3^{1.85}}{(17,076)(100)^{1.85}(1 \text{ ft})^{4.87}}$$

$$= 0.00000117 Q_3^{1.85}$$

Using trial and error, test several values of z_B, and solve for $\sum Q = Q_1 - Q_2 - Q_3 = 0$.

Because z_A is 110 ft and z_C is 50 ft, z_B must be between 50 ft and 110 ft. Assume z_B is 60 ft, 90 ft, then 65 ft. For each of these assumptions, the sum of the flows is not equal to zero. Therefore, for the next iteration, assume z_B is 70 ft, then solve for Q_1, Q_2, and Q_3.

For pipe 1,

$$0.00000117 Q_1^{1.85} = 110 \text{ ft} - z_B$$

$$= 110 \text{ ft} - 70 \text{ ft}$$

$$= 40 \text{ ft}$$

$$Q_1^{1.85} = 34,188,034$$

$$Q_1 = 11,813 \text{ gal/min}$$

For pipe 2,

$$0.00000117 Q_2^{1.85} = z_B - 50 \text{ ft}$$

$$= 70 \text{ ft} - 50 \text{ ft}$$

$$= 20 \text{ ft}$$

$$Q_2^{1.85} = 17,094,017$$

$$Q_2 = 8122 \text{ gal/min}$$

For pipe 3,

$$0.00000117 Q_3^{1.85} = z_B - 65 \text{ ft}$$

$$= 70 \text{ ft} - 65 \text{ ft}$$

$$= 5 \text{ ft}$$

$$Q_3^{1.85} = 4,273,504$$

$$Q_3 = 3838 \text{ gal/min}$$

By the continuity equation,

$$\sum Q = Q_1 - Q_2 - Q_3 = 0 \text{ gal/min}$$

$$= 11,813 \, \frac{\text{gal}}{\text{min}} - 8112 \, \frac{\text{gal}}{\text{min}} - 3838 \, \frac{\text{gal}}{\text{min}}$$

$$= -137 \text{ gal/min} \quad [-137 \text{ gal/min} \approx 0, \text{ OK}]$$

Therefore,

$$Q_1 = 11{,}813 \text{ gal/min}$$

$$Q_2 = 8112 \text{ gal/min}$$

$$Q_3 = 3838 \text{ gal/min}$$

10. MINOR LOSSES

Minor head losses caused by turbulence at gates, valves, bends, and tees in a pipeline may be estimated by the *velocity head equation*, Eq. 2.20, or the equivalent pipe method. Minor head losses, h_L, are proportional to the velocity head, $v^2/2g$.

$$h_L = \frac{Kv^2}{2g} \qquad\qquad 2.20$$

To work with the *friction loss coefficient, K*, in Eq. 2.20, velocity is given in feet per second. Values of the coefficient, K, are given in Table 2.2.

Minor head losses may also be estimated by the *equivalent pipe method*. This method models minor head losses from gates, valves, bends, and tees as an equivalent length of pipe that experiences the same head loss from pipe friction. Table 2.3 gives typical estimates of equivalent lengths for several diameters of steel pipe, as calculated by the Darcy-Weisbach equation.

Table 2.2 Minor Friction Loss Coefficients

friction source	loss coefficient, K
globe valve, fully open	10.0
angle valve, fully open	5.0
swing check valve, fully open	2.5
gate valve, fully open	0.2
90° pipe elbow	0.9
45° pipe elbow	0.4
tee	1.8
square entrance	0.5
contraction	0.4
exit	1.0

Table 2.3 Typical Equivalent Lengths for Various Sources of Friction in Steel Pipe

	equivalent length (ft)		
	pipe diameter		
friction source	1 in	2 in	4 in
globe valve	29	54	110
angle valve	14	18	18
swing check valve	11	19	38
gate valve	0.8	1.5	2.5
90° pipe elbow	5.2	8.5	13
45° pipe elbow	1.3	2.7	5.5
tee	3.2	7.7	17

Example 2.9

The following illustration shows a 24 in diameter iron pipe system. The flow rate from reservoir A to reservoir B is 10 ft³/sec. The pipeline contains one swing check valve (open), one gate valve, one tee, three 90° elbows, and two 45° elbows. Determine minor friction head losses through the pipe system.

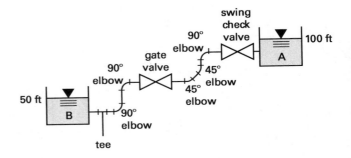

Solution

Calculate the velocity using Eq. 2.3.

$$v = \frac{Q}{A}$$

$$= \frac{10 \ \dfrac{\text{ft}^3}{\text{sec}}}{\pi \left(\dfrac{24 \text{ in}}{(2)\left(12 \ \frac{\text{in}}{\text{ft}}\right)} \right)^2}$$

$$= 3.18 \text{ ft/sec}$$

The head loss coefficient, K, is the sum of the head loss coefficients, given in Table 2.2, for all the valves, elbows, and tees in the system.

$$K = 2.5 + 0.2 + 1.8 + (3)(0.9) + (2)(0.4) = 8$$

Use Eq. 2.20.

$$h_L = \frac{Kv^2}{2g} = \frac{(8)\left(3.18 \ \dfrac{\text{ft}}{\text{sec}}\right)^2}{(2)\left(32.2 \ \dfrac{\text{ft}}{\text{sec}^2}\right)}$$

$$= 1.3 \text{ ft}$$

Example 2.10

In the 2 in steel pipe system shown in the illustration, the flow rate from tank A to tank B is 0.2 ft³/sec. Determine the total equivalent pipe length of the fittings.

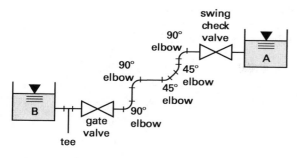

Solution

From Table 2.3, the total equivalent pipe lengths of the sources of friction in the illustration are summarized in the following table.

source of friction	(no.)(pipe length (ft))	equivalent pipe length (ft)
gate valve	(1)(1.5)	1.5
swing check valve	(1)(19)	19
tee	(1)(7.7)	7.7
45° elbow	(2)(2.7)	5.4
90° elbow	(3)(8.5)	25.5
total		59.1

The total equivalent pipe length is 59.1 ft.

11. PUMP APPLICATION AND ANALYSIS

Pumping stations usually contain one of two types of pump: axial or centrifugal. *Axial flow pumps* generate lower pressures at higher flow rates by expelling the fluid inline through the pipe, as air is expelled through the propeller of an airplane. *Centrifugal flow pumps* generate pressure by accelerating the flow with an impeller, which gives velocity to the fluid through centrifugal force. As shown in Fig. 2.5, the *total dynamic pumping head*, TDH, is described by Eq. 2.21 as the sum of Δz, the total static head (the vertical distance between fluid water levels from the pump centerline to the fluid water level); h_L, the friction head losses from the Darcy-Weisbach or Hazen-Williams equations, including the minor head losses from the velocity head equation for gates, valves, bends, and tees.

$$\text{TDH} = h_L + \Delta z + h_v \qquad 2.21$$

Figure 2.5 Total Dynamic Head Schematic

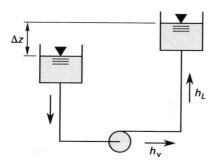

In Eq. 2.22, power, P_s, supplied by a pump is calculated by a variation of the energy equation.

$$\frac{-P_s}{\gamma Q} = \frac{v_2^2 - v_1^2}{2g} - \frac{p_2 - p_1}{\gamma} + z_2 - z_1 + h_L \qquad 2.22$$

Power in pipe systems can be defined by pump head, pump power, and brake head, as depicted in Fig. 2.6. Power can be converted to horsepower using the unit conversion of 1 hp = 550 ft-lbf/sec.

The *pump efficiency* is η. The *brake power* input is

$$P_n = \frac{P_s}{\eta} \qquad 2.23$$

Figure 2.6 Pump and Pipe Losses

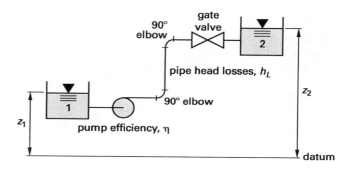

Example 2.11

How much horsepower does a pump need to increase water pressure by 30 lbf/in² in a 6 in diameter level pipe with a flow rate of 1.0 ft³/sec and no friction losses?

Solution

By continuity, the flow rate and the pipe diameter are constant. Therefore, the velocity along the pipe is constant.

$$\frac{v_2^2 - v_1^2}{2g} = 0$$

Both ends of the pipe are at the same elevation.

$$z_2 - z_1 = 0$$

The pump power is given by Eq. 2.22. Convert foot-pounds per second to horsepower.

$$\frac{-P_s}{\gamma Q} = \frac{v_2^2 - v_1^2}{2g} - \frac{p_2 - p_1}{\gamma}$$
$$+ z_2 - z_1 + h_L$$

$$\frac{-P_s\left(550 \; \dfrac{\text{ft-lbf}}{\text{hp-sec}}\right)}{\left(62.4 \; \dfrac{\text{lbf}}{\text{ft}^3}\right)\left(1.0 \; \dfrac{\text{ft}^3}{\text{sec}}\right)} = 0 \text{ ft}$$

$$- \frac{\left(30 \; \dfrac{\text{lbf}}{\text{in}^2} - 0 \; \dfrac{\text{lbf}}{\text{in}^2}\right)\left(12 \; \dfrac{\text{in}}{\text{ft}}\right)^2}{62.4 \; \dfrac{\text{lbf}}{\text{ft}^3}}$$

$$+ 0 \text{ ft} + 0 \text{ ft}$$

$$P_s = 7.9 \text{ hp}$$

Example 2.12

What pump power is necessary for a flow rate of 20 ft³/sec of 50°F water through a 2000 ft, 24 in diameter, ductile iron pipe from reservoir 1 (elevation 100 ft above msl) to reservoir 2 (elevation 200 ft above msl), as shown in the illustration?

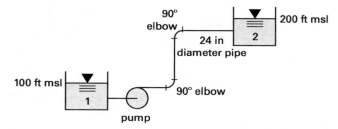

Solution

The water surfaces in both reservoirs are still and open to the atmosphere. Therefore, velocity and gage pressure are equal to zero.

$$v_1 = v_2 = 0 \text{ ft/sec}$$
$$p_2 = p_1 = 0 \text{ lbf/in}^2$$

Starting with the Darcy-Weisbach expression for the head loss from pipe friction, find the velocity, v, of the pipe.

$$v = \frac{Q}{A} = \frac{20 \; \dfrac{\text{ft}^3}{\text{sec}}}{\pi\left(\dfrac{24 \text{ in}}{(2)\left(12 \; \dfrac{\text{in}}{\text{ft}}\right)}\right)^2}$$

$$= 6.37 \text{ ft/sec}$$

Calculate the Reynolds number using Eq. 2.10, taking ν, the kinematic viscosity of water, to be approximately 1.41×10^{-5} ft²/sec at 50°F.

$$\text{Re} = \frac{Dv}{\nu} = \frac{(24 \text{ in})\left(6.37 \; \dfrac{\text{ft}}{\text{sec}}\right)}{\left(1.41 \times 10^{-5} \; \dfrac{\text{ft}^2}{\text{sec}}\right)\left(12 \; \dfrac{\text{in}}{\text{ft}}\right)}$$

$$= 9.04 \times 10^5$$

Calculate the relative roughness of the pipe. From the Moody friction factor chart (see Fig. 2.2), the roughness of the interior iron pipe surface, e, is 0.00015 ft.

$$\frac{e}{D} = \frac{0.00015 \text{ ft}}{2 \text{ ft}} = 0.000075$$

From the Moody friction factor chart for a relative roughness of 0.000075 and a Re $= 9.04 \times 10^5$, the friction factor, f, is 0.013.

$K = 0.9$ for 90° elbows. Calculate the pump power using Eq. 2.22.

$$\frac{-P_s}{\gamma Q} = \frac{v_2^2 - v_1^2}{2g} - \frac{p_2 - p_1}{\gamma} + z_2 - z_1 + h_L$$
$$= \frac{v_2^2 - v_1^2}{2g} - \frac{p_2 - p_1}{\gamma} + z_2 - z_1$$
$$+ \left(\frac{fL}{D} + K_{90°} + K_{90°}\right)\frac{v^2}{2g}$$

$$\frac{-P_s\left(550 \; \dfrac{\text{ft-lbf}}{\text{sec}}{\text{hp}}\right)}{\left(62.4 \; \dfrac{\text{lbf}}{\text{ft}^3}\right)} = 0 \; \frac{\text{ft}}{\text{sec}} + 0 \; \frac{\text{lbf}}{\text{in}^2} + 200 \text{ ft} - 100 \text{ ft}$$

$$\times \left(20 \; \frac{\text{ft}^3}{\text{sec}}\right)$$

$$+ \left(\frac{(0.013)(2000 \text{ ft})}{\dfrac{24 \text{ in}}{12 \; \dfrac{\text{in}}{\text{ft}}}} + 0.9 + 0.9\right)$$

$$\times \frac{\left(6.37 \; \dfrac{\text{ft}}{\text{sec}}\right)^2}{(2)\left(32.2 \; \dfrac{\text{ft}}{\text{sec}^2}\right)}$$

$$-P_s = 246 \text{ hp}$$

A 246 hp pump is needed.

Example 2.13

Water is pumped from a creek at elevation 70 ft msl through a 1000 ft long, 12 in diameter ductile iron pipe to a reservoir at elevation 170 ft msl. The flow rate is 3 ft³/sec, and the pipe contains an open gate valve, as shown in the illustration. Determine the head loss through the pipe using the Hazen-Williams equation.

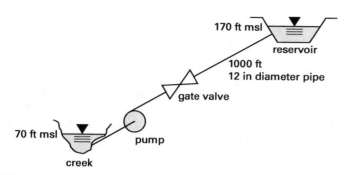

Solution

Determine the pipe area and velocity.

$$A = \pi\left(\frac{D}{2}\right)^2 = \pi\left(\frac{12 \text{ in}}{(2)\left(12 \frac{\text{in}}{\text{ft}}\right)}\right)^2$$

$$= 0.785 \text{ ft}^2$$

$$\mathrm{v} = \frac{Q}{A} = \frac{3 \frac{\text{ft}^3}{\text{sec}}}{0.785 \text{ ft}^2}$$

$$= 3.82 \text{ ft/sec}$$

Table 2.1 gives 130 as the roughness coefficient, C, for new ductile iron pipe. Use the Hazen-Williams equation.

$$h_{L,\text{pipe}} = \frac{3.022 L_{\text{ft}} \mathrm{v}_{\text{ft/sec}}^{1.85}}{C^{1.85} D_{\text{ft}}^{1.17}} = \frac{(3.022)(1000 \text{ ft})\left(3.82 \frac{\text{ft}}{\text{sec}}\right)^{1.85}}{(130)^{1.85}(1 \text{ ft})^{1.17}}$$

$$= 4.43 \text{ ft}$$

The head loss from the open gate valve is given by Eq. 2.20.

$$h_{L,\text{valve}} = \frac{K\mathrm{v}^2}{2g} = \frac{(0.2)\left(3.82 \frac{\text{ft}}{\text{sec}}\right)^2}{(2)\left(32.2 \frac{\text{ft}}{\text{sec}^2}\right)}$$

$$= 0.0453 \text{ ft}$$

The total head loss is

$$h_{L,\text{total}} = h_{L,\text{pipe}} + h_{L,\text{valve}} = 4.43 \text{ ft} + 0.0453 \text{ ft}$$

$$= 4.48 \text{ ft}$$

Example 2.14

The following illustration shows data for total dynamic head versus discharge for three pumps. Select a pump for a total capacity of 1000 gal/min at a head of 50 ft.

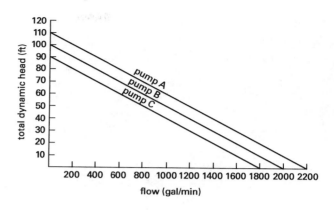

Solution

Pump B will provide 1000 gal/min at a head of 50 ft. Therefore, select pump B.

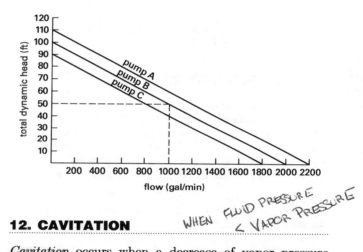

12. CAVITATION

WHEN FLUID PRESSURE < VAPOR PRESSURE

Cavitation occurs when a decrease of vapor pressure causes internal water pressures to drop, resulting in the collapse of the pump or pipe system. The *net positive suction head* (NPSH) is used to design pumps to avoid cavitation. The net positive suction head available (NPSHA) must not be less than the net positive suction head required (NPSHR).

APP. 2A/B

$$\text{NPSHA} \geq \text{NPSHR}$$

$$= \frac{p_{\text{atm}} + p_{\text{vap}}}{\gamma} - \Delta z - h_L \qquad \textit{2.24}$$

The head loss, h_L, on the suction pipe between locations 1 and 2 is found by means of the Hazen-Williams equation.

Example 2.15

The required net positive suction head for a pump is 25 ft. It pumps water through a 100 ft long pipe with a diameter of 12 in at a flow rate of 1000 gal/min. C is 130 for a new ductile iron pipe. Assume p_{vap} is 0.5 lbf/in^2 at 80°F. At the pipe inlet, what height above the water surface should the pump be located to avoid cavitation?

Solution

Set NPSHA equal to NPSHR. Use the Hazen-Williams equation to calculate the head loss due to friction.

$$\text{NPSHA} = \text{NPSHR} = \frac{p_{\text{atm}} - p_{\text{vap}}}{\gamma} - \Delta z - h_L$$

$$\Delta z = \frac{p_{\text{atm}} - p_{\text{vap}}}{\gamma} - \frac{L_{\text{ft}} Q_{\text{gpm}}^{1.85}}{17{,}076\, C^{1.85} D_{\text{ft}}^{4.87}} - \text{NPSHR}$$

$$= \frac{\left(14.7\, \dfrac{\text{lbf}}{\text{in}^2} - 0.5\, \dfrac{\text{lbf}}{\text{in}^2}\right)\left(12\, \dfrac{\text{in}}{\text{ft}}\right)^2}{62.4\, \dfrac{\text{lbf}}{\text{ft}^3}}$$

$$- \frac{(100\text{ ft})\left(1000\, \dfrac{\text{gal}}{\text{min}}\right)^{1.85}}{(17{,}076)(130)^{1.85}(1\text{ ft})^{4.87}} - 25\text{ ft}$$

$$= 7.5\text{ ft}$$

The pump should be placed no higher than 7.5 ft above the inlet end of the pipe.

13. PIPE NETWORK ANALYSIS

Pipe network analyses for closed conduits under pressure use the *Hardy-Cross method* to balance the head losses in each pipe. Pipe networks are designed with the assumption that the sum of the head losses in any pipe loop must equal zero. This is the fundamental method for designing water supply distribution networks.

According to the continuity or conservation of mass equation, the flow entering a particular node must equal the flow leaving it. For instance, if pipe A with a flow of 1000 ft^3/sec enters node 1, then the sum of flows leaving node 1 through pipes B and C must equal 1000 ft^3/sec.

The Hardy-Cross method is one of the more complex calculations in water resources engineering and is often remembered as an iterative, five-step process.

step 1: Number each of the various pipe loops by identifying the junctions (nodes) by alphabet or number.

step 2: Assume a flow direction and assume a flow through each pipe. Clockwise flow through a pipe loop is assumed to be positive (+), and counterclockwise flow is assumed to be negative (−).

step 3: Calculate the head loss through each pipe in a loop using the Hazen-Williams equation.

step 4: Sum the head losses in all of the pipes in each of the loops, being careful to use the correct positive or negative signs, depending on the flow direction within the loop.

step 5: If the sum of the calculated head losses is reasonably small (e.g., less than 1 ft (0.3 m)), then the assumptions for flow and direction through the pipes are correct, and the calculation is complete.

If the sum of the head losses is not reasonably small, then employ a flow correction factor, q. Calculate the correction factor using Eq. 2.25.

$$q_{\text{loop}} = \frac{-\sum_{\text{loop}} h_{Li}}{\sum_{\text{loop}} 1.85 \left| \dfrac{h_L}{Q} \right|} \qquad 2.25$$

The sum of the head losses in the pipe loops is $\sum h_{Li}$, given in feet. The head loss in each pipe is h_{Li}. The assumed flow rate through each pipe is Q, in gallons per minute. Add the flow correction factor to the assumed flow for each pipe and repeat the calculation starting with step 2. Continue the calculation cycle until the sum of head losses in each loop is less than 1.

Example 2.16

Referring to the water supply network depicted in the following illustration and assuming a Hazen-Williams roughness coefficient of 130 for new, ductile iron pipe, calculate the flow through each pipe in loop ABC using one iteration of the Hardy-Cross method.

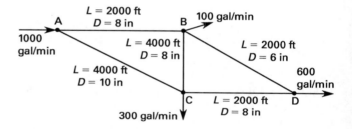

Solution

step 1: The various pipe loops are defined with nodes A, B, C, and D.

step 2: Assume a flow direction through each loop, clockwise (+) and counterclockwise (−). For instance, the flow through pipes AB and BC is clockwise (+) in loop ABC. The flow through pipe AC is counterclockwise (−) in loop ABC. Assume a flow through each pipe as depicted in the illustration.

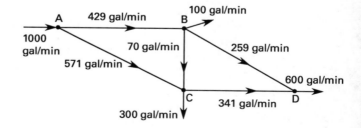

step 3: Calculate the head losses through each pipe using the Hazen-Williams equation. For pipe AB,

$$h_{L,AB} = \frac{L_{ft} Q_{gpm}^{1.85}}{17{,}076\, C^{1.85} D_{ft}^{4.87}}$$

$$= \frac{(2000\ \text{ft})\left(429\ \dfrac{\text{gal}}{\text{min}}\right)^{1.85}}{(17{,}076)(130)^{1.85}(0.67\ \text{ft})^{4.87}}$$

$$= +7.49\ \text{ft}$$

For pipe BC,

$$h_{L,BC} = \frac{L_{ft} Q_{gpm}^{1.85}}{17{,}076\, C^{1.85} D_{ft}^{4.87}}$$

$$= \frac{(4000\ \text{ft})\left(70\ \dfrac{\text{gal}}{\text{min}}\right)^{1.85}}{(17{,}076)(130)^{1.85}(0.67\ \text{ft})^{4.87}}$$

$$= +0.524\ \text{ft}$$

For pipe AC,

$$h_{L,AC} = \frac{L_{ft} Q_{gpm}^{1.85}}{17{,}076\, C^{1.85} D_{ft}^{4.87}}$$

$$= \frac{(4000\ \text{ft})\left(571\ \dfrac{\text{gal}}{\text{min}}\right)^{1.85}}{(17{,}076)(130)^{1.85}(0.83\ \text{ft})^{4.87}}$$

$$= -8.97\ \text{ft}$$

step 4: Sum the head losses in all pipes in loop ABC.

$$\sum h_{Li} = h_{L,AB} + h_{L,BC} + h_{L,AC}$$

$$= 7.49\ \text{ft} + 0.524\ \text{ft} - 8.97\ \text{ft}$$

$$= -0.96\ \text{ft}$$

step 5: If the sum of the head losses is greater than 1 ft, calculate the flow correction factor. For this example, step 5 is not needed because the sum of head losses in loop ABC is less than 1 ft. However, to demonstrate the concept of the flow correction factor, step 5 is performed as follows.

$$q_{\text{loop}} = \frac{-\sum_{\text{loop}} h_{Li}}{\sum_{\text{loop}} 1.85 \left| \dfrac{h_{Li}}{Q} \right|}$$

$$= \frac{-(-0.96\ \text{ft})}{(1.85)\left|\dfrac{7.49\ \text{ft}}{429\ \dfrac{\text{gal}}{\text{min}}}\right| + (1.85)\left|\dfrac{0.524\ \text{ft}}{70\ \dfrac{\text{gal}}{\text{min}}}\right| + (1.85)\left|\dfrac{8.97\ \text{ft}}{571\ \dfrac{\text{gal}}{\text{min}}}\right|}$$

$$= 12.8\ \text{gal/min}$$

Add 12.8 gal/min to the flow through each pipe, taking the flow direction into consideration. Repeat calculations starting in step 2 until the sum of head losses in all pipes in each loop is reasonably small.

$$Q_{AB} = 429\ \frac{\text{gal}}{\text{min}} + 12.8\ \frac{\text{gal}}{\text{min}}$$

$$= 441.8\ \text{gal/min} \quad (442\ \text{gal/min})$$

$$Q_{BC} = 70\ \frac{\text{gal}}{\text{min}} + 12.8\ \frac{\text{gal}}{\text{min}}$$

$$= 82.8\ \text{gal/min} \quad (83\ \text{gal/min})$$

$$Q_{CA} = 571\ \frac{\text{gal}}{\text{min}} - 12.8\ \frac{\text{gal}}{\text{min}}$$

$$= 558.2\ \text{gal/min} \quad (558\ \text{gal/min})$$

14. WATER HAMMER

Water hammer (also known as *fluid hammer*) occurs when a gate or valve in a piped system suddenly closes, resulting in an instantaneous decrease in flow velocity and a substantial increase in pressure. Unless controlled by closing valves, the shock wave caused by water hammer can cause water main breaks and pipe failure. Water hammer is affected by the pressure wave velocity, the length of time that the pressure is constant, and the resulting pressure increase after a valve closes.

The pressure wave velocity after a valve closes, C_p, is given by Eq. 2.26. E is the *modulus of elasticity*, equal to 300,000 lbf/in² (2068.4 MPa) in water, and E_p is the pipe modulus of elasticity. Table 2.4 gives approximate modulus of elasticity values for various materials at room temperature. D is the pipe diameter, and t_h is the pipe thickness.

$$C_p = C \sqrt{\frac{1}{1 + \dfrac{ED}{E_p t_h}}} \qquad 2.26$$

The pressure wave velocity in water or the speed of sound in water, C, is given by Eq. 2.27. The value of C is 4720 ft/sec (1440 m/s) at 50°F (10°C).

$$C = \sqrt{\frac{E}{\rho}} \qquad \text{[SI]} \quad 2.27(a)$$

$$C = \sqrt{\frac{Eg}{\gamma}} \qquad \text{[U.S.]} \quad 2.27(b)$$

The critical length of time pressure is constant at the valve and is found from Eq. 2.28 from the length of the pipe, L, and the pressure wave velocity, C.

$$t = \frac{2L}{C} \qquad 2.28$$

Table 2.4 *Approximate Modulus of Elasticity for Various Materials at Room Temperature*

material	lbf/in²	GPa
aluminum alloys	10–11 × 10⁶	70–80
brass	15–16 × 10⁶	100–110
cast iron	15–22 × 10⁶	100–150
cast iron, ductile	22–25 × 10⁶	150–170
cast iron, malleable	26–27 × 10⁶	180–190
copper alloys	17–18 × 10⁶	110–112
glass	7–12 × 10⁶	50–80
magnesium alloys	6.5 × 10⁶	45
molybdenum	47 × 10⁶	320
nickel alloys	26–30 × 10⁶	180–210
steel, hard*	30 × 10⁶	210
steel, soft*	29 × 10⁶	200
steel, stainless	28–30 × 10⁶	190–210
titanium	15–17 × 10⁶	100–110

*Common values given.
(Multiply lbf/in² by 6.89×10^{-6} to obtain GPa.)

Reprinted with permission from Michael R. Lindeburg, PE, *Civil Engineering Reference Manual*, 12th ed., copyright © 2011, by Professional Publications, Inc.

After the valve closes, flow velocity, v, decreases to zero, causing pressure to increase. This change in water pressure, Δp, after the valve closes and creates water hammer can be determined using Eq. 2.29.

$$\Delta p = \rho C_p \Delta v \qquad \text{[SI]} \qquad 2.29(a)$$

$$\Delta p = \frac{\gamma}{g} C_p \Delta v \qquad \text{[U.S.]} \qquad 2.29(b)$$

Example 2.17

Water flows at 3 ft/sec through a 5000 ft soft steel pipe with a wall thickness of 0.75 in and a pipe diameter of 24 in. What is the change in water pressure if a valve completely closes in 3 sec?

Solution

Use Eq. 2.26 to determine the pressure wave velocity after the valve closes.

$$C_p = C \sqrt{\frac{1}{1 + \dfrac{ED}{E_p t_h}}}$$

$$= \left(4720 \ \frac{\text{ft}}{\text{sec}}\right) \sqrt{\frac{1}{1 + \dfrac{\left(3 \times 10^5 \ \dfrac{\text{lbf}}{\text{in}^2}\right)(24 \ \text{in})}{\left(29 \times 10^6 \ \dfrac{\text{lbf}}{\text{in}^2}\right)(0.75 \ \text{in})}}}$$

$$= 4108 \ \text{ft/sec}$$

Use Eq. 2.29 to find the change in water pressure.

$$\Delta p = \frac{\gamma}{g} C_p \Delta v$$

$$= \frac{\left(\dfrac{62.4 \ \dfrac{\text{lbf}}{\text{ft}^3}}{32.2 \ \dfrac{\text{ft}}{\text{sec}^2}}\right)\left(4108 \ \dfrac{\text{ft}}{\text{sec}}\right)\left(3 \ \dfrac{\text{ft}}{\text{sec}}\right)}{\left(12 \ \dfrac{\text{in}}{\text{ft}}\right)^2}$$

$$= 165.9 \ \text{lbf/in}^2$$

15. FLOW MEASURING DEVICES

Flow through pressurized, closed-conduit pipes may be measured by orifice meters, flow nozzles, venturi meters, or pitot-static gauges (see Fig. 2.7). Orifice meters, flow nozzles, and venturi meters are examples of *obstruction meters*, which rely on a decrease in static pressure to measure the flow velocity.

Figure 2.7 *Flow Measuring Devices*

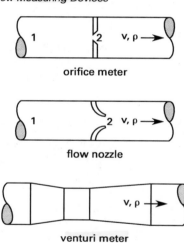

orifice meter

flow nozzle

venturi meter

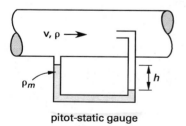

pitot-static gauge

The flow rate, Q, can be calculated for orifice meters, flow nozzles, and venturi meters using Eq. 2.30.

$$Q = C_f A_2 \sqrt{\frac{2g(p_1 - p_2)}{\gamma}} \qquad 2.30$$

Water Resources

C_f is the unitless *flow coefficient* and is calculated as the ratio of the actual flow rate to the theoretical flow rate.

$$C_f = \frac{Q_{\text{actual}}}{Q_{\text{theoretical}}} \qquad 2.31$$

A *pitot-static gauge* is a type of velocity probe that combines a pitot tube and a static pressure tap to measure flow velocity. In a pitot-static gauge, one end of a manometer is acted on by the static pressure, while the other end experiences the total pressure. The difference in elevation between the two ends determines the velocity of flow. The velocity for a pitot-static gauge is calculated using Eq. 2.32.

$$v = \sqrt{\frac{2gh(\rho_m - \rho)}{\rho}} \qquad 2.32$$

Example 2.18

Water ($\rho = 62.4$ lbm/ft^3) is flowing through a pipe. A pitot-static gauge measures 5 in of mercury ($\rho = 848.6$ lbm/ft^3). What is the velocity of the water in the pipe?

Solution

Use Eq. 2.32 to calculate the velocity.

$$v = \sqrt{\frac{2gh(\rho_m - \rho)}{\rho}} = \sqrt{\frac{(2)\left(32.2 \frac{\text{ft}}{\text{sec}^2}\right)(5 \text{ in})}{\times \left(848.6 \frac{\text{lbm}}{\text{ft}^3} - 62.4 \frac{\text{lbm}}{\text{ft}^3}\right)}{\left(62.4 \frac{\text{lbm}}{\text{ft}^3}\right)\left(12 \frac{\text{in}}{\text{ft}}\right)}}$$

$$= 18.4 \text{ ft/sec}$$

PRACTICE PROBLEMS

1. How many days are required to drain a 2 billion gallon reservoir with an outlet pipe capacity of 100 ft^3/sec?

2. Calculate the head loss in a 1000 ft long, 12 in diameter new ductile iron pipe with a flow rate of 1000 gal/min.

3. For water pumped through a 100 ft long new ductile iron pipe with a diameter of 12 in, the allowable net positive suction head for a pump is 50 ft. The flow rate is 1000 gal/min, and the roughness coefficient, C, is 130. Assume $p_{\text{vap}} = 0.5$ lbf/in^2 at 80°F. At what height above the pipe inlet water surface should the pump be located to avoid cavitation?

4. What is the flow through a 24 in pipe with a velocity of 5.00 ft/sec?

(A) 15.7 ft^3/sec

(B) 31.4 ft^3/sec

(C) 62.8 ft^3/sec

(D) 127 ft^3/sec

5. If the head loss is 0.5 ft, what is the flow rate through a 2 ft diameter, 2000 ft long new ductile iron pipe?

(A) 1000 gal/min

(B) 1500 gal/min

(C) 1800 gal/min

(D) 3400 gal/min

PITOT TUBE IN OPEN CHANNEL

$$V = C\sqrt{2gh_v}$$

SOLUTIONS

1. The time required to drain the reservoir is

$$t_{\text{days}} = \frac{V}{Q} = \frac{2 \times 10^9 \text{ gal}}{\left(100 \ \frac{\text{ft}^3}{\text{sec}}\right)\left(7.48 \ \frac{\text{gal}}{\text{ft}^3}\right)\left(60 \ \frac{\text{sec}}{\text{min}}\right)}$$

$$\times \left(60 \ \frac{\text{min}}{\text{hr}}\right)\left(24 \ \frac{\text{hr}}{\text{day}}\right)$$

$$= 31 \text{ days}$$

2. Use the Hazen-Williams equation. From Table 2.1, the roughness coefficient is 130.

$$h_L = \frac{L_{\text{ft}} \, Q_{\text{gpm}}^{1.85}}{17{,}076 \, C^{1.85} D_{\text{ft}}^{4.87}}$$

$$= \frac{(1000 \text{ ft})\left(1000 \ \frac{\text{gal}}{\text{min}}\right)^{1.85}}{(17{,}076)(130)^{1.85}(1 \text{ ft})^{4.87}}$$

$$= 2.6 \text{ ft}$$

3. Find the head loss using the Hazen-Williams equation.

$$h_L = \frac{L_{\text{ft}} \, Q_{\text{gpm}}^{1.85}}{17{,}076 \, C^{1.85} D_{\text{ft}}^{4.87}}$$

$$= \frac{(100 \text{ ft})\left(1000 \ \frac{\text{gal}}{\text{min}}\right)^{1.85}}{(17{,}076)(130)^{1.85}(1 \text{ ft})^{4.87}}$$

$$= 0.255 \text{ ft}$$

Solve Eq. 2.24 for Δz.

$$\text{NPSHA} \geq \text{NPSHR} = \frac{p_{\text{atm}} - p_{\text{vap}}}{\gamma} - \Delta z - h_L$$

$$\Delta z = \frac{p_{\text{atm}} - p_{\text{vap}}}{\gamma} - h_L - \text{NPSHR}$$

$$= \frac{\left(14.7 \ \frac{\text{lbf}}{\text{in}^2} - 0.5 \ \frac{\text{lbf}}{\text{in}^2}\right)\left(12 \ \frac{\text{in}}{\text{ft}}\right)^2}{62.4 \ \frac{\text{lbf}}{\text{ft}^3}}$$

$$- 0.255 \text{ ft} - 50 \text{ ft}$$

$$= -17.49 \text{ ft}$$

The pump should be placed no higher than 17.49 ft below the inlet end of the pipe.

4. Calculate the area.

$$A = \pi r^2 = \pi \left(\frac{D}{2}\right)^2$$

$$= \pi \left(\frac{24 \text{ in}}{(2)\left(12 \ \frac{\text{in}}{\text{ft}}\right)}\right)^2$$

$$= 3.14 \text{ ft}^2$$

Calculate the flow rate.

$$Q = vA$$

$$= \left(5.00 \ \frac{\text{ft}}{\text{sec}}\right)(3.14 \text{ ft}^2)$$

$$= 15.7 \text{ ft}^3/\text{sec}$$

The answer is (A).

5. From Table 2.1, the Hazen-Williams roughness coefficient for ductile iron pipe is 130. Rearrange Eq. 2.12 and solve for flow.

$$h_L = \frac{L_{\text{ft}} \, Q_{\text{gpm}}^{1.85}}{17{,}076 \, C^{1.85} D_{\text{ft}}^{4.87}}$$

$$Q^{1.85} = \frac{h_L (17{,}076) \, C^{1.85} D^{4.87}}{L}$$

$$= \frac{(0.50 \text{ ft})(17{,}076)(130)^{1.85}(2.0 \text{ ft})^{4.87}}{2000 \text{ ft}}$$

$$Q = 1766 \text{ gal/min} \quad (1800 \text{ gal/min})$$

The answer is (C).

Open Channel Hydraulics

Nomenclature

A	area	ft^2	m^2
b	throat width	ft	m
B	surface width	ft	m
c	coefficient	–	–
C	weir flow coefficient	–	–
C_c	contraction coefficient	–	–
C_d	discharge coefficient	–	–
C_e	expansion coefficient	–	–
d	flow depth	ft	m
d_1	flow depth upstream from hydraulic jump	ft	m
d_2	flow depth downstream from hydraulic jump	ft	m
d_a	upstream flow depth	ft	m
d_b	downstream flow depth	ft	m
d_c	critical depth	ft	m
d_H	depth of water surface flowing over a weir	ft	m
d_{HW}	depth of headwater upstream	ft	m
d_n	normal depth	ft	m
d_{TW}	depth of tailwater downstream	ft	m
D	diameter	in	mm
e	roughness of interior pipe surface	ft	m
f	friction factor	–	–
Fr	Froude number	–	–
g	gravitational acceleration	ft/sec^2	m/s^2
h	difference between upstream and downstream WSEL	ft	m
h	standard step parameter accounting for losses	ft	m
h	head	ft	m
H	height (weirs)	ft	m
i	intensity	in/hr	cm/h
I	inflow rate	ft^3/sec	m^3/s
K	conveyance	ft^3/sec	m^3/s
K	friction loss coefficient	–	–
K_e	culvert entrance loss coefficient	–	–
K_t	average conveyance for reach	ft^3/sec	m^3/s
K_{td}	conveyance in downstream sections of reach	ft^3/sec	m^3/s
K_{tu}	conveyance in upstream sections of reach	ft^3/sec	m^3/s
L	length of pipe or channel	ft	m
n	Manning roughness coefficient	–	–
n	Parshall flume exponent	–	–
N	number	–	–
O	outflow rate	ft^3/sec	m^3/s
p	pressure	lbf/in^2	kPa
p_{atm}	atmospheric pressure	lbf/in^2	kPa
p_{vap}	vapor pressure of water	lbf/in^2	kPa
q	specific discharge	ft^3/sec-ft	m^3/s·m
Q	flow rate	ft^3/sec	m^3/s
r	geometric radius of pipe	ft	m
R	hydraulic radius	ft	m
S	longitudinal slope of a pipe or channel	ft/ft	m/m
S_L	longitudinal slope of street	%	%
S_n	storage volume of basin or reservoir	ft^3	m^3
S_x	cross slope	%	%
t	time	sec	s
T	top gutter width	ft	m
TDH	total dynamic head	ft	m
v	velocity	ft/sec	m/s
V	volume	gal	gal
W	width	ft	m
W_o	gutter width of combination inlet	ft	m
WP	wetted perimeter	ft	m
WSEL	starting water surface elevation	ft	m
x	horizontal dimension of sideslope	ft	m
z	elevation	ft	m

Symbols

α	velocity distribution coefficient	

Subscripts

CHAN	channel
i	incremental
L	lanes or loss
LOB	left overbank
o	other losses

ROB   right overbank
t total

1. INTRODUCTION

Open channel hydraulic engineering applies the study of hydraulics to the design of stormwater systems, flood control channels, and roadway culverts and bridges. *Open channel flow*, which includes streams, rivers and floodplains, stormwater collection systems, and sanitary sewers, is open to the atmosphere. That is, open channel flow is only under atmospheric pressure. One of the most important equations for open channel flow is the *Bernoulli equation.*

2. BERNOULLI ENERGY EQUATION

In open channel flow calculations, the water is exposed to the atmosphere everywhere. Therefore, the atmospheric pressure term appears on both sides of the energy equation and cancels out. For open channels, the *Bernoulli energy equation* is

ALSO
CERM p.19-7

$$\frac{v_1^2}{2g} + d_1 + z_1 = \frac{v_2^2}{2g} + d_2 + z_2 \qquad 3.1$$

In the open channel flow regime shown in Fig. 3.1, flow is open to the atmosphere at section 1 and section 2.

Figure 3.1 *Open Channel Flow Water Surface Profile*

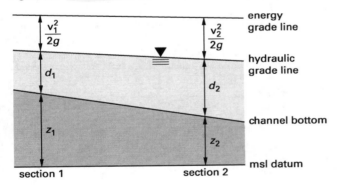

The tank in Fig. 3.2 is open to the atmosphere at the top water surface (elevation 1) and at the orifice below (elevation 2). The atmospheric pressure, p_{atm}, at elevation 1 and elevation 2 are equal. The water level d_1 is assumed to be constant. Because there is no variable channel bottom to consider, z_1 and z_2 are relative to the same datum. Therefore, $\Delta z = 0$. The Bernoulli equation reduces to

$$\frac{v_1^2}{2g} + d_1 = \frac{v_2^2}{2g} + d_2 \qquad 3.2$$

Figure 3.2 *Velocity of Flow Exiting a Tank*

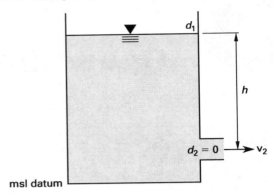

The velocity, v_1, at the water surface in the tank is 0 ft/sec. The equation becomes

$$d_1 = \frac{v_2^2}{2g} + d_2 \qquad 3.3$$

The distance, h, between d_1 and d_2 is $d_1 - d_2$. Therefore,

$$h = \frac{v_2^2}{2g} \qquad 3.4$$

The velocity at the outlet is

$$v_2 = \sqrt{2gh} \qquad 3.5$$

Example 3.1

Refer to the open channel in Fig. 3.1. At section 1, the velocity is 5 ft/sec, and the water surface elevation is 100 ft above mean sea level (msl). The elevation at section 2 is 99 ft msl. Calculate the velocity at section 2.

Solution

The total surface elevations are

$$d_1 + z_1 = 100 \text{ ft msl}$$
$$d_2 + z_2 = 99 \text{ ft msl}$$

Use Eq. 3.1 to find the velocity at section 2.

$$\frac{v_1^2}{2g} + d_1 + z_1 = \frac{v_2^2}{2g} + d_2 + z_2$$

$$\frac{\left(5 \, \frac{\text{ft}}{\text{sec}}\right)^2}{(2)\left(32.2 \, \frac{\text{ft}}{\text{sec}^2}\right)} + 100 \text{ ft} = \frac{v_2^2}{(2)\left(32.2 \, \frac{\text{ft}}{\text{sec}^2}\right)} + 99 \text{ ft}$$

$$v_2 = 9.5 \text{ ft/sec}$$

Example 3.2

For the tank in Fig. 3.2, the water level is kept constant at 10 ft above the centerline of the orifice outlet. The diameter of the orifice is 6 in. Ignoring friction losses, calculate the velocity and flow rate at the outlet of the tank.

Solution

Using Eq. 3.5,

$$v_2 = \sqrt{2gh}$$
$$= \sqrt{(2)\left(32.2 \ \frac{ft}{sec^2}\right)(10 \ ft)}$$
$$= 25.4 \ ft/sec$$

The area of the orifice is

$$A = \pi r^2 = \pi\left(\frac{D}{2}\right)^2$$
$$= \pi\left(\frac{6 \ in}{(2)\left(12 \ \frac{in}{ft}\right)}\right)^2$$
$$= 0.196 \ ft^2$$

The flow rate from the orifice at the bottom of the tank is

$$Q = vA$$
$$= \left(25.4 \ \frac{ft}{sec}\right)(0.196 \ ft^2)$$
$$= 4.98 \ ft^3/sec \quad (5 \ ft^3/sec)$$

3. CONTINUITY IN OPEN CHANNEL FLOW

The same *continuity equation* from Chap. 2 that applies to open channel flow also applies to flow through conduits.

$$Q_1 = Q_2 = Q_3 = Q_n$$
$$A_1v_1 = A_2v_2 = A_3v_3 = A_nv_n \qquad \textbf{3.6}$$

Example 3.3

Streamflow through a creek at section 1 has a cross-sectional area of 20 ft² and a velocity of 4 ft/sec. Downstream at section 2, the velocity is measured at 2 ft/sec. What is the flow through the stream? What is the area at section 2?

Solution

Using Eq. 3.6, the flows at section 1 and section 2 are

$$Q_1 = A_1v_1$$
$$= (20 \ ft^2)\left(4 \ \frac{ft}{sec}\right)$$
$$= 80 \ ft^3/sec$$

Continuity requires that

$$Q = Q_1 = Q_2$$
$$= A_1v_1 = A_2v_2$$

At section 2, the velocity is 2 ft/sec. Therefore, the area is

$$A_2 = \frac{Q}{v_2}$$
$$= \frac{80 \ \frac{ft^3}{sec}}{2 \ \frac{ft}{sec}}$$
$$= 40 \ ft^2$$

4. FLOW REGIMES

There are four general *flow regimes* in hydraulic engineering.

1. *Steady flow:* The depth of flow does not change over time. The peak flow of a storm is used in a steady flow regime to design storm sewers, culverts, and open channels. The peak flow is taken to be constant, and the design characteristics are calculated to accommodate it. For example, runoff equations such as the rational method (see Sec. 1.11) are used to calculate peak flows in a steady flow regime.

2. *Unsteady Flow:* The depth of flow changes with time. Unsteady flow is depicted in a hydrograph with increasing and decreasing rainfall volume. The flow rate increases to a peak, then decreases after the storm (see Fig. 1.11).

3. *Uniform flow:* The depth and velocity of flow along the channel is uniform for a pipeline or an engineered trapezoidal water supply canal (see Fig. 3.3). The shape and slope of the channel is uniform along the entire length. Most engineered open channels are examples of uniform flow.

4. *Nonuniform, varied flow:* The depth and velocity changes along the channel as the sections and channel slopes vary. Nonuniform flow occurs along natural rivers and streams. The channel sections have different shapes and areas, while longitudinal slopes range between mild and steep depending on the stream reach (see Fig. 3.4).

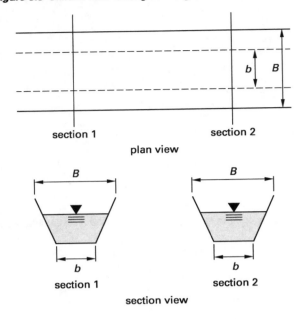

Figure 3.3 *Uniform Flow Through an Engineered Canal*

plan view

section view

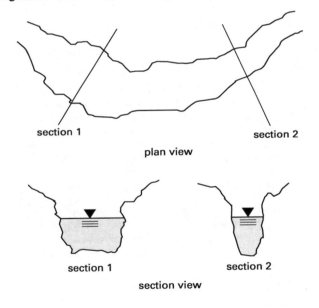

Figure 3.4 *Nonuniform Flow Through a Natural River System*

plan view

section view

5. OPEN CHANNEL FLOW

Open channel flow through streams, rivers, storm sewers, canals, and other conveyances is channel flow that is open to the atmosphere and not under pressure. The velocity and flow rate in open channels are calculated from the hydraulic radius, R, and the longitudinal slope of the channel, S, using the *Manning equation*. (The time of concentration calculations in Sec. 1.11 use the

Manning equation with *TR-55* Worksheet 3 to calculate the channel flow component.)

$$v = \left(\frac{1.00}{n}\right) R^{2/3}\sqrt{S} \qquad \text{[SI]} \qquad 3.7(a)$$

$$v = \left(\frac{1.49}{n}\right) R^{2/3}\sqrt{S} \qquad \text{[U.S.]} \qquad 3.7(b)$$

The *Manning roughness coefficient*, n, for different channel types is obtained from Table 3.1.

Table 3.1 *Manning Roughness Coefficients*

channel type	Manning roughness coefficient, n
asphalt	0.016
brass	0.011
brick	0.015
cast iron	0.013
clay tile	0.014
concrete	0.013
corrugated metal	0.022
earth channel	0.022
farm	0.035
finished concrete	0.012
heavy brush	0.075
masonry, stones	0.025
meadow	0.050
natural streams	0.030
polyvinyl chloride (PVC)	0.010
riprap, rock	0.035
steel	0.012
trees	0.150
unfinished concrete	0.014

Reprinted from *Urban Drainage Design Manual*, 3rd ed., Federal Highway Administration, 2009.

The cross-sectional area of the channel is A. The wetted perimeter, WP, is the length of the channel cross-sectional perimeter that is wetted by the flow. The *hydraulic radius*, R, is

$$R = \frac{A}{\text{WP}} \qquad 3.8$$

Flow is the product of velocity and area.

$$Q = vA \qquad 3.9$$

Therefore, the Manning equation can be multiplied by area to calculate the flow rate in an open channel as

$$Q = \left(\frac{1.00}{n}\right) A R^{2/3}\sqrt{S} \qquad \text{[SI]} \qquad 3.10(a)$$

$$Q = \left(\frac{1.49}{n}\right) A R^{2/3}\sqrt{S} \qquad \text{[U.S.]} \qquad 3.10(b)$$

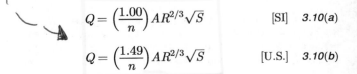

Several of the parameters (the hydraulic radius, the slope, the area, and the wetted perimeter) in the Manning equations, Eq. 3.7 and Eq. 3.10, are based on geometric considerations, which are described in the following subsections.

Cross-Sectional Area, *A*

The area of rectangular, triangular, or trapezoidal channels can be computed geometrically by summing the area of the triangles and rectangles. It is important to sketch out and mark the triangles and squares on the cross section to calculate the area and wetted perimeter.

The area of a *rectangular channel*, with a width of *b* and a depth of flow of *d* (see Fig. 3.5), is

$$A = bd \qquad 3.11$$

Figure 3.5 *Rectangular Open Channel*

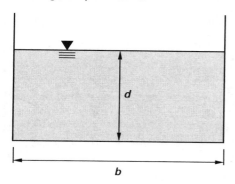

The area of a *triangular channel* (see Fig. 3.6) is

$$A = \tfrac{1}{2}bd \qquad 3.12$$

Figure 3.6 *Triangular Open Channel*

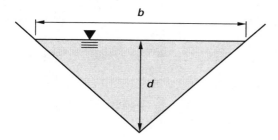

The area of a *trapezoidal channel* is

$$A = bd + xd = (b + x)d \qquad 3.13$$

The first term on the right-hand side of Eq. 3.13 represents the rectangle in the middle of the trapezoid. The second term represents the right triangles on the sides. The parameters *x* and *d* are the lengths of the sides of those triangles (see Fig. 3.7).

Figure 3.7 *Trapezoidal Open Channel*

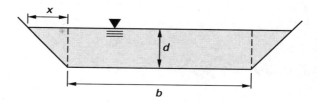

The cross-sectional area of a *circular pipe flowing full* is

$$A = \pi r^2 \qquad 3.14$$

The area of a *circular pipe flowing half full* is

$$A = \frac{\pi r^2}{2} \qquad 3.15$$

For pipes flowing at some level other than half full, the area of flow can be found in Table 3.2, using the variables defined in Fig. 3.8. Table 3.2 simplifies the calculation of the area of the pipe.

Table 3.2 *Hydraulic Properties of Circular Pipes*   (APP. 16A IN CERM)

d/D	A/D^2	WP/D	R/D
0.05	0.0147	0.4510	0.0326
0.10	0.0409	0.6435	0.0635
0.15	0.0739	0.7954	0.0929
0.20	0.1118	0.9273	0.1206
0.25	0.1535	1.0472	0.1466
0.30	0.1982	1.1593	0.1709
0.35	0.2450	1.2661	0.1935
0.40	0.2934	1.3694	0.2142
0.45	0.3428	1.4706	0.2331
0.50	0.3927	1.5708	0.2500
0.55	0.4426	1.6710	0.2649
0.60	0.4920	1.7722	0.2776
0.65	0.5404	1.8755	0.2881
0.70	0.5872	1.9823	0.2962
0.75	0.6318	2.0944	0.3017
0.80	0.6736	2.2143	0.3042
0.85	0.7115	2.3462	0.3033
0.90	0.7445	2.4981	0.2980
0.95	0.7707	2.6906	0.2864
1.00	0.7854	3.1416	0.2500

Figure 3.8 *Circular Open Channel*

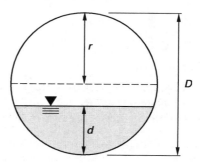

Example 3.4

What is the area of a pipe with a 4 ft diameter flowing 1 ft deep?

Solution

The area of the pipe is

$$\frac{d}{D} = \frac{1 \text{ ft}}{4 \text{ ft}} = 0.25$$

Using Table 3.2,

$$\frac{A}{D^2} = 0.1535$$

$$A = 0.1535 D^2$$

$$= (0.1535)(4 \text{ ft})^2$$

$$= 2.46 \text{ ft}^2$$

Example 3.5

What is the area of a 3 ft deep × 10 ft top width triangular channel?

Solution

The area of the triangular channel is

$$A = \tfrac{1}{2}bd$$

$$= \left(\tfrac{1}{2}\right)(3 \text{ ft})(10 \text{ ft})$$

$$= 15 \text{ ft}^2$$

Example 3.6

What is the area of a 3 ft deep trapezoidal channel with 1 ft horizontal to 1 ft vertical sideslopes and a 10 ft bottom?

Solution

From Eq. 3.13, the area of the channel is

$$A = bd + xd$$

$$= (10 \text{ ft})(3 \text{ ft}) + (3 \text{ ft})(3 \text{ ft})$$

$$= 39 \text{ ft}^2$$

Wetted Perimeter

The *wetted perimeter*, WP, of rectangular, triangular, and trapezoidal channels can be calculated using geometry. The wetted perimeter of a rectangular channel is

$$\text{WP} = 2d + b \qquad\qquad 3.16$$

Using the Pythagorean theorem, the wetted perimeter of a triangular channel is

$$\text{WP} = 2\sqrt{d^2 + \left(\frac{b}{2}\right)^2} \qquad\qquad 3.17$$

The wetted perimeter of a trapezoidal channel is

$$\text{WP} = b + 2\sqrt{d^2 + x^2} \qquad\qquad 3.18$$

The vertical dimension of the sideslope is d (see Fig. 3.7). The horizontal dimension of the sideslope is x.

The wetted perimeter of circular pipes flowing full is

$$\text{WP} = 2\pi r \qquad\qquad 3.19$$

The wetted perimeter of circular pipes flowing half full is

$$\text{WP} = \frac{2\pi r}{2} = \pi r \qquad\qquad 3.20$$

For circular pipes flowing greater than or less than half full, the wetted perimeter can be calculated using Table 3.2.

Example 3.7

What is the wetted perimeter of a 3 ft deep × 10 ft wide rectangular channel?

Solution

From Eq. 3.17, the wetted perimeter is

$$\text{WP} = 2d + b$$

$$= (2)(3 \text{ ft}) + 10 \text{ ft}$$

$$= 16 \text{ ft}$$

Example 3.8

What is the wetted perimeter of a 3 ft deep × 10 ft top width triangular channel?

Solution

From Eq. 3.17, the wetted perimeter is

$$\text{WP} = 2\sqrt{d^2 + \left(\frac{b}{2}\right)^2}$$

$$= (2)\sqrt{(3 \text{ ft})^2 + \left(\frac{10 \text{ ft}}{2}\right)^2}$$

$$= 11.7 \text{ ft}$$

Example 3.9

What is the wetted perimeter of a 3 ft deep × 10 ft bottom width trapezoidal channel with 1 ft horizontal to 1 ft vertical sideslopes?

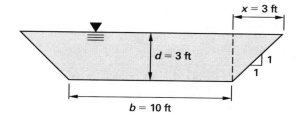

Solution

From Eq. 3.18, the wetted perimeter is

$$WP = b + 2\sqrt{d^2 + x^2}$$
$$= 10 \text{ ft} + (2)\sqrt{(3 \text{ ft})^2 + (3 \text{ ft})^2}$$
$$= 18.5 \text{ ft}$$

Example 3.10

What is the wetted perimeter of a 4 ft diameter pipe flowing 1 ft deep?

Solution

Calculate the d/D ratio of the pipe.

$$\frac{d}{D} = \frac{1 \text{ ft}}{4 \text{ ft}} = 0.25$$

For a d/D of 0.25 and using Table 3.2, the wetted perimeter is

$$\frac{WP}{D} = 1.0472$$
$$WP = (1.0472)(4 \text{ ft})$$
$$= 4.19 \text{ ft}$$

Channel Slope

The slope of a channel or pipe is

$$S = \frac{z_1 - z_2}{L} \qquad\qquad 3.21$$

The *thalweg* is the lowest elevation in a stream channel or waterway. The *invert* is the lowest elevation in a pipe section. z_1 is the elevation of the channel thalweg or pipe invert at section 1. The elevation of the channel thalweg or pipe invert at section 2 is z_2. The horizontal distance between the two sections is L.

Example 3.11

What is the slope of a stream channel in which the thalweg elevation at section 1, 2100 ft upstream from a bridge, is 100 ft above sea level, and the thalweg elevation at section 2, 2200 ft upstream from a bridge, is 99 ft?

Solution

Use Eq. 3.21.

$$S = \frac{z_1 - z_2}{L}$$
$$= \frac{100 \text{ ft} - 99 \text{ ft}}{2200 \text{ ft} - 2100 \text{ ft}}$$
$$= 0.01 \text{ ft/ft}$$

Example 3.12

A rectangular concrete channel is 5 m deep and 10 m wide. The channel slope decreases 1 m over the next 1 km. Calculate the flow through the channel.

Solution

The wetted perimeter is calculated using Eq. 3.16.

$$WP = 2d + b$$
$$= (2)(5 \text{ m}) + 10 \text{ m}$$
$$= 20 \text{ m}$$

The hydraulic radius is found from Eq. 3.8.

$$R = \frac{A}{WP} = \frac{bd}{WP}$$
$$= \frac{(10 \text{ m})(5 \text{ m})}{20 \text{ m}}$$
$$= 2.5 \text{ m}$$

The slope is found from Eq. 3.21.

$$S = \frac{z_1 - z_2}{L}$$
$$= \frac{1 \text{ m}}{(1 \text{ km})\left(1000 \ \frac{\text{m}}{\text{km}}\right)}$$
$$= 0.001 \text{ m/m}$$

Table 3.1 gives the Manning roughness coeffficient for concrete as 0.013. The flow rate is determined by using the Manning equation, Eq. 3.10.

$$Q = \left(\frac{1.00}{n}\right) A R^{2/3} \sqrt{S}$$
$$= \left(\frac{1.00}{0.013}\right)(50 \text{ m}^2)(2.5 \text{ m})^{2/3} \sqrt{0.001 \ \frac{\text{m}}{\text{m}}}$$
$$= 224 \text{ m}^3/\text{s}$$

Example 3.13

The trapezoidal earth channel shown has a 20 ft bottom width, sideslopes at 1 ft horizontal to 1 ft vertical (1:1), and a depth of flow of 5 ft. The difference in the channel bottom elevation is 10 ft over 1000 ft of length. Calculate the flow through the channel.

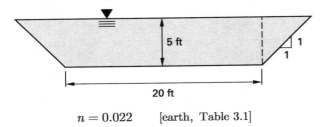

$$n = 0.022 \qquad [\text{earth, Table 3.1}]$$

Solution

From Eq. 3.13, the area of the trapezoidal channel is

$$
\begin{aligned}
A &= bd + xd \\
&= (20 \text{ ft})(5 \text{ ft}) + (5 \text{ ft})(5 \text{ ft}) \\
&= 125 \text{ ft}^2
\end{aligned}
$$

From Eq. 3.18, the wetted perimeter is

$$
\begin{aligned}
\text{WP} &= b + 2\sqrt{d^2 + x^2} \\
&= 20 \text{ ft} + (2)\sqrt{(5 \text{ ft})^2 + (5 \text{ ft})^2} \\
&= 34.1 \text{ ft}
\end{aligned}
$$

From Eq. 3.8, the hydraulic radius is

$$
\begin{aligned}
R &= \frac{A}{\text{WP}} = \frac{125 \text{ ft}^2}{34.1 \text{ ft}} \\
&= 3.66 \text{ ft}
\end{aligned}
$$

From Eq. 3.21, the slope is

$$
\begin{aligned}
S &= \frac{z_1 - z_2}{L} \\
&= \frac{10 \text{ ft}}{1000 \text{ ft}} \\
&= 0.01 \text{ ft/ft}
\end{aligned}
$$

From Table 3.1, the Manning roughness coefficient for an earth channel is 0.022. Using Eq. 3.10, find the flow through the channel.

$$
\begin{aligned}
Q &= \left(\frac{1.49}{n}\right) A R^{2/3} \sqrt{S} \\
&= \left(\frac{1.49}{0.022}\right)(125 \text{ ft}^2)(3.66 \text{ ft})^{2/3} \sqrt{0.01 \; \frac{\text{ft}}{\text{ft}}} \\
&= 2011 \text{ ft}^3/\text{sec}
\end{aligned}
$$

Example 3.14

A trapezoidal natural channel has a bottom width of 20 ft at elevation 92 ft msl and sideslopes at 5 ft horizontal to 1 ft vertical. The longitudinal slope of the stream channel is 0.002 ft/ft. Calculate the flow depth necessary to convey the 10 year flow of 175 ft³/sec through the channel.

Solution

The solution is most efficiently calculated by trial and error, as summarized in the *Trapezoidal Channel Flow Depth Calculation* table.

Use the Manning equation and solve for depth. From Table 3.1, the roughness coefficient for a natural channel is 0.03. The first term of the Manning equation is

$$
\begin{aligned}
\frac{1.49}{n} &= \frac{1.49}{0.03} \\
&= 49.7
\end{aligned}
$$

At elevation 92 ft, the depth of flow is zero. Therefore, A, WP, v, and Q are all zero.

At elevation 93 ft,

$$d = 93 \text{ ft} - 92 \text{ ft} = 1 \text{ ft}$$

Calculate the area of the channel at a depth of 1 ft using Eq. 3.13.

$$
\begin{aligned}
A &= bd + xd \\
&= (20 \text{ ft})(1 \text{ ft}) + (5 \text{ ft})(1 \text{ ft}) \\
&= 25 \text{ ft}^2
\end{aligned}
$$

Calculate the wetted perimeter at a depth of 1 ft using Eq. 3.18.

$$
\begin{aligned}
\text{WP} &= b + 2\sqrt{d^2 + x^2} \\
&= 20 \text{ ft} + (2)\sqrt{(1 \text{ ft})^2 + (5 \text{ ft})^2} \\
&= 30.2 \text{ ft}
\end{aligned}
$$

Calculate the hydraulic radius using Eq. 3.8.

$$
\begin{aligned}
R &= \frac{A}{\text{WP}} \\
&= \frac{25 \text{ ft}^2}{30.2 \text{ ft}} \\
&= 0.828 \text{ ft}
\end{aligned}
$$

Trapezoidal Channel Flow Depth Calculation

elevation (ft)	depth (ft)	$\dfrac{1.49}{n}$	$\sqrt{S}$ (ft/ft)	A (ft²)	WP (ft)	$R = \dfrac{A}{\text{WP}}$ (ft)	$R^{2/3} = \left(\dfrac{A}{\text{WP}}\right)^{2/3}$	$v = \left(\dfrac{1.49}{n}\right)R^{2/3}\sqrt{S}$ (ft/sec)	$Q = vA$ (ft³/sec)
92	0	49.7	0.0447	0	0.0	0.000	0.0000	0.00	0
93	1	49.7	0.0447	25	30.2	0.828	0.882	1.96	49
94	2	49.7	0.0447	60	40.4	1.49	1.30	2.89	173

The velocity at a depth of 1 ft is calculated using Eq. 3.7.

$$v = \left(\frac{1.49}{n}\right)R^{2/3}\sqrt{S}$$
$$= \left(\frac{1.49}{0.03}\right)(0.828\ \text{ft})^{2/3}\sqrt{0.002\ \frac{\text{ft}}{\text{ft}}}$$
$$= 1.96\ \text{ft/sec}$$

From Eq. 3.9, the flow rate at a depth of 1 ft is

$$Q = vA$$
$$= \left(1.96\ \frac{\text{ft}}{\text{sec}}\right)(25\ \text{ft}^2)$$
$$= 49\ \text{ft}^3/\text{sec} \quad [<Q_{10\,\text{yr}} = 175\ \text{ft}^3/\text{sec}]$$

At an elevation of 94 ft, the flow depth is

$$d = 94\ \text{ft} - 92\ \text{ft} = 2\ \text{ft}$$

Calculate the area of the channel at 2 ft deep using Eq. 3.13.

$$A = bd + xd$$
$$= (20\ \text{ft})(2\ \text{ft}) + (10\ \text{ft})(2\ \text{ft})$$
$$= 60\ \text{ft}^2$$

Calculate the wetted perimeter at a depth of 2 ft using Eq. 3.18.

$$\text{WP} = b + 2\sqrt{d^2 + x^2}$$
$$= 20\ \text{ft} + (2)\sqrt{(2\ \text{ft})^2 + (10\ \text{ft})^2}$$
$$= 40.4\ \text{ft}$$

Calculate the hydraulic radius at a depth of 2 ft using Eq. 3.8.

$$R = \frac{A}{\text{WP}}$$
$$= \frac{60\ \text{ft}^2}{40.4\ \text{ft}}$$
$$= 1.49\ \text{ft}$$

The velocity at a depth of 2 ft is calculated using Eq. 3.7.

$$v = \left(\frac{1.49}{n}\right)R^{2/3}\sqrt{S}$$
$$= \left(\frac{1.49}{0.03}\right)(1.49\ \text{ft})^{2/3}\sqrt{0.002\ \frac{\text{ft}}{\text{ft}}}$$
$$= 2.89\ \text{ft/sec}$$

From Eq. 3.9, the flow rate at a depth of 2 ft is

$$Q = vA$$
$$= \left(2.89\ \frac{\text{ft}}{\text{sec}}\right)(60\ \text{ft}^2)$$
$$= 173\ \text{ft}^3/\text{sec} \quad [= Q_{10\,\text{yr}},\ \text{therefore OK}]$$

The depth of flow is 2 ft, calculated for a 10 yr flow rate of 175 ft³/sec. The 10 yr flow elevation is

$$92\ \text{ft} + 2\ \text{ft} = 94\ \text{ft}$$

Example 3.15

Calculate the channel slope necessary to convey a flow of 100 ft³/sec through a 20 ft wide × 5 ft deep concrete rectangular channel.

Solution

Table 3.1 gives the Manning roughness coefficient for concrete.

$$n = 0.013$$

The channel is rectangular, so use Eq. 3.11 to find the area.

$$A = bd = (20\ \text{ft})(5\ \text{ft})$$
$$= 100\ \text{ft}^2$$

The wetted perimeter is found using Eq. 3.16 for a rectangular channel.

$$\text{WP} = 2d + b = (2)(5 \text{ ft}) + 20 \text{ ft}$$
$$= 30 \text{ ft}$$

From Eq. 3.8, the hydraulic radius is

$$R = \frac{A}{\text{WP}} = \frac{100 \text{ ft}^2}{30 \text{ ft}}$$
$$= 3.33 \text{ ft}$$

Use Eq. 3.10 and solve for the slope.

$$Q = \left(\frac{1.49}{n}\right) AR^{2/3}\sqrt{S}$$
$$100 \; \frac{\text{ft}^3}{\text{sec}} = \left(\frac{1.49}{0.013}\right)(100 \text{ ft}^2)(3.33 \text{ ft})^{2/3}\sqrt{S}$$
$$S = 0.0000153 \text{ ft/ft} \quad (0.0015\% \text{ slope})$$

Example 3.16

A 48 in circular concrete pipe as shown conveys flow at a depth of half full. If the longitudinal slope of the pipe is 0.02 ft/ft, calculate the velocity and flow rate.

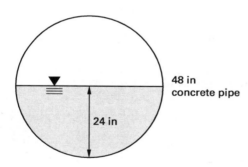

48 in concrete pipe

24 in

Solution

Table 3.1 gives the Manning roughness coefficient for concrete.

$$n = 0.013$$

The area of the flow is half the area of the circular pipe. Use Eq. 3.15.

$$A = \frac{\pi r^2}{2} = \frac{\pi \left(\dfrac{24 \text{ in}}{12 \; \frac{\text{in}}{\text{ft}}}\right)^2}{2}$$
$$= 6.28 \text{ ft}^2$$

Calculate the wetted perimeter using Eq. 3.19.

$$\text{WP} = \pi r = \pi \left(\frac{24 \text{ in}}{12 \; \frac{\text{in}}{\text{ft}}}\right)$$
$$= 6.28 \text{ ft}$$

From Eq. 3.8, the hydraulic radius is

$$R = \frac{A}{\text{WP}} = \frac{6.28 \text{ ft}^2}{6.28 \text{ ft}}$$
$$= 1 \text{ ft}$$

Substitute the parameters into the Manning equation, Eq. 3.7.

$$v = \left(\frac{1.49}{n}\right) R^{2/3}\sqrt{S}$$
$$= \left(\frac{1.49}{0.013}\right)(1 \text{ ft})^{2/3}\sqrt{0.02 \; \frac{\text{ft}}{\text{ft}}}$$
$$= 16.2 \text{ ft/sec}$$

From Eq. 3.9, the flow rate is

$$Q = vA = \left(16.2 \; \frac{\text{ft}}{\text{sec}}\right)(6.3 \text{ ft}^2)$$
$$= 102 \text{ ft}^3/\text{sec}$$

6. SUBCRITICAL AND SUPERCRITICAL FLOW

From an energy standpoint, *critical flow* is the most efficient flow in an open channel. The specific discharge, q, in a rectangular channel with flow rate, Q, and flow width, b, is

$$q = \frac{Q}{b} \qquad \text{3.22}$$

The *critical depth*, d_c, is

$$d_c = \frac{q^{2/3}}{g^{1/3}} \qquad \text{3.23}$$

If the *normal depth*, d_n, calculated with the Manning equation is greater than the critical depth, d_c, then the flow is *subcritical* and the velocity is <u>low</u>.

$$d_n > d_c \quad [\text{subcritical flow depth}] \qquad \text{3.24}$$

If the normal depth, d_n, is equal to the critical depth, d_c, then critical flow exists (e.g., as at a hydraulic jump at the base of a dam).

$$d_n = d_c \quad [\text{critical flow depth}] \qquad \text{3.25}$$

If the normal depth, d_n, is less than the critical depth, d_c, then the flow is *supercritical* and the velocity is high (e.g., as at the riffles or rapids section (the whitewater) of a stream).

$$d_n < d_c \quad \text{[supercritical flow depth]} \qquad 3.26$$

Example 3.17

Calculate the critical depth in a rectangular channel in which the flow rate is 50 ft³/sec and the flow width is 10 ft.

Solution

Using Eq. 3.22 and Eq. 3.23, calculate the critical depth for a rectangular channel.

$$q = \frac{Q}{b} = \frac{50 \ \dfrac{\text{ft}^3}{\text{sec}}}{10 \ \text{ft}}$$

$$= 5 \ \text{ft}^3/\text{sec-ft}$$

$$d_c = \frac{q^{2/3}}{g^{1/3}} = \frac{\left(5 \ \dfrac{\text{ft}^3}{\text{sec-ft}}\right)^{2/3}}{\left(32.2 \ \dfrac{\text{ft}}{\text{sec}^2}\right)^{1/3}}$$

$$= 0.9 \ \text{ft}$$

7. GRADUALLY VARIED FLOW

Gradually varied flow occurs in open channels in which the flow depth and velocity vary slowly from section to section. In contrast, rapidly varied flow undergoes swift changes in depth and velocity, such as the flow at a hydraulic jump over a spillway. Water surface depth profiles can be estimated for gradually varied flow using the normal depth, d_n, as calculated by the Manning equation; using the critical depth, d_c, from Eq. 3.23; using Eq. 3.30 for the Froude number, Fr; or using the standard step method discussed in Sec. 3.10.

The *Froude number* in a rectangular channel can be estimated by

$$\text{Fr} = \frac{q}{\sqrt{gd_1^3}} \qquad 3.27$$

CERH
p.19-18/19
MORE FR
EQUATIONS

The *specific discharge* (flow rate), q, in a rectangular channel is calculated using Eq. 3.22. The depth of flow is d, and g is the gravitational acceleration.

Figure 3.9 describes typical water surface profiles for mild and steep channel slopes with gradually varied flow. Mild channel slope profiles M1, M2, and M3 are controlled by downstream water surface elevations. Steep channel slopes S1, S2, and S3 are controlled by upstream water surface elevations.

Example 3.18

The flow in a 10 ft wide, concrete, rectangular channel is 1900 ft³/sec. The flow depth upstream is 12 ft, and the downstream flow is critical. The channel slope is 0.002 ft/ft. Determine the normal and critical depths, and estimate the type of water surface profile.

Solution

Determine the normal depth using trial and error and the Manning equation. Assume a normal depth, d_n, of 15 ft. Table 3.1 gives the Manning roughness coefficient of concrete as 0.013. From Eq. 3.7,

$$Q = \left(\frac{1.49}{n}\right) A R^{2/3} \sqrt{S}$$

$$= \left(\frac{1.49}{0.013}\right)(15 \ \text{ft})(10 \ \text{ft})\left(\frac{(15 \ \text{ft})(10 \ \text{ft})}{15 \ \text{ft} + 10 \ \text{ft} + 15 \ \text{ft}}\right)^{2/3}$$

$$\times \sqrt{0.002 \ \frac{\text{ft}}{\text{ft}}}$$

$$= 1855 \ \text{ft}^3/\text{sec}$$

This is close enough to the known flow rate, so use a normal depth, d_n, of 15 ft.

Determine the critical depth from Eq. 3.22 and Eq. 3.23.

$$d_c = \frac{q^{2/3}}{g^{1/3}} = \frac{\left(\dfrac{Q}{b}\right)^{2/3}}{g^{1/3}} = \frac{\left(\dfrac{1900 \ \dfrac{\text{ft}^3}{\text{sec}}}{10 \ \text{ft}}\right)^{2/3}}{\left(32.2 \ \dfrac{\text{ft}}{\text{sec}^2}\right)^{1/3}}$$

$$= 10.4 \ \text{ft}$$

The normal depth of 15 ft is greater than the critical depth of 10.4 ft. Therefore, the flow is subcritical and mild. The depth at the upstream end, d_1, is 12 ft, which is greater than critical depth and less than normal depth. Therefore, $d_n > d_1 > d_c$, and the water surface profile type is mild at M2, as shown in Fig. 3.9.

8. HYDRAULIC JUMP

Hydraulic jump occurs when supercritical flow at high velocity abruptly transitions to subcritical flow at low velocity. This occurs when a steeply-sloped stream channel abruptly changes to a mild slope. The high velocity flow at lower depth transitions quickly to a low velocity flow at higher depth. Hydraulic jump commonly occurs at the base of dams or spillways, or at rapids or waterfalls along rivers and streams. Figure 3.10 describes the characteristics of hydraulic jump.

To calculate the parameters of hydraulic jump through a rectangular channel, three equations are necessary: the critical depth equation (see Eq. 3.23), the Froude

Figure 3.9 *Gradually Varied Flow Water Surface Profiles*

slope/profile	flow depth	Froude number, Fr
mild, M1	$d > d_n > d_c$	< 1
mild, M2	$d_n > d > d_c$	< 1
mild, M3	$d_n > d_c > d$	> 1
steep, S1	$d > d_c > d_n$	< 1
steep, S2	$d_c > d > d_n$	> 1
steep, S3	$d_c > d_n > d$	> 1

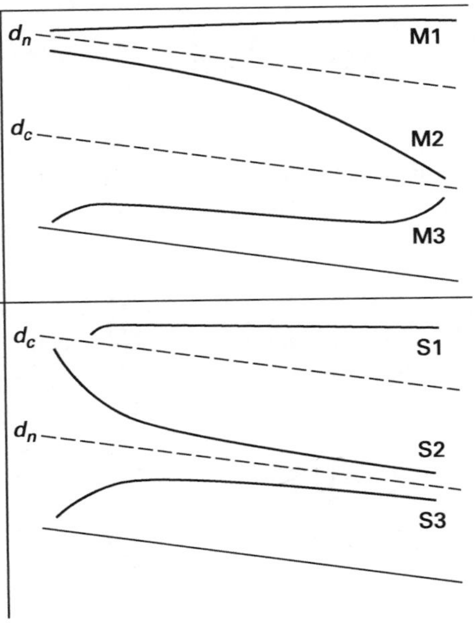

Figure 3.10 *Hydraulic Jump*

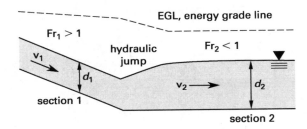

number equation (see Eq. 3.27) upstream from the hydraulic jump, and the hydraulic jump equation (see Eq. 3.28).

RECTANGULAR CHANNEL

$$\frac{d_2}{d_1} = \frac{1}{2}\left(\sqrt{1 + 8(\mathrm{Fr})^2} - 1 \right) \qquad 3.28$$

Upstream from the hydraulic jump, where supercritical flow is at high velocity, the depth, d_1, is lower, and the Froude number is greater than 1. Downstream from the hydraulic jump, the flow is subcritical at lower velocity, the depth, d_2, is higher, and the Froude number is less than 1. Table 3.3 summarizes the various facets of hydraulic jump.

Table 3.3 *Facets of Hydraulic Jump*

parameter	section 1	section 2
location	upstream	downstream
flow regime	supercritical	subcritical
velocity, v	higher	lower
flow depth, d	lower	higher
Froude number, Fr	> 1	< 1

Example 3.19

Calculate the depth downstream from a hydraulic jump in a 10 ft wide rectangular channel in which the flow is 100 ft³/sec and the depth upstream from the jump is 1 ft. Calculate the upstream and downstream velocities.

Solution

Calculate the specific discharge in a rectangular channel using Eq. 3.22.

$$q = \frac{Q}{b}$$

$$= \frac{100 \ \dfrac{\text{ft}^3}{\text{sec}}}{10 \ \text{ft}}$$

$$= 10 \ \text{ft}^3/\text{sec-ft}$$

Calculate the Froude number upstream from the hydraulic jump at section 1. From Eq. 3.27,

$$\mathrm{Fr} = \frac{q}{\sqrt{gd_1^3}}$$

$$= \frac{10 \ \dfrac{\text{ft}^3}{\text{sec-ft}}}{\sqrt{\left(32.2 \ \dfrac{\text{ft}}{\text{sec}^2}\right)(1 \ \text{ft})^3}}$$

$$= 1.76$$

Calculate the flow depth, d_2, downstream from the hydraulic jump using the hydraulic jump equation, Eq. 3.28.

$$\frac{d_2}{d_1} = \frac{1}{2}\left(\sqrt{1 + 8(\text{Fr})^2} - 1\right)$$

$$d_2 = \frac{1}{2}\left(\sqrt{1 + 8(\text{Fr})^2} - 1\right)d_1$$

$$= \left(\tfrac{1}{2}\right)\left(\sqrt{1 + (8)(1.76)^2} - 1\right)(1 \text{ ft})$$

$$= 2.04 \text{ ft}$$

The upstream velocity, v_1, is

$$Q = A_1 v_1$$

$$100 \ \frac{\text{ft}^3}{\text{sec}} = (10 \text{ ft})(1 \text{ ft})v_1$$

$$v_1 = \frac{100 \ \dfrac{\text{ft}^3}{\text{sec}}}{10 \text{ ft}^2}$$

$$= 10 \text{ ft/sec}$$

Solve for the downstream velocity, v_2.

$$A_2 = bd_2 = (10 \text{ ft})(2.04 \text{ ft})$$

$$= 20.4 \text{ ft}^2$$

$$v_2 = \frac{Q}{A_2} = \frac{100 \ \dfrac{\text{ft}^3}{\text{sec}}}{20.4 \text{ ft}^2}$$

$$= 4.9 \text{ ft/sec}$$

9. ENERGY DISSIPATION

Hydraulic jump is a transition between rapid, turbulent flow at high velocity and placid, laminar flow at low velocity (see Fig. 3.11). It often forms downstream from spillways that cause high velocity flow. When a hydraulic jump occurs, some form of energy dissipation is required to prevent erosion of the channel. The hydraulic jump equation for a <u>rectangular</u> channel is

$$d_2 = -\tfrac{1}{2}d_1 + \sqrt{\frac{2v_1^2 d_1}{g} + \frac{d_1^2}{4}} \quad \begin{bmatrix} \text{rectangular} \\ \text{channels} \end{bmatrix} \qquad 3.29$$

d_1 is flow depth upstream from the hydraulic jump, and d_2 is the flow depth downstream from the hydraulic jump. The velocity upstream from the hydraulic jump is v_1 in feet per second. The gravitational acceleration, g, is 32.2 ft/sec^2 (9.81 m/s^2).

The required energy dissipation below a spillway depends on the depth downstream from the hydraulic jump, d_2, compared to the tailwater depth, d_{TW}. If the tailwater depth is greater than d_2, the tailwater submerges the hydraulic jump and flow continues at high

energy along the channel floor, requiring some form of energy dissipation. The most common forms of energy dissipation downstream from spillways are stilling basins and scour pools, both of which allow the tailwater depth to exceed and drown out the hydraulic jump. Table 3.4 lists the various types of energy dissipation or channel scour (erosion) protection for the various forms of hydraulic jump.

Figure 3.11 *Hydraulic Jump Downstream from Spillway*

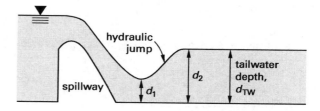

10. SURFACE WATER PROFILE

The *standard step method* is commonly used to estimate water surface profiles (flow elevation) and to delineate floodplains for streams and open channels. United States Army Corps of Engineers (USACE) data and calculations for Colorado's Red Fox River in Fig. 3.12 illustrates the method.

Representative data for cross sections of the river are provided in Fig. 3.12 and Fig. 3.13. Figure 3.12 gives the plan view and locates the sections. Figure 3.13 plots the elevations for two of the stream sections.

Table 3.5 gives the standard step backwater calculations for the Red Fox River. In this calculation, a water surface profile is calculated for a river discharge of 5000 ft^3/sec. A starting water surface elevation of 5709.0 ft is assumed. Expansion and contraction coefficients are assumed to be 0.3 and 0.1, respectively. The water surface profile is solved correctly when the assumed water surface elevation in column 2 is within $\pm$ 0.05 ft of the calculated water surface elevation in column 3.

The calculations follow the form. Column 2 and column 4 through column 12 are used to solve the Manning equation to obtain the energy losses caused by friction. Column 13 and column 14 contain calculations for the velocity distributions across the sections. Column 15 through column 17 contain the average kinetic energies. Column 18 contains calculations for other losses (expansion and contraction), and column 19 contains the computed changes in water surface elevation.

Column 1 is the cross section ID number. Cross sections are labeled sequentially, starting at the farthest point downstream and working upstream. For example, in Fig. 3.12, cross section 1 is the farthest point downstream, and cross section 4 is the farthest point upstream.

Table 3.4 *Forms of Energy Dissipation and Scour Protection*

class of flow	depth (d_2) vs. tailwater (d_{TW})	hydraulic jump	scour protection
1	$d_2 = d_{TW}$	toe of spillway	concrete apron with walls
2	$d_1^* > d_2$	jump submerged by tailwater, little energy dissipated	sloping concrete apron or concrete deflector bucket at base of spillway
3	$d_2 > d_1$	toe of spillway	stilling pool or stilling basin to increase tailwater depth

$^*d_1$ is the upstream flow depth.

Table 3.5 *Standard Step Water Surface Elevation (WSEL) Profile Calculations*

1	2	3	4	5	6	7	8	9	10	11	12	13	14	15	16	17	18	19
cross section no.^{a}	assumed WSEL (ft)	computed WSEL (ft)	A (ft^2)	R (ft)^b	$R^{2/3}$ (ft)	n^c	K^d	K_t^e	S_f^f (ft/ft) $\times 10^{-3}$	L (ft)	h_L (ft)^g	K^5/A^2 (ft^5/sec^3) $\times 10^6$	α^h	v (ft/sec) Q/A_t	$\dfrac{\alpha v^2}{2g}$ (ft)	$\Delta \dfrac{\alpha v^2}{2g}$ (ft)	h_o (ft)^i	ΔWSEL (ft)
1		5709.0	355	6.1	3.35	0.03	58,900					1568	1.0	14.1	3.08			
2	5712.6		360	6.1	3.34	0.03	59,719											
			80	1.6	1.37	0.05	3266					5.44						
		5713.0	440				62,100	60,943	6.83	500	3.37	1648	1.10	11.4	2.22	+1.18	0.118	4.67

Project: Red Fox River
$Q = 5000$ ft^3/sec, $C_e = 0.3$, and $C_c = 0.1$
^{a}cross section 1 is downstream from cross section 2
^{b}see Eq. 3.8
^{c}see Table 3.1
^{d}see Eq. 3.30
^{e}see Eq. 3.31
^{f}see Eq. 3.32
^{g}see Eq. 3.33
^{h}see Eq. 3.34
^{i}see Eq. 3.35 or Eq. 3.36
Adapted from *Water Surface Profiles*. Volume 6. Hydrologic Engineering Center. Davis, California. 1976. USACE.

The starting water surface elevations may be obtained from the normal depth calculated with the Manning equation.

Column 3 is the rating curve value for the first section. For the following sections of the waterway, column 3 is calculated by adding ΔWSEL to the calculated water surface elevation for the previous section.

Column 4 is the cross-sectional area in square feet. If the section is complex and has been subdivided into several parts (e.g., left overbank, channel, and right overbank), use one line of the form for each section and sum to get the total cross-sectional area, A_t.

Column 5 is the hydraulic radius, R, in feet, which equals the area divided by the wetted perimeter (see Eq. 3.8).

Figure 3.14 plots both the cross-sectional areas and hydraulic radii as a function of the elevations. Elevations are given on the y-axis, and two x-axes are calibrated—the top for the hydraulic radius and the bottom for the area. Curves for hydraulic radii and areas are labeled for each of the four sections shown in Fig. 3.12.

The wetted perimeter is given in feet. Column 6 is the hydraulic radius to the 2/3 power, as in the Manning equation. Column 7 is the Manning roughness coefficient.

Column 8 is the *conveyance*, K, defined as

$$K = \frac{1.49 A R^{2/3}}{n} \qquad \text{[U.S.]} \qquad \textbf{3.30}$$

Column 9 is the average conveyance, K_t, for the reach.

$$K_t = 0.5(K_{td} + K_{tu}) \qquad \textbf{3.31}$$

K_{td} is the conveyance in the downstream, and K_{tu} is the conveyance in the upstream sections of the reach.

Column 10 is the average friction slope, S_f, in feet per foot, through the reach.

$$S_f = \left(\frac{Q}{K_t}\right)^2 \qquad \textbf{3.32}$$

Column 11 is the distance, L, in unit length, between sections.

Column 12 is the head loss, h_L, caused by friction.

$$h_L = S_f L = \left(\frac{Q}{K_t}\right)^2 L \qquad \textbf{3.33}$$

Figure 3.12 *Plan View of the Red Fox River, Colorado*

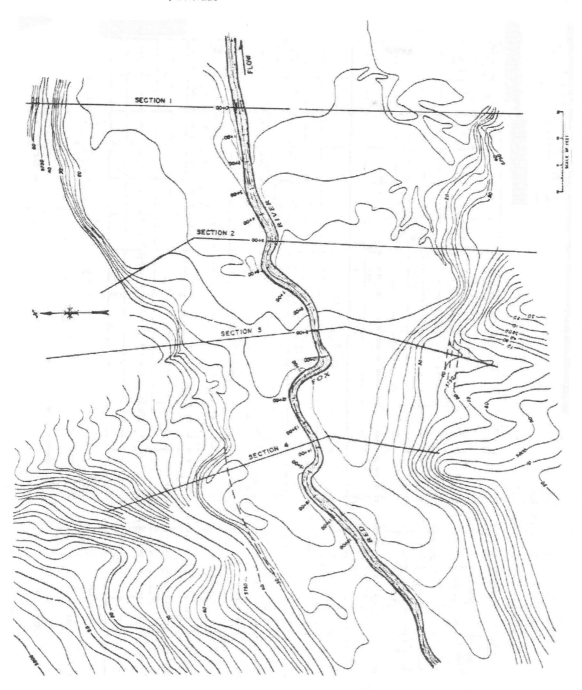

Reprinted from *Water Surface Profiles*. Volume 6. Hydrologic Engineering Center. Davis, California. 1976. USACE.

Water Resources

Figure 3.13 *Stream Cross Sections Along the Red Fox River, Colorado*

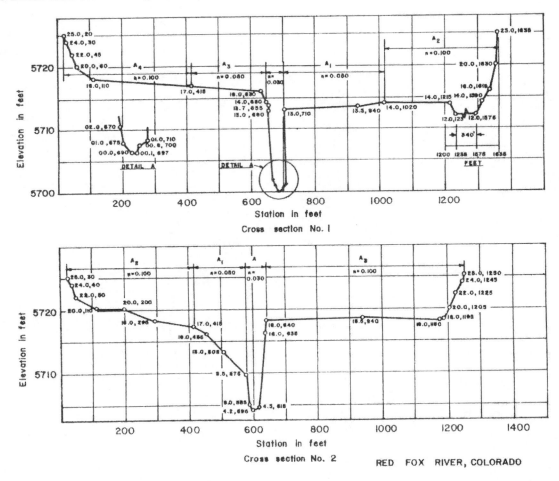

Reprinted from *Water Surface Profiles*. Volume 6. Hydrologic Engineering Center. Davis, California. 1976. USACE.

Column 13 is K^5/A^2, which relates the range of flow velocities distributed across the section to the average velocity in the section. Because of friction effects along the boundaries, velocities vary across the section. This term averages the velocities for the section.

Column 14 is the velocity distribution coefficient, α.

$$\alpha = \frac{A_t^2 \sum_i \left(\dfrac{K^3}{A_i^2}\right)}{K_t^3} \qquad 3.34$$

K_i and A_i are the incremental values of K and A, and the total values are K_t and A_t.

Column 15 is the average velocity in feet per second, calculated as Q/A_t.

Column 16 is $\alpha v^2/2g$, the average velocity head corrected for flow distribution.

Column 17 is $\Delta(\alpha v^2/2g)$, the difference between the velocity heads at the upstream and downstream sections. A positive value indicates the velocity is increasing in the

downstream direction and sets a *contraction coefficient*, C_c, equal to 0.1. A negative value indicates that the velocity is decreasing in the downstream direction and sets an *expansion coefficient*, C_e, equal to 0.3.

Column 18 accounts for other head losses, h_o. When the quantity $\Delta(\alpha v^2/2g)$ is positive, and velocity is increasing,

$$h_o = C_c \Delta\left(\frac{\alpha v^2}{2g}\right) \qquad 3.35$$

When $\Delta(\alpha v^2/2g)$ is negative and velocity is decreasing,

$$h_o = C_e \Delta\left(\frac{\alpha v^2}{2g}\right) \qquad 3.36$$

Column 19 is the change in water surface elevation from the previous section, ΔWSEL, in feet, and is found by adding columns 12, 17, and 18.

Example 3.20

Complete the next row of standard step profile calculations according to the instructions for the profile table

Figure 3.14 *Area and Hydraulic Radius Curves for Cross Section No. 1, Red Fox River, Colorado*

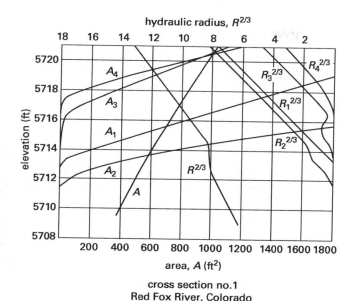

cross section no.1
Red Fox River, Colorado

Reprinted from *Water Surface Profiles*. Volume 6. Hydrologic Engineering Center. Davis, California. 1976. USACE.

begun in Table 3.5. The flow is 5000 ft³/sec. Assume the expansion coefficient is 0.3 and the contraction coefficient is 0.1. The starting water surface elevation at section 1 is 5709.0 ft. The starting water surface elevation for section 2 is 5712.6 ft.

Solution

The calculations describing section 2 are entered in the extension of the standard step surface profile calculations table which follows them. Column 3 is calculated by adding ΔWSEL to the calculated water surface elevation for the previous section 1.

For column 4, the areas are

$$A_{\text{channel}} = 360 \text{ ft}^2 \quad [\text{Fig. 3.14}]$$
$$A_{\text{floodplain}} = 80 \text{ ft}^2 \quad [\text{Fig. 3.14}]$$
$$A_{\text{channel+floodplain}} = 360 \text{ ft}^2 + 80 \text{ ft}^2$$
$$= 440 \text{ ft}^2$$

For column 5, the hydraulic radii are

$$R_{\text{channel}} = 6.1 \text{ ft}$$
$$R_{\text{floodplain}} = 1.6 \text{ ft}$$

For column 6,

$$R_{\text{channel}}^{2/3} = 3.34 \text{ ft}$$
$$R_{\text{floodplain}}^{2/3} = 1.37 \text{ ft}$$

For column 7, the Manning coefficients are

$$n_{\text{channel}} = 0.03$$
$$n_{\text{floodplain}} = 0.05$$

For column 8, the conveyances are given by Eq. 3.30.

$$K_{\text{channel}} = \frac{1.49 A R^{2/3}}{n}$$
$$= \frac{(1.49)(360 \text{ ft}^2)(3.34 \text{ ft})}{0.03}$$
$$= 59{,}719 \text{ ft}^3/\text{sec}$$

$$K_{\text{floodplain}} = \frac{1.49 A R^{2/3}}{n}$$
$$= \frac{(1.49)(80 \text{ ft}^2)(1.37 \text{ ft})}{0.05}$$
$$= 3266 \text{ ft}^3/\text{sec}$$

$$K_{\text{channel+floodplain}} = 59{,}719 \, \frac{\text{ft}^3}{\text{sec}} + 3266 \, \frac{\text{ft}^3}{\text{sec}}$$
$$= 62{,}985 \text{ ft}^3/\text{sec}$$

For column 10,

$$K_{\text{tu}} = 58{,}900 \text{ ft}^3/\text{sec} \quad [\text{from section 1}]$$

The average conveyance is calculated using Eq. 3.31.

$$K_t = 0.5(K_{\text{td}} + K_{\text{tu}})$$
$$= (0.5)\left(62{,}985 \, \frac{\text{ft}^3}{\text{sec}} + 58{,}900 \, \frac{\text{ft}^3}{\text{sec}}\right)$$
$$= 60{,}943 \text{ ft}^3/\text{sec}$$

For column 11, the distance between section 1 (0 ft) and section 2 (5 ft) is

$$L = 500 \text{ ft} - 0 \text{ ft} = 500 \text{ ft}$$

For column 12, the head loss caused by friction is calculated using Eq. 3.33.

$$h_L = \left(\frac{Q}{K_t}\right)^2 L = \left(\frac{5000 \, \dfrac{\text{ft}^3}{\text{sec}}}{60{,}943 \, \dfrac{\text{ft}^3}{\text{sec}}}\right)^2 (500 \text{ ft})$$
$$= 3.37 \text{ ft}$$

For column 13, calculate the ratios K^3/A^2 for averaging the velocities.

$$\left(\frac{K^3}{A^2}\right)_{channel} = \frac{\left(59{,}719\ \frac{ft^3}{sec}\right)^3}{(360\ ft^2)^2}$$

$$= 1643 \times 10^6\ ft^5/sec^3$$

$$\left(\frac{K^3}{A^2}\right)_{floodplain} = \frac{\left(3266\ \frac{ft^3}{sec}\right)^3}{(80\ ft^2)^2}$$

$$= 5.44 \times 10^6\ ft^5/sec^3$$

$$\left(\frac{K^3}{A^2}\right)_{channel+floodplain} = 1643 \times 10^6\ \frac{ft^5}{sec^3}$$

$$+\ 5.44 \times 10^6\ \frac{ft^5}{sec^3}$$

$$= 1648 \times 10^6\ ft^5/sec^3$$

For column 14, calculate the velocity distribution coefficient using Eq. 3.34.

$$\alpha = \frac{A_t^2 \sum_i \left(\dfrac{K^3}{A_i^2}\right)}{K_t^3}$$

$$= \frac{(440\ ft^2)^2 \left(\dfrac{\left(62{,}985\ \frac{ft^3}{sec}\right)^3}{(440\ ft^2)^2}\right)}{\left(60{,}943\ \frac{ft^3}{sec}\right)^3}$$

$$= 1.10$$

For column 15, calculate the average velocity.

$$v = \frac{Q}{A_t} = \frac{5000\ \frac{ft^3}{sec}}{440\ ft^2}$$

$$= 11.4\ ft/sec$$

For column 16, calculate the average velocity head corrected for flow distribution.

$$\frac{\alpha v^2}{2g} = \frac{(1.10)\left(11.4\ \frac{ft}{sec}\right)^2}{(2)\left(32.2\ \frac{ft}{sec^2}\right)}$$

$$= 2.22\ ft$$

For column 17, calculate the average velocity head at section 1 minus the average velocity head at section 2.

$$\Delta\left(\frac{\alpha v^2}{2g}\right) = \left(\frac{\alpha v^2}{2g}\right)_{section\ 1} - \left(\frac{\alpha v^2}{2g}\right)_{section\ 2}$$

$$= \frac{(1.10)\left(14.1\ \frac{ft}{sec}\right)^2}{(2)\left(32.2\ \frac{ft}{sec^2}\right)} - 2.22\ ft$$

$$= 1.18\ ft \quad [\text{positive}]$$

The positive value of 1.18 ft indicates that the velocity is increasing in the downstream direction, and the contraction coefficient, C_c, of 0.1 should be used. Calculate the other head losses using Eq. 3.35.

$$h_o = C_c\Delta\left(\frac{\alpha v^2}{2g}\right) = (0.1)(1.18\ ft)$$

$$= 0.118\ ft$$

For column 19, the change in the water surface elevation, $\Delta WSEL$, is found by adding columns 12, 17, and 18.

$$\Delta WSEL = 3.37\ ft + 1.18\ ft + 0.118\ ft$$

$$= 4.67\ ft$$

For column 3, calculate the water surface elevation at section 2.

$$WSEL_{section\ 2} = WSEL_{section\ 1} + \Delta WSEL$$

$$= 5709.0\ ft + 4.67\ ft$$

$$= 5714\ ft$$

At section 2, the assumed and calculated water surface elevations are not within ± 0.05 ft. Therefore, perform a second iteration of calculations with an assumed water surface elevation of 5714 ft. Repeat iterations for each assumed water surface elevation until the assumed and calculated values are within ± 0.05 ft.

11. STORMWATER COLLECTION

Storm sewers are usually designed for the 10 yr design flow, $Q_{10\,yr}$. The Manning equation can be used to calculate design pipe diameters and slopes. Figure 3.15 provides a convenient nomograph for the design of pipe flow using the Manning equation. The range from minimum to maximum pipe velocities for storm sewers should be from 2 ft/sec to 10 ft/sec (0.61 m/s to 3.04 m/s). The minimum velocity criterion is to prevent sediment and debris from settling in the storm sewer, and the maximum velocity criterion is to minimize turbulence and scouring.

Storm sewer design can be accomplished using the following seven-step process.

step 1: Sketch out the plan and profile of the storm sewer system. Manholes should be spaced at junctions of the storm sewer pipes no more than 400 ft to 500 ft (121 m to 152 m) apart to allow access to the system for maintenance. Inlets should be located at intersections and at low points in the gutter. At least 3 ft (0.91 m) of cover should be placed over the crown of the pipe for structural purposes.

step 2: Calculate the 10 yr design flow for each storm sewer section using the rational method (see Sec. 1.11).

step 3: Estimate invert elevations of the pipes at each manhole or inlet, assuming first that the pipes are parallel to the ground surface. Calculate the changes in invert elevations.

step 4: Measure the length of the pipe between each node. Calculate the slope of each pipe by dividing the change in invert elevation by the pipe length.

step 5: Estimate the Manning roughness value of the storm sewer. n is 0.013 for concrete.

step 6: Estimate the diameter of each pipe using the Manning equation nomograph (see Fig. 3.15). Round the design pipe size to the nearest standard 6 in (15.2 cm) diameter increment. For instance, if the nomograph indicates a 43 in pipe is needed, select a 48 in diameter pipe.

step 7: Check the selected pipe size for the recommended velocity criteria of 2 ft/sec (0.61 m/s) minimum and 10 ft/sec (3.05 m/s) maximum.

Example 3.21

Calculate the 10 yr design flow, $Q_{10\,yr}$, for a drainage area of 50 ac of single family residential use. The runoff coefficient, c, is 0.4, and the rainfall intensity is given by the following table.

recurrence interval (yr)	intensity (in/hr)
2	4
10	5
25	6
50	7
100	8

Solution

Calculate the 10 yr design flow, $Q_{10\,yr}$, using the rational method equation, Eq. 1.9 (Note that A is in acres, so Q will be in ac-in/hr. However, Q is more commonly taken

as ft³/sec since the conversion factor between these two units is 1.008.)

$$Q_{10\,yr} = ciA = (0.4)\left(5\ \frac{\text{in}}{\text{hr}}\right)(50\ \text{ac})$$
$$= 100\ \text{ac-in/hr} \quad (100\ \text{ft}^3/\text{sec})$$

Example 3.22

For the plan and profile shown, the 10 yr design flow is 60 ft³/sec at station 2+00, 40 ft³/sec at station 4+00, and 20 ft³/sec at station 5+00. Calculate the slope of a 48 in reinforced concrete circular pipe and the change in invert elevation between station 0+00 and station 2+00.

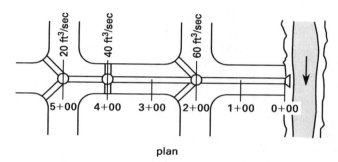

plan

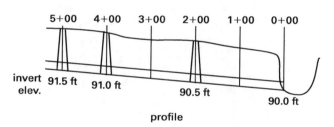

profile

Solution

Refer to the following table, which summarizes the pipe values between station 0+00 and station 2+00.

station	Q (ft³/sec)	invert elevation (ft)	Δ invert elevation (ft)	L (ft)	S (ft/ft)
0+00	0	90.0	–	–	–
2+00	60	90.5	0.5	200	0.0025

At station 2+00, the $Q_{10\,yr}$ is 60 ft³/sec.

The change in invert elevation between station 0+00 and station 2+00 is 90.5 ft − 90.0 ft = 0.5 ft.

The length of the pipe is 200 ft. The slope of the pipe is

$$S = \frac{0.5\ \text{ft}}{200\ \text{ft}} = 0.0025\ \text{ft/ft} \quad (0.25\%)$$

Figure 3.15 *Manning Equation Nomograph for Pipe Flow*

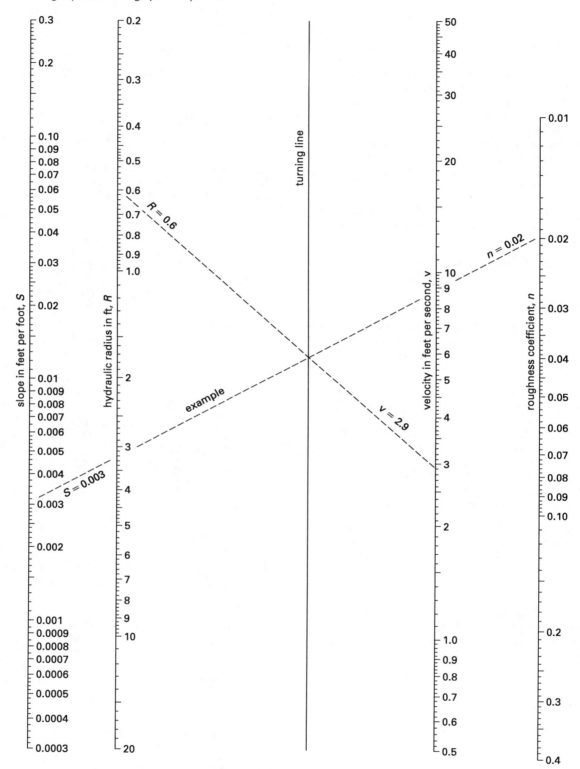

HEC22(4-2) $K_U = 0.56$ $n = Manning's$

OPEN CHANNEL HYDRAULICS 3-21

$$Q = \frac{K_U}{n} S_x^{1.67} S_L^{0.5} T^{2.67}$$

Gutter Flow *cfs*

An important aspect of stormwater collection is gutter flow. Stormwater from urban and suburban neighborhoods flows in street gutters and enters the storm sewer system through grated inlets. Roads are designed with a cross slope and longitudinal slope that directs flow to a triangular shaped gutter. Stormwater flows down the gutter and is discharged into inlets to the sewer system. Figure 3.16 depicts a typical curb and gutter stormwater system.

Figure 3.16 *Curb and Gutter Stormwater Dimensions*

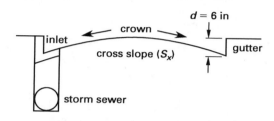

Gutter flow is calculated from Eq. 3.37, where n is the Manning roughness coefficient (0.013 for asphalt and concrete pavement), S_x is the cross slope, d is the depth at curb (6 in (15.2 cm) maximum), and S_L is the longitudinal slope of the street.

$$Q = 0.4 \left(\frac{1}{nS_x} \right) d^{8/3} \sqrt{S_L} \quad \text{[U.S. only]} \qquad 3.37$$

Generally, at least half of the street should be free of water. Therefore, the gutter longitudinal slope should exceed 0.5%, and the street cross slope should exceed 2%. The gutter flow should be a maximum of 6 in (15.2 cm) deep and at least 2 ft (0.61 m) wide, with the depth not exceeding the top of the curb. The depth of flow, d, and the top width of the gutter, T, are related by Eq. 3.38 and illustrated in Fig. 3.17.

$$d = TS_x \qquad 3.38$$

The top gutter width is the product of the number of lanes and the gutter width of the lanes.

$$T = N_L W_L \qquad 3.39$$

Figure 3.17 *Flow Depth and Gutter Top Width*

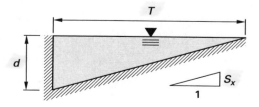

Example 3.23

Determine inlet spacing for an asphalt road designed for four 12 ft lanes (two lanes going in each direction) with a cross slope of 2% and a longitudinal slope of 1% in an area with a rainfall intensity of 5 in/hr.

Solution

Since the roadway is four lanes and half the road must be free of water, from Eq. 3.39, the top width is

$$
\begin{aligned}
T &= \frac{N_L W_L}{2} \\
&= \frac{(4 \text{ lanes})\left(12 \ \frac{\text{ft}}{\text{lane}}\right)}{2} \\
&= 24 \text{ ft}
\end{aligned}
$$

From Eq. 3.38, the depth at the curb is

$$d = TS_x = (24 \text{ ft})(0.02) = 0.48 \text{ ft}$$

From Eq. 3.37, the gutter flow rate is

$$
\begin{aligned}
Q &= 0.4 \left(\frac{1}{nS_x} \right) d^{8/3} \sqrt{S_L} \\
&= (0.4)\left(\frac{1}{(0.013)(0.02)} \right) (0.48 \text{ ft})^{8/3} \sqrt{0.01} \\
&= 21.73 \text{ ft}^3/\text{sec}
\end{aligned}
$$

Since intensity, i, is 5 in/hr, the drainage area, A, is

$$
\begin{aligned}
A &= \frac{Q}{i} \\
&= \frac{\left(21.73 \ \frac{\text{ft}^3}{\text{sec}}\right)\left(12 \ \frac{\text{in}}{\text{ft}}\right)\left(60 \ \frac{\text{min}}{\text{hr}}\right)\left(60 \ \frac{\text{sec}}{\text{min}}\right)}{5 \ \frac{\text{in}}{\text{hr}}} \\
&= 187{,}747 \text{ ft}^2
\end{aligned}
$$

Each lane of the four-lane roadway contributes runoff to a gutter. The width of each contributing lane is 12 ft. The length between inlets is

$$
\begin{aligned}
L &= \frac{A}{24 \text{ ft}} = \frac{187{,}747 \text{ ft}^2}{24 \text{ ft}} \\
&= 7823 \text{ ft}
\end{aligned}
$$

Stormwater Inlets

Stormwater inlets are usually specified as curb, grate, or combination inlets, as illustrated in Fig. 3.18.

Figure 3.18 *Stormwater Inlet Types*

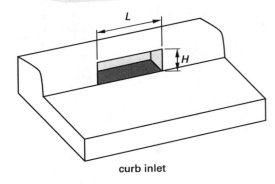

curb inlet

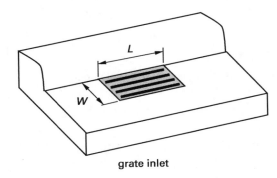

grate inlet

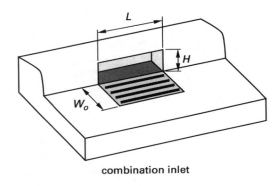

combination inlet

Curb inlets convey stormwater through an opening in the curb. The flow to a curb inlet is

(HORIZ. OPENING)

$$Q = 1.27 L H^{1.5} \quad \text{[U.S. only]} \qquad 3.40$$

Grate inlets are gutter openings covered by a grate. The capacity of a grate inlet is calculated using Eq. 3.41.

(VERTICAL OPENING)
$$Q = 1.66(2W + 2L)d^{1.5} \quad \text{[U.S. only]} \qquad 3.41$$

Flow into a combination inlet is the sum of Eq. 3.40 and Eq. 3.41.

Example 3.24

Calculate the flow capacity of a 3 ft × 2 ft grate inlet if the maximum flow depth at the gutter is 6 in.

Solution

Use Eq. 3.41.

$$Q = 1.66(2W + 2L)d^{1.5}$$

$$= (1.66)\Big((2)(2 \text{ ft}) + (2)(3 \text{ ft})\Big)\left(\frac{6 \text{ in}}{12 \dfrac{\text{in}}{\text{ft}}}\right)^{1.5}$$

$$= 5.9 \text{ ft}^3/\text{sec}$$

12. SPILLWAY CAPACITY

Flow through spillways from dams, reservoirs, and stormwater detention basins may be estimated using the weir flow equations. Generally, the following three types of weirs are used in water resources engineering.

- rectangular weir
- V-notch (triangular) weir
- trapezoidal weir

Rectangular Weir

Rectangular weirs are used to model flow over a dam or road. The rectangular weir flow equation is

$$Q = C b d_H^{3/2} \qquad 3.42$$

The flow through the weir, Q, is given in cubic feet per second. The weir flow coefficient, C, ranges from 3.0 to 4.0, depending on the depth of water flowing over the weir. Usually the assumed value is 3.5 for a rectangular weir. The width, b, of the weir and the depth, d_H, of the water surface flowing over the weir (see Fig. 3.19) are given in feet. The weir flow coefficient converts the units to cubic feet per second.

Figure 3.19 *Rectangular Weir*

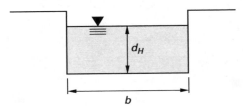

V-Notch (Triangular) Weir

Triangular weirs are frequently used in wastewater or water treatment measurements. The flow through a V-notch (triangular) weir is

$$Q = C(0.5)\tan\frac{\theta}{2} d_H^{5/2}\sqrt{2g} \qquad 3.43$$

The V-notch weir flow coefficient, C, is 0.6. The angle of the V-notch weir is θ. The depth, d_H, of the water surface flowing over the weir (see Fig. 3.20) is given in feet, and g is the gravitational acceleration.

Figure 3.20 *V-Notch or Triangular Weir*

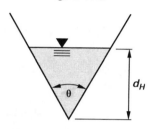

C=0.6 for Broad-crested also

Trapezoidal Weir

Trapezoidal weirs are often used as reservoir overflow spillways. If the velocity of approach is insignificant, flow is given by Eq. 3.44.

$$Q = Cbd_H^{3/2} \qquad 3.44$$

The trapezoidal weir flow coefficient, C, is 3.3. The width, b, of the weir and the depth, d_H, of the water surface (see Fig. 3.21) flowing over the weir are given in feet.

Trapezoidal weirs are essentially rectangular weirs with a triangular weir on either end. Flow through a trapezoidal weir is given as

$$Q = \tfrac{2}{3}Cbd_H^{3/2}\sqrt{2g} \qquad 3.45$$

Figure 3.21 *Trapezoidal Weir*

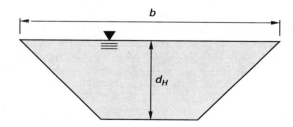

Example 3.25

Calculate the flow through a rectangular weir in which the width, b, is 20 ft, and the depth, d_H, of flow over the weir is 3 ft.

Solution

Use the rectangular weir flow formula, Eq. 3.42. The rectangular weir flow coefficient is 3.5.

$$Q = Cbd_H^{3/2} = (3.5)(20\text{ ft})(3\text{ ft})^{3/2}$$
$$= 364\text{ ft}^3/\text{sec}$$

Example 3.26

Estimate the rectangular weir length necessary at the outlet of a detention basin in which the water level is at elevation 101 ft msl, the crest of the weir is set at elevation 100 ft msl, and the outlet flow from the basin is 100 ft³/sec.

Solution

The depth of the weir is the difference between the water level of the detention basin and the crest of the weir.

$$d_H = 101\text{ ft} - 100\text{ ft} = 1\text{ ft}$$

Use Eq. 3.42, and solve for the weir length.

$$Q = Cbd_H^{3/2}$$
$$100\ \frac{\text{ft}^3}{\text{sec}} = (3.5)b(1\text{ ft})^{3/2}$$
$$b = \frac{100\ \frac{\text{ft}^3}{\text{sec}}}{(3.5)(1\text{ ft})^{3/2}}$$
$$= 28.6\text{ ft}$$

Example 3.27

Calculate the flow through a V-notch weir in which θ is 90° and the flow through the weir is 5 ft deep.

Solution

Use the V-notch weir formula, Eq. 3.43.

$$Q = C(0.5)\tan\frac{\theta}{2} d_H^{5/2}\sqrt{2g}$$
$$= (0.6)(0.5)\left(\tan\frac{90°}{2}\right)(5\text{ ft})^{5/2}\sqrt{(2)\left(32.2\ \frac{\text{ft}}{\text{sec}^2}\right)}$$
$$= 134\text{ ft}^3/\text{sec}$$

Example 3.28

Calculate the flow through a trapezoidal weir of length 20 ft. The depth of flow over the weir is 3 ft.

Solution

Use the trapezoidal weir flow formula from Eq. 3.44.

$$Q = Cbd_H^{3/2} = (3.3)(20 \text{ ft})(3 \text{ ft})^{3/2}$$
$$= 343 \text{ ft}^3/\text{sec}$$

13. FLOW MEASUREMENT DEVICES

Flow, or discharge, can be metered using measurement devices such as weirs and flumes. Open channel flow through water treatment and wastewater plants is often measured using a *Parshall flume* (see Fig. 3.22).

Figure 3.22 *Parshall Flume*

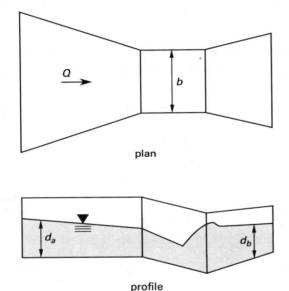

The flow equation for the Parshall flume is

$$Q = 4bd_a^n \quad \text{[U.S. only]} \qquad 3.46$$

b is the width in unit length of the flume throat, and d_a is the upstream flow depth in unit length. The Parshall flume exponent, n, is

$$n = 1.5b^{0.03} \quad \text{[U.S. only]} \qquad 3.47$$

Example 3.29

Calculate the flow in a 4 ft wide Parshall flume in which the upstream flow depth is 2 ft.

Solution

Use Eq. 3.47 to find the Parshall flume exponent.

$$n = 1.5b^{0.03}$$
$$= (1.5)(4 \text{ ft})^{0.03}$$
$$= 1.56$$

Use Eq. 3.46.

$$Q = 4bd_a^n$$
$$= (4)(4 \text{ ft})(2 \text{ ft})^{1.56}$$
$$= 47.2 \text{ ft}^3/\text{sec}$$

14. STORMWATER DETENTION/RETENTION PONDS

Many jurisdictions in the United States have regulations that control the quality and quantity of stormwater runoff from new development by requiring stormwater detention and retention ponds. *Detention ponds* hold enough volume to store runoff and slowly release the flow to the waterway or storm sewer system, primarily through an outlet structure such as a pipe or weir (see Fig. 3.23). *Retention ponds* store the runoff volume, then release the runoff through infiltration and evaporation, not through an outlet structure.

Figure 3.23 *Typical Stormwater Detention Pond Outlet*

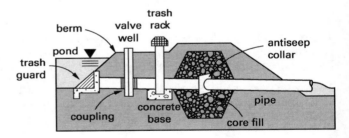

Reprinted from *Ponds—Planning, Design, Construction*, Agriculture Handbook 590, Natural Resources Conservation Services, 1997.

Stormwater facilities control water quality by settling out pollutants in runoff and by bio-removal in vegetated basins. Stormwater ponds for quality control are often sized to store the first 1 in to 2 in (2.5 cm to 5.1 cm) of runoff, known as the pollutant-laden first flush, for the 1 yr or 2 yr design storm. Table 3.6 summarizes the pollutant removal efficiencies of various stormwater *best management practices*, such as detention and retention ponds, stormwater wetlands, and infiltration basins.

Many regulations require stormwater quantity control by specifying that the peak postdevelopment stormwater runoff should not exceed the peak predevelopment runoff. This criterion is met by sizing stormwater detention or retention ponds with sufficient volume to store much of the runoff, then to slowly release the flow though a restricted outlet structure. Stormwater ponds are often sized to release the runoff volume within 48 hr to 72 hr after the end of the design storm. Depending on the jurisdiction, stormwater ponds for quantity control are often sized for the 100 yr design storm. The maximum depth of stormwater ponds should not exceed 4 ft to 6 ft (1.22 m to 1.83 m) for safety reasons. To prevent soil erosion and accommodate mowing and maintenance equipment, the pond sideslopes should not exceed a ratio of 4 ft (4 m) horizontal to 1 ft (1 m) vertical (4:1).

Table 3.6 *Stormwater Best Management Practice Pollutant Removal Efficiencies (%)*

pollutant	detention ponds	retention ponds	stormwater wetlands	filters wetlands	infiltration basins
total phosphorus	19	51	49	59	70
total nitrogen	25	33	30	38	51
TSS[*]	47	80	76	86	95
Cu	26	57	40	49	–
Zn	26	66	44	88	99

[*]Total Suspended Solids

Adapted from *Stormwater Best Management Practice Design Guide*, Volume I, EPA/600/R-04/121, USEPA, 2004.

15. POND AND RESERVOIR ROUTING

The *pond and reservoir routing technique* is used to design the storage area and outlet characteristics of stormwater basins and ponds. The reservoir routing technique is defined by the following equation, which is a variation of the conservation of mass principle (continuity principle) that inflow, I, must equal outflow, O, plus the change in pond storage, ΔS_n.

$$\frac{S_{n2} - S_{n1}}{t} = \frac{I_1 + I_2}{2} - \frac{O_1 + O_2}{2} \qquad 3.48$$

The storage of the basin, S_n, is given in unit volume, the inflow and outflow are given in unit volume per second, and t is the time step of the hydrograph in seconds.

The pond and reservoir routing technique is performed according to the following six-step process.

step 1: Calculate, tabulate, and graph the design storm (i.e., 100 yr) predevelopment and post development condition hydrographs using the rational method synthetic hydrograph or the *TR-55* procedures outlined in Chap. 1.

step 2: Define the outflow rating curve at 0.5 ft intervals of depth, d_H, for a weir or culvert pipe outlet structure. For instance, if a rectangular weir were proposed as an outlet structure, the outflow rating curve would be calculated according to the weir flow equation, $Q = Cbd_H^{3/2}$.

step 3: Define the surface area-volume relationship for the stormwater pond using the average end area method at depth increments of 1 ft. When the area, A, of a reservoir is known at two different depths, and the difference in the depths is given by Δd_H (see Fig. 3.24), the reservoir volume, V, between those depths is estimated as the average area.

$$V = \Delta d_H \left(\frac{A_1 + A_2}{2} \right) \qquad 3.49$$

When d_H is given in feet, A_1 and A_2 are given in square feet, and volume, V, is given in cubic feet.

Figure 3.24 *Average End Area Method*

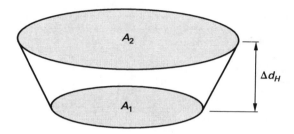

step 4: Tabulate and plot a rating curve with outflow, O, on the x-axis versus $2S/t + O$ on the y-axis.

step 5: Route the flows through the stormwater pond or reservoir by setting up a table in the following column number format, using units of cubic feet per second for all flow quantities. The subscript 1 refers to the time interval in the current row. The subscript 2 refers to the next time interval.

(i)   time interval, t (hr)

(ii) post development inflow to the pond, I

(iii) inflow from the current time interval (1) plus inflow from the next time interval (2)

$$I_1 + I_2$$

(iv) two times the storage divided by the time interval minus outflow from the current time interval

$$\frac{2S_n}{t} - O_1$$

(v)   two times the storage divided by the time interval plus outflow from the next time interval

$$\frac{2S_n}{t} + O_2$$

(vi) outflow from the stormwater pond, O

step 6: Plot hydrographs delineating the post develop-ment inflow, I, to the stormwater pond and the outflow from the outlet structure, O.

Example 3.30 outlines stormwater pond design using the pond and reservoir routing technique and the average end method.

Example 3.30

The following tables provide data for the design of a stormwater detention pond. The first table gives the 100 yr storm predevelopment and post development hydrograph data. The second table relates the pond elevation to area. Assuming a weir flow outlet structure, and assuming that the peak post development discharge shall not exceed the peak predevelopment discharge, design a stormwater detention pond.

Stormwater Hydrographs for Predevelopment and Post Development Conditions

time (hr)	predevelopment (ft^3/sec)	post development (ft^3/sec)
10	0	0
11	2	9
12	10	42
13	25	89 (peak)
14	38 (peak)	28
15	23	19
16	17	15
17	14	12
18	10	11
19	8	10
20	6	8
21	4	7
22	2	0
23	1	0
24	0	0

Pond Elevation-Surface Area-Volume Relationship

depth, d_H (ft)	surface area, A (ft^2)
0.0	40,000
1.0	100,000
2.0	200,000
3.0	300,000
4.0	400,000

Solution

Use the six-step process.

step 1: Plot a graph of the 100 yr storm predevelopment and post development condition hydrographs using the data in the problem statement's first table.

As shown in *Stormwater Hydrographs for Pre-development and Post Development Conditions*,

the peak predevelopment discharge is 38 ft^3/sec. Therefore, a stormwater detention pond should be designed with a peak 100 yr outflow of 38 ft^3/sec.

step 2: Define the outflow rating curve at 0.5 ft inter-vals of depth, d_H, for a weir outlet structure. The calculations for the outflow rating curve, using the rectangular weir coefficient, the given weir length, b, and half foot increments in depth, d_H, are organized in the following table of weir outflow rating curve calculations. The rectangular weir flow coefficient, C, is 3.5. The maximum depth of the storm water pond should not exceed 4 ft to 6 ft for safety pur-poses. The weir length should be designed to pond no more than 4 ft with a maximum peak pond outflow of 38 ft^3/sec. Assume an initial weir length of 1.7 ft to pond the water at near 4 ft depth and an outflow of 38 ft^3/sec.

Using Eq. 3.42, calculate the weir outflow, Q, rating curve calculations up to a pond elevation of 4 ft.

elevation (ft)	C	b (ft)	d_H (ft)	$d_H^{3/2}$ (ft)	$Q = Cbd_H^{3/2}$
0.0	3.5	1.7	0.0	0.00	0
0.5	3.5	1.7	0.5	0.35	2.1
1.0	3.5	1.7	1.0	1.00	6
1.5	3.5	1.7	1.5	1.80	11
2.0	3.5	1.7	2.0	2.80	17
2.5	3.5	1.7	2.5	3.95	24
3.0	3.5	1.7	3.0	5.20	31
3.5	3.5	1.7	3.5	6.50	39 (peak)
4.0	3.5	1.7	4.0	8.00	48

step 3: Define the surface area-volume relationship for the stormwater pond using the average end area method at depth increments of 1 ft, as shown in the following table. For each depth, incremental storage, cumulative storage, weir outflow, and the rating curve quantity $2S_n/t + O$ are calculated.

At a depth of 1 ft, the surface area (column 2) is

$$A = 100,000 \text{ ft}^2 \quad \text{[given]}$$

The average area (column 3) is

$$\frac{A_1 + A_2}{2} = \frac{40,000 \text{ ft}^2 + 100,000 \text{ ft}^2}{2}$$
$$= 70,000 \text{ ft}^2$$

The difference in depth, Δd_H (column 4), is 1.0 ft.

The incremental storage (column 5) is the depth times the average area.

$$\Delta d_H \left(\frac{A_1 + A_2}{2} \right) = (1.0 \text{ ft})(70,000 \text{ ft}^2)$$
$$= 70,000 \text{ ft}^3$$

The cumulative storage (column 6) is

$$S = 0 \text{ ft}^3 + 70,000 \text{ ft}^3 = 70,000 \text{ ft}^3$$

According to the weir outflow rating curve calculations in the table for step 2, the weir outflow at the depth, d_H, of 1 ft (column 7) is 6 ft^3/sec.

The rating curve calculation (column 8) is

$$\frac{2S_n}{t} + O = \frac{(2)(70,000 \text{ ft}^3)}{3600 \text{ sec}} + 6 \frac{\text{ft}^3}{\text{sec}}$$
$$= 44.9 \text{ ft}^3/\text{sec}$$

At a depth of 2 ft, the surface area (column 2) is

$$A = 200,000 \text{ ft}^2 \quad \text{[given]}$$

The average area (column 3) is

$$\frac{A_1 + A_2}{2} = \frac{100,000 \text{ ft}^2 + 200,000 \text{ ft}^2}{2}$$
$$= 150,000 \text{ ft}^2$$

The difference in depth, Δd_H (column 4), is 1.0 ft.

The incremental storage (column 5) is the depth times the average area.

$$\Delta d_H \left(\frac{A_1 + A_2}{2} \right) = (1.0 \text{ ft})(150,000 \text{ ft}^2)$$
$$= 150,000 \text{ ft}^3$$

The cumulative storage (column 6) is

$$S = 70,000 \text{ ft}^3 + 150,000 \text{ ft}^3 = 220,000 \text{ ft}^3$$

According to the weir outflow rating curve calculations in the table for step 2, the weir outflow at the depth, d_H, of 2 ft (column 7) is 17 ft^3/sec.

The rating curve calculation (column 8) is

$$\frac{2S_n}{t} + O = \frac{(2)(220,000 \text{ ft}^3)}{3600 \text{ sec}} + 17 \frac{\text{ft}^3}{\text{sec}}$$
$$= 139 \text{ ft}^3/\text{sec}$$

The calculations for pond depths of 3 ft and 4 ft are entered in the following table.

Surface Area-Volume Relationship for the Stormwater Pond

depth, d_H (ft)	surface area, A (ft^2)	average area (ft^2)	Δd_H (ft)	incremental storage, S_n (ft^3)	cumulative storage, S_n (ft^3)	weir outflow, O (ft^3/sec)	$2S/t$ $+O^*$ (ft^3/sec)
0.0	40,000	-	0	0	0	0	0
1.0	100,000	70,000	1.0	70,000	70,000	6	45
2.0	200,000	150,000	1.0	150,000	220,000	17	139
3.0	300,000	250,000	1.0	250,000	470,000	31	292
4.0	400,000	350,000	1.0	350,000	820,000	48	503

$^*t = 1 \text{ hr} = 3600 \text{ sec}$

step 4: Plot a rating curve with weir outflow, O, on the x axis versus $2S/t + O$ on the y-axis in the following illustration, using the data from the table of surface area-volume relationships completed in step 3.

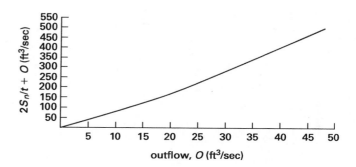

step 5: Inflow and Outflow Curves From the Stormwater Pond routes the flow of water through the stormwater pond.

Inflow and Outflow Curves From the Stormwater Pond

time (hr)	post development inflow, I (ft^3/sec)	$I_1 + I_2$ (ft^3/sec)	$\frac{2S_n}{t} - O_2$ [a] (ft^3/sec)	$\frac{2S_n}{t} + O_2$ [b] (ft^3/sec)	outflow [c], O (ft^3/sec)
10	0	9	0	–	0
11	9	51	0	–	0
12	42	131	35 C	51 A	8 B
13	89	117	126	166	20
14	28	47	183	243	30
15	19	34	174	230	28
16	15	27	160	208	24
17	12	23	145	187	21
18	11	21	128	168	20
19	10	18	113	149	18
20	8	15	100	131	15
21	7	7	93	115	11
22	0	0	80	100	10
23	0	0	62	80	9
24	0	0	48	62	7

[a] column 4 = column 5 − 2(column 6)
[b] column 3 + column 4 = column 5 of net interval
[c] column 6 obtained from O versus $2S_n/t + O$ rating curve

step 6: Plot hydrographs delineating the inflow from the developed site into the stormwater pond, and the outflow from the weir outlet structure. Tabular and graphical hydrographs delineating the inflow from the developed site into the stormwater pond and the outflow from the weir outlet structure are shown in the following table and illustration.

time (hr)	post development inflow (ft³/sec)	outflow from weir (ft³/sec)
10	0	0
11	9	0
12	42	8
13	89 (peak)	20
14	28	30 (peak)
15	19	28
16	15	24
17	12	21
18	11	20
19	10	18
20	8	15
21	7	11
22	0	10
23	0	9
24	0	7

- - - - - inflow post development (ft³/sec)
——— outflow weir (ft³/sec)

The peak inflow into the storm water pond is 89 ft³/sec. The peak outflow from the weir outlet structure is 30 ft³/sec. The maximum surface area of the stormwater pond at a depth of 3.0 ft is 300,000 ft² (6.9 ac). The volume of the stormwater pond is 470,000 ft³ (10.8 ac-ft).

16. CULVERT DESIGN

Culverts are large enclosed channels or pipes that convey flow under roads, railroads, embankments, and other structures. Culverts are usually composed of concrete or corrugated metal in standard circular, rectangular, oval, and arch shapes and sizes. Culvert design should incorporate the following hydraulic criteria.

- Culverts under main highways and railroads should be designed to convey the 100 yr flow.

- Culverts under secondary roads, driveways, and bike paths should be designed to convey the 50 yr flow.

- Design the culvert to flow partially full, to provide freeboard, and to allow sufficient capacity for debris to flow through the culvert and avoid clogging. The headwater, or depth of flow at the entrance to the culvert, should not exceed 90% of the culvert diameter or rise, 0.9D.

- At least 1 ft of freeboard should be included between the design headwater elevation and the roof of the culvert.

Culvert hydraulics behave under two flow regimes: inlet control and outlet control. *Inlet control* occurs when the barrel of the culvert remains unsubmerged by the flow, and the outlet of the culvert flows partially full, as depicted in Fig. 3.25. Under inlet control, the hydraulic opening and contraction coefficient at the entrance are the limiting factors in conveying the flow through the culvert. The entrance to the culvert may be partly full (if the culvert was designed by the 0.9D criterion), or the inlet may be submerged. The culvert equation for inlet control is

$$Q = C_d A \sqrt{2gh} \qquad 3.50$$

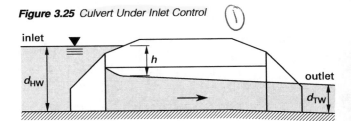

Figure 3.25 *Culvert Under Inlet Control*

The culvert discharge coefficient, C_d, is 0.6 for square edge or unbeveled edge entrance conditions, and 1.0 for rounded or beveled edge entrance conditions. The cross-sectional area, A, of the culvert is given in square feet. The difference, h, between the headwater depth, d_{HW}, upstream from the culvert and the tailwater depth, d_{TW}, downstream from the culvert is given in feet. The culvert coefficient of contraction converts the units of Eq. 3.50 into cubic feet per second.

Outlet control occurs when the barrel and the culvert outlet remain fully submerged (see Fig. 3.26). The hydraulic opening of the culvert and the friction losses caused by the barrel of the culvert and/or the tailwater downstream from the culvert are the limiting factors in the hydraulic capacity of the culvert. The outlet control culvert equation is

$$h = \left(K_e + 1 + \frac{29n^2 L}{R^{4/3}} \right) \frac{v^2}{2g} \qquad 3.51$$

K_e is the entrance loss coefficient, which is 0.5 for a square edged entrance and 0.05 for a rounded entrance. The Manning roughness coefficient of the culvert material is n. The length, L, of the culvert barrel is given in feet. R is the hydraulic radius, in feet, found by dividing the culvert cross-sectional area, A, by the wetted perimeter, WP. The velocity of flow through the culvert, in feet per second, is Q/A.

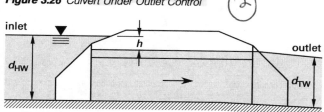

Figure 3.26 *Culvert Under Outlet Control*

Example 3.31

A 10 ft wide × 5 ft rise concrete box culvert is under inlet control and has rounded, beveled edges at the entrance. The headwater depth is 6 ft, the tailwater depth is 4 ft, and the length of the culvert is 50 ft. Estimate the flow capacity of the culvert.

Solution

Use the inlet control culvert equation, Eq. 3.50.

$$C_d = 1.0 \quad \text{[for rounded edge entrances]}$$
$$A = bd = (10 \text{ ft})(5 \text{ ft}) = 50 \text{ ft}^2$$
$$h = d_{\text{HW}} - d_{\text{TW}} = 6 \text{ ft} - 4 \text{ ft} = 2 \text{ ft}$$

$$Q = C_d A \sqrt{2gh}$$
$$= (1.0)(50 \text{ ft}^2)\sqrt{(2)\left(32.2 \frac{\text{ft}}{\text{sec}^2}\right)(2 \text{ ft})}$$
$$= 567 \text{ ft}^3/\text{sec}$$

Example 3.32

Estimate the flow capacity, Q, for a 50 ft long × 10 ft wide × 5 ft rise box culvert under outlet control. The culvert is flowing full, with a headwater depth of 10 ft and a tailwater depth of 9 ft.

Solution

Use the outlet control culvert equation, Eq. 3.51. To find h,

$$h = d_{\text{HW}} - d_{\text{TW}} = 10 \text{ ft} - 9 \text{ ft} = 1 \text{ ft}$$

The entrance loss coefficient, K_e, is 0.05 for a rounded entrance. From Table 3.1, n is 0.013 for concrete.

Calculate the geometrical properties and the flow. From Eq. 3.16,

$$\text{WP} = 2d + 2b = (2)(5 \text{ ft}) + (2)(10 \text{ ft})$$
$$= 30 \text{ ft}$$

From Eq. 3.8,

$$R = \frac{A}{\text{WP}} = \frac{50 \text{ ft}^2}{30 \text{ ft}}$$
$$= 1.7 \text{ ft}$$

$$\text{v} = \frac{Q}{A}$$
$$= \frac{Q}{50 \text{ ft}^2}$$

Substitute into the outlet control culvert equation, Eq. 3.51.

$$h = \left(K_e + 1 + \frac{29n^2 L}{R^{4/3}}\right)\frac{\text{v}^2}{2g}$$

$$1 \text{ ft} = \left(0.05 + 1 + \frac{(29)(0.013)^2(50 \text{ ft})}{(1.7 \text{ ft})^{4/3}}\right)$$
$$\times \left(\frac{\left(\dfrac{Q}{50 \text{ ft}^2}\right)^2}{(2)\left(32.2 \dfrac{\text{ft}}{\text{sec}^2}\right)}\right)$$

$$= (0.05 + 1 + 0.1208)\left(\frac{Q^2}{161{,}000}\right)$$

$$Q^2 = \frac{161{,}000}{1.17}$$
$$= 137{,}607$$
$$Q = 371 \text{ ft}^3/\text{sec}$$

17. VELOCITY CONTROL

Erosion along open channels can be avoided by properly sizing the cross-sectional area, the sideslopes, and the longitudinal slope for the given channel material. The velocity in open channels can be estimated using the Manning equation, Eq. 3.7.

Table 3.7 lists the maximum allowable velocities along open channels for various channel materials. Table 3.8

Table 3.7 *Maximum Allowable Velocities Along Open Channels*

channel material	maximum velocity (ft/sec)
sand	1.5
silt	2.0
pebbles	2.5
clay	3.0
gravel	4.0
cobbles	5.0
concrete	40

(Multiply ft/sec by 0.3 to obtain m/s.)
Reprinted from *Design Charts for Open Channel Flow*, Federal Highway Administration, App. A, Table 2.

Table 3.8 Minimum Channel Sideslopes

channel material	minimum sideslope (horizontal:vertical)
rock	$\frac{1}{2}$:1
firm soil	1:1
gravel, loam	$1\frac{1}{2}$:1
sand	3:1
grass turf, mowed	4:1

summarizes the minimum sideslopes for a given channel material.

Example 3.33

The trapezoidal gravel channel depicted in the illustration has a bottom width of 10 ft, 3:1 sideslopes, and a depth of flow of 3 ft. Define the maximum longitudinal channel slope necessary to prevent erosion of gravel banks.

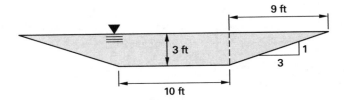

Solution

The minimum channel longitudinal slope, S, is calculated using the Manning equation for velocity, Eq. 3.7.

$$n = 0.022 \quad \text{[earth, Table 3.1]}$$

$$v = 4 \text{ ft/sec} \quad \text{[gravel, from Table 3.7]}$$

From Eq. 3.13, the area of the trapezoidal channel is

$$A = bd + xd$$
$$= (10 \text{ ft})(3 \text{ ft}) + (9 \text{ ft})(3 \text{ ft})$$
$$= 57 \text{ ft}^2$$

From Eq. 3.18, the wetted perimeter of a trapezoid is

$$\text{WP} = b + 2\sqrt{d^2 + x^2} = 10 \text{ ft} + 2\sqrt{(3 \text{ ft})^2 + (9 \text{ ft})^2}$$
$$= 29 \text{ ft}$$

The hydraulic radius is calculated using Eq. 3.8.

$$R = \frac{A}{\text{WP}} = \frac{57 \text{ ft}^2}{29 \text{ ft}}$$
$$= 1.97 \text{ ft}$$

Substitute into Eq. 3.7 and solve for channel slope, S.

$$v = \left(\frac{1.49}{n}\right) R^{2/3}\sqrt{S}$$

$$4 \frac{\text{ft}}{\text{sec}} = \left(\frac{1.49}{0.022}\right)(1.97 \text{ ft})^{2/3}\sqrt{S}$$

$$\sqrt{S} = 0.038 \frac{\text{ft}^2}{\text{ft}^2}$$

$$S = \left(0.038 \frac{\text{ft}^2}{\text{ft}^2}\right)^2$$

$$= 0.0014 \text{ ft/ft}$$

18. FLOODPLAIN AND FLOODWAY ANALYSIS

Many jurisdictions seek to prevent flood damage by enacting floodplain ordinances. Most ordinances seek to regulate or prevent development in the 100 yr floodway and floodplain of waterways. *Floodplains* are delineated by calculating the 100 yr flood depth ($d_{100 \text{ yr}}$) using the Manning equation, Eq. 3.10, and the standard step backwater calculations (see Sec. 3.7). Once the flood depth is known, it is plotted on topographic maps to delineate the floodplain. The *floodway* is the portion of the floodplain that includes the stream channel and the adjacent area that carries the greatest portion of the floodwaters. The bottom of the channel is called the thalweg. Figure 3.27 depicts the floodway and floodplain on a stream section. Figure 3.28 delineates a floodplain overlaid on a street map for the Federal Emergency Management Agency (FEMA) National Insurance Program.

The floodplain is delineated by calculating the flood elevation, then plotting the elevation on aerial photographs and topographic maps as depicted in Fig. 3.28. The flood elevation is calculated using the Manning equation, Eq. 3.10, or the standard step method (see Sec. 3.10). Because the floodplain condition often varies from the channel roughness, several values of Manning coefficients, n, are often used in a composite Manning equation. Hydraulic computer models such as the USACE HEC-RAS model are often used to delineate floodplains for the FEMA flood insurance studies.

Figure 3.27 The Floodplain and Floodway

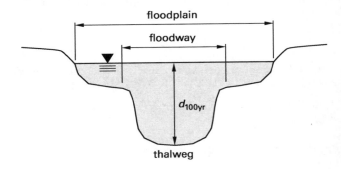

Figure 3.28 *Floodplain Mapping Along the Christina River in Newark, Delaware*

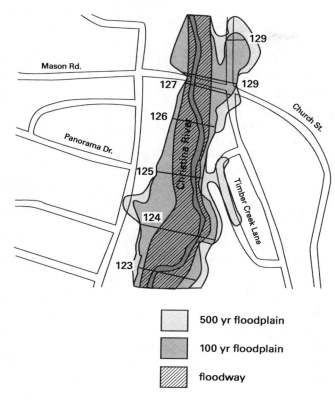

☐ 500 yr floodplain

▨ 100 yr floodplain

▧ floodway

Example 3.34

For the floodplain section shown in Fig. 3.28, the 100 yr flood elevation is 100 ft msl. The left and right overbanks of the floodplain are forest, and the channel is a natural stream. The thalweg of the stream section is at elevation 88 ft msl, 1000 ft downstream. The geometric slope is the same for the main channel and floodplain. Calculate the 100 yr flood flow for the floodplain section.

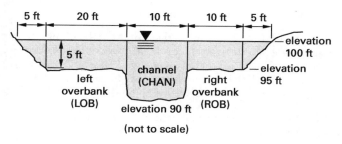

(not to scale)

Solution

For the left overbank (LOB),

$$n = 0.150 \quad \text{[trees, Table 3.1]}$$

The area of the LOB is estimated from the area of a triangle plus the area of a rectangle, as delineated in the illustration.

$$A_{\text{LOB}} = (5 \text{ ft})\left(\frac{100 \text{ ft} - 95 \text{ ft}}{2}\right) + (5 \text{ ft})(20 \text{ ft})$$
$$= 112.5 \text{ ft}^2$$

The wetted perimeter of the LOB, WP_{LOB}, has contributions from the bottom of the rectangle and the hypotenuse of the triangle.

$$\text{WP}_{\text{LOB}} = b + \sqrt{d^2 + x^2}$$
$$= 20 \text{ ft} + \sqrt{(5 \text{ ft})^2 + (5 \text{ ft})^2}$$
$$= 27.1 \text{ ft}$$

The hydraulic radius of the LOB, R_{LOB}, is

$$R_{\text{LOB}} = \frac{A_{\text{LOB}}}{\text{WP}_{\text{LOB}}} = \frac{112.5 \text{ ft}^2}{27.1 \text{ ft}}$$
$$= 4.15 \text{ ft}$$

The longitudinal slope is

$$S = \frac{90 \text{ ft} - 88 \text{ ft}}{1000 \text{ ft}}$$
$$= 0.002 \text{ ft/ft}$$

Use Eq. 3.10 to find the flow rate through the LOB.

$$Q_{\text{LOB}} = \left(\frac{1.49}{n}\right) A R^{2/3} \sqrt{S}$$
$$= \left(\frac{1.49}{0.150}\right)(112.5 \text{ ft}^2)(4.15 \text{ ft})^{2/3}\sqrt{0.002 \frac{\text{ft}}{\text{ft}}}$$
$$= 129 \text{ ft}^3/\text{sec}$$

For the right overbank (ROB),

$$n = 0.150 \quad \text{[trees, Table 3.1]}$$

The area of the ROB is estimated from the area of a triangle plus the area of a rectangle, as delineated in the illustration.

$$A_{\text{ROB}} = (5 \text{ ft})\left(\frac{100 \text{ ft} - 95 \text{ ft}}{2}\right) + (5 \text{ ft})(10 \text{ ft})$$
$$= 62.5 \text{ ft}^2$$

The wetted perimeter of the ROB, WP_{ROB}, has contributions from the bottom of the rectangle and the hypotenuse of the triangle.

$$\text{WP}_{\text{ROB}} = b + \sqrt{d^2 + x^2}$$
$$= 10 \text{ ft} + \sqrt{(5 \text{ ft})^2 + (5 \text{ ft})^2}$$
$$= 17.1 \text{ ft}$$

The hydraulic radius of the ROB, R_{ROB}, is

$$R_{\text{ROB}} = \frac{A_{\text{ROB}}}{\text{WP}_{\text{ROB}}} = \frac{62.5 \text{ ft}^2}{17.1 \text{ ft}}$$

$$= 3.65 \text{ ft}$$

Use Eq. 3.10 to find the flow rate through the ROB.

$$Q_{\text{ROB}} = \left(\frac{1.49}{n}\right) A R^{2/3} \sqrt{S}$$

$$= \left(\frac{1.49}{0.150}\right) (62.5 \text{ ft}^2)(3.65 \text{ ft})^{2/3} \sqrt{0.002 \frac{\text{ft}}{\text{ft}}}$$

$$= 66 \text{ ft}^3/\text{sec}$$

For the channel (CHAN),

$$n = 0.030 \quad [\text{streams, Table 3.1}]$$

The area of the channel, A_{CHAN}, and the wetted perimeter, WP_{CHAN}, are estimated from the illustration.

$$A_{\text{CHAN}} = (10 \text{ ft})(100 \text{ ft} - 90 \text{ ft}) = 100 \text{ ft}^2$$

$$\text{WP}_{\text{CHAN}} = 10 \text{ ft} + 5 \text{ ft} + 5 \text{ ft} = 20 \text{ ft}$$

The hydraulic radius of the channel, R_{CHAN}, is

$$R_{\text{CHAN}} = \frac{A_{\text{CHAN}}}{\text{WP}_{\text{CHAN}}} = \frac{100 \text{ ft}^2}{20 \text{ ft}} = 5 \text{ ft}$$

Use Eq. 3.10 to find the flow rate through the channel.

$$Q_{\text{CHAN}} = \left(\frac{1.49}{n}\right) A R^{2/3} \sqrt{S}$$

$$= \left(\frac{1.49}{0.030}\right) (100 \text{ ft}^2)(5 \text{ ft})^{2/3} \sqrt{0.002 \frac{\text{ft}}{\text{ft}}}$$

$$= 649 \text{ ft}^3/\text{sec}$$

The 100 yr flood flow in the stream section is the sum of the 100 yr flood flows in the right and left overbanks and the channel.

$$Q_{100} = Q_{\text{LOB}} + Q_{\text{ROB}} + Q_{\text{CHAN}}$$

$$= 129 \frac{\text{ft}^3}{\text{sec}} + 66 \frac{\text{ft}^3}{\text{sec}} + 649 \frac{\text{ft}^3}{\text{sec}}$$

$$= 844 \text{ ft}^3/\text{sec}$$

PRACTICE PROBLEMS

1. What is the flow in a creek in million gallons per day if a stream gauge estimates the flow as 100 ft^3/sec?

2. Calculate the flow in a concrete rectangular channel with a 20 ft bottom width, a depth of flow of 5 ft, and a difference in the channel bottom elevation of 20 ft over 1000 ft in length.

3. *SI units:* How many days will it take to drain a 200 million liter reservoir with an outlet pipe capacity of 0.1 m^3/s? *Customary U.S. units:* What is the time in days required to drain a 2 billion gallon reservoir with an outlet pipe capacity of 100 ft^3/sec?

4. Calculate the flow through a rectangular weir that has a width of 10 ft. The depth of water flow over the weir is 1 ft.

5. Calculate the flow through a V-notch weir in which the angle is 90° and the flow through the weir is 1 ft deep.

6. Calculate the flow through a trapezoidal weir that has a width of 10 ft. The depth of water flow over the weir is 1 ft.

7. Calculate the velocity at section 2 in an open channel with a velocity of 4 ft/sec, a water surface elevation of 200 ft at section 1, and a water surface elevation at section 2 of 199 ft.

8. Calculate the flow in an earthen trapezoidal channel with a 20 ft bottom width and 1:1 sideslopes, a depth of flow of 5 ft, and a difference in the channel bottom elevation of 20 ft over 1000 ft in length.

9. Calculate the critical depth in a rectangular channel in which the discharge is 100 ft^3/sec and the flow width is 10 ft.

10. Calculate the depth downstream from a hydraulic jump in a 20 ft wide rectangular channel in which the flow is 200 ft^3/sec and the depth upstream is 1 ft. Calculate the upstream velocity and the downstream velocity.

11. Calculate the flow in a 5 ft wide Parshall flume in which the upstream flow depth is 1 ft.

12. A 12 ft wide × 6 ft rise concrete box culvert is under inlet control, with rounded, beveled edges at the entrance. The headwater depth is 12 ft, and the tailwater depth is 10 ft. The length of the culvert is 100 ft. Estimate the flow capacity for the culvert.

13. What is the flow through a 24 in pipe in which the velocity is measured at 5 ft/sec?

(A) 16 ft^3/sec

(B) 30 ft^3/sec

(C) 63 ft^3/sec

(D) 130 ft^3/sec

14. What is the velocity through a 10 ft wide concrete rectangular channel, flowing at a depth of 5 ft with a slope of 2%?

(A) 5.0 ft/sec

(B) 14 ft/sec

(C) 30 ft/sec

(D) 40 ft/sec

15. What diameter concrete pipe can convey 200 ft^3/sec when full, at a slope of 0.02 ft/ft?

(A) 24 in

(B) 36 in

(C) 42 in

(D) 48 in

16. If the critical depth is 1.1 ft through a 10 ft wide concrete channel, what is the discharge?

(A) 25 ft^3/sec

(B) 65 ft^3/sec

(C) 75 ft^3/sec

(D) 100 ft^3/sec

17. What is the flow through a 10 ft wide rectangular weir at a depth of 1 ft?

(A) 6.0 ft^3/sec

(B) 25 ft^3/sec

(C) 35 ft^3/sec

(D) 40 ft^3/sec

18. A 12 ft wide × 5 ft rise concrete box culvert with beveled edges is under inlet control. The headwater depth is 6 ft, and tailwater depth is 4 ft. Estimate the flow capacity of the culvert.

(A) 570 ft^3/sec

(B) 680 ft^3/sec

(C) 960 ft^3/sec

(D) 1200 ft^3/sec

19. The maximum allowable velocity for gravel in an open channel is

(A) 3.0 ft/sec

(B) 3.5 ft/sec

(C) 4.0 ft/sec

(D) 7.0 ft/sec

20. A 48 in circular concrete pipe conveys flow at a depth of 3 ft. If the longitudinal slope of the pipe is 0.02 ft/ft, what are the velocity and flow rate?

SOLUTIONS

1. The creek's flow in million gallons per day (MGD) is

$$Q = \left(100 \ \frac{\text{ft}^3}{\text{sec}}\right)\left(60 \ \frac{\text{sec}}{\text{min}}\right)\left(60 \ \frac{\text{min}}{\text{hr}}\right)$$
$$\times \left(24 \ \frac{\text{hr}}{\text{day}}\right)\left(7.48 \ \frac{\text{gal}}{\text{ft}^3}\right)$$
$$= 64{,}627{,}200 \ \text{gal/day} \quad (65 \ \text{MGD})$$

2. From Eq. 3.11, the area of a rectangular channel is

$$A = bd$$
$$= (20 \ \text{ft})(5 \ \text{ft})$$
$$= 100 \ \text{ft}^2$$

From Eq. 3.16, the wetted perimeter of a rectangular channel is

$$\text{WP} = 2d + b$$
$$= (2)(5 \ \text{ft}) + 20 \ \text{ft}$$
$$= 30 \ \text{ft}$$

From Eq. 3.8, the hydraulic radius is

$$R = \frac{A}{\text{WP}}$$
$$= \frac{100 \ \text{ft}^2}{30 \ \text{ft}}$$
$$= 3.33 \ \text{ft}$$

The longitudinal slope is calculated from the given difference in channel elevations, $z_1 - z_2$, over L feet. From Eq. 3.21,

$$S = \frac{z_1 - z_2}{L}$$
$$= \frac{20 \ \text{ft}}{1000 \ \text{ft}}$$
$$= 0.02 \ \text{ft/ft}$$

The Manning roughness coefficient of concrete is

$$n = 0.013$$

Use Eq. 3.10(b).

$$Q = \left(\frac{1.49}{n}\right) A R^{2/3} \sqrt{S}$$
$$= \left(\frac{1.49}{0.013}\right)(100 \ \text{ft}^2)(3.33 \ \text{ft})^{2/3}\sqrt{0.02 \ \frac{\text{ft}}{\text{ft}}}$$
$$= 3615 \ \text{ft}^3/\text{sec}$$

3. *SI units:*

$$t = \frac{V}{Q} = \frac{(200 \text{ ML})\left(1\,000\,000\ \frac{\text{L}}{\text{ML}}\right)}{\left(0.1\ \frac{\text{m}^3}{\text{s}}\right)\left(1000\ \frac{\text{L}}{\text{m}^3}\right)\left(60\ \frac{\text{s}}{\text{min}}\right)}$$

$$\times \left(60\ \frac{\text{min}}{\text{h}}\right)\left(24\ \frac{\text{h}}{\text{d}}\right)$$

$$= 23 \text{ d}$$

Customary U.S. units:

$$t = \frac{V}{Q} = \frac{(2 \text{ BG})\left(1 \times 10^9\ \frac{\text{gal}}{\text{BG}}\right)}{\left(100\ \frac{\text{ft}^3}{\text{sec}}\right)\left(7.48\ \frac{\text{gal}}{\text{ft}^3}\right)\left(60\ \frac{\text{sec}}{\text{min}}\right)}$$

$$\times \left(60\ \frac{\text{min}}{\text{hr}}\right)\left(24\ \frac{\text{hr}}{\text{day}}\right)$$

$$= 31 \text{ days}$$

4. The rectangular weir flow coefficient, C, is 3.5. Equation 3.42 describes rectangular weir flow.

$$Q = Cbd_H^{3/2}$$
$$= (3.5)(10 \text{ ft})(1 \text{ ft})^{3/2}$$
$$= 35 \text{ ft}^3/\text{sec}$$

5. The V-notch weir flow coefficient, C, is 0.6. The V-notch weir flow formula is Eq. 3.43.

$$Q = C(0.5)\tan\frac{\theta}{2} d_H^{5/2}\sqrt{2g}$$
$$= (0.6)(0.5)\left(\tan\frac{90°}{2}\right)(1 \text{ ft})^{5/2}\sqrt{(2)\left(32.2\ \frac{\text{ft}}{\text{sec}^2}\right)}$$
$$= 2.4 \text{ ft}^3/\text{sec}$$

6. The trapezoidal weir flow coefficient, C, is 3.3. The trapezoidal weir flow formula is Eq. 3.44.

$$Q = Cbd_H^{3/2}$$
$$= (3.3)(10 \text{ ft})(1 \text{ ft})^{3/2}$$
$$= 33 \text{ ft}^3/\text{sec}$$

7.

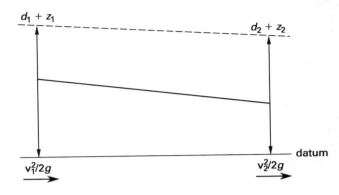

Rearrange the Bernoulli energy equation, Eq. 3.1.

$$\frac{\text{v}_1^2}{2g} + d_1 + z_1 = \frac{\text{v}_2^2}{2g} + d_2 + z_2$$

$$\frac{\left(4\ \frac{\text{ft}}{\text{sec}}\right)^2}{(2)\left(32.2\ \frac{\text{ft}}{\text{sec}^2}\right)} + 200 \text{ ft} = \frac{\text{v}_2^2}{(2)\left(32.2\ \frac{\text{ft}}{\text{sec}^2}\right)} + 199 \text{ ft}$$

$$\text{v}_2^2 = 80.4\ \frac{\text{ft}^2}{\text{sec}^2}$$

$$\text{v}_2 = 9.0 \text{ ft/sec}$$

8. From Eq. 3.13, the area of a trapezoidal channel is

$$A = bd + xd$$
$$= (20 \text{ ft})(5 \text{ ft}) + (5 \text{ ft})(5 \text{ ft})$$
$$= 125 \text{ ft}^2$$

From Eq. 3.18, the wetted perimeter of a trapezoidal channel is

$$\text{WP} = b + 2\sqrt{d^2 + x^2}$$
$$= 20 \text{ ft} + (2)\sqrt{(5 \text{ ft})^2 + (5 \text{ ft})^2}$$
$$= 34.1 \text{ ft}$$

From Eq. 3.8, the hydraulic radius is

$$R = \frac{A}{\text{WP}} = \frac{125 \text{ ft}^2}{34.1 \text{ ft}}$$
$$= 3.67 \text{ ft}$$

The longitudinal slope is calculated from the given difference in channel elevations, $z_1 - z_2$, over L feet. From Eq. 3.21,

$$S = \frac{z_1 - z_2}{L}$$
$$= \frac{20 \text{ ft}}{1000 \text{ ft}}$$
$$= 0.02 \text{ ft/ft}$$

The Manning roughness coefficient is

$$n = 0.022 \quad \text{[earth, Table 3.1]}$$

Substitute into Eq. 3.10(b).

$$Q = \left(\frac{1.49}{n}\right) A R^{2/3} \sqrt{S}$$
$$= \left(\frac{1.49}{0.022}\right)(125 \text{ ft}^2)(3.67 \text{ ft})^{2/3}\sqrt{0.02 \frac{\text{ft}}{\text{ft}}}$$
$$= 2849 \text{ ft}^3/\text{sec}$$

9. From Eq. 3.22, the specific flow is

$$q = \frac{Q}{b}$$
$$= \frac{100 \frac{\text{ft}^3}{\text{sec}}}{10 \text{ ft}}$$
$$= 10 \text{ ft}^3/\text{sec-ft}$$

Use Eq. 3.23.

$$d_c = \frac{(q^2)^{1/3}}{g^{1/3}} = \frac{\left(\left(10 \frac{\text{ft}^3}{\text{sec-ft}}\right)^2\right)^{1/3}}{\left(32.2 \frac{\text{ft}}{\text{sec}^2}\right)^{1/3}}$$
$$= \frac{\left(100 \frac{\text{ft}^3}{\text{sec-ft}}\right)^{1/3}}{\left(32.2 \frac{\text{ft}}{\text{sec}^2}\right)^{1/3}}$$
$$= 1.46 \text{ ft}$$

10. Calculate the specific discharge in a rectangular channel using Eq. 3.22.

$$q = \frac{Q}{b}$$
$$= \frac{200 \frac{\text{ft}^3}{\text{sec}}}{20 \text{ ft}}$$
$$= 10 \text{ ft}^3/\text{sec-ft}$$

Calculate the Froude number upstream from the hydraulic jump at section 1 using Eq. 3.27.

$$\text{Fr} = \frac{q}{\sqrt{g d_1^3}}$$
$$= \frac{10 \frac{\text{ft}^3}{\text{sec-ft}}}{\sqrt{\left(32.2 \frac{\text{ft}}{\text{sec}^2}\right)(1 \text{ ft})^3}}$$
$$= 1.76$$

Calculate the flow depth downstream from the hydraulic jump using Eq. 3.28.

$$\frac{d_2}{d_1} = \frac{1}{2}\left(\sqrt{1 + 8(\text{Fr})^2} - 1\right)$$
$$d_2 = \left(\frac{1}{2}\right)\left(\sqrt{1 + (8)(1.76)^2} - 1\right)(1 \text{ ft})$$
$$= 2.04 \text{ ft}$$

The downstream depth is 2.04 ft.

Use the continuity equation, Eq. 3.6, to determine the upstream velocity.

$$Q = A_1 v_1 = A_2 v_2$$
$$200 \frac{\text{ft}^3}{\text{sec}} = (20 \text{ ft})(1 \text{ ft})v_1$$
$$v_1 = \frac{200 \frac{\text{ft}^3}{\text{sec}}}{20 \text{ ft}^2}$$
$$= 10 \text{ ft/sec}$$

The upstream velocity is 10 ft/sec.

Use the continuity equation to determine the downstream velocity.

$$Q = A_2 v_2$$
$$200 \frac{\text{ft}^3}{\text{sec}} = (20 \text{ ft})(2.04 \text{ ft})v_2$$
$$v_2 = \frac{200 \frac{\text{ft}^3}{\text{sec}}}{40.8 \text{ ft}^2}$$
$$= 4.90 \text{ ft/sec}$$

11. Calculate the Parshall flume exponent, n.

$$n = 1.5 b^{0.03}$$
$$= (1.5)(5 \text{ ft})^{0.03}$$
$$= 1.57$$

Use the Parshall flume equation.

$$Q = 4bd_a^n = (4)(5 \text{ ft})(1 \text{ ft})^{1.57}$$
$$= 20 \text{ ft}^3/\text{sec}$$

12. The culvert coefficient of contraction for a rounded edge entrance is

$$C_d = 1.0$$

The area of the culvert is the product of the given width and rise.

$$A = bd = (12 \text{ ft})(6 \text{ ft})$$
$$= 72 \text{ ft}^2$$

The difference between the headwater and tailwater depth is the difference in the given values.

$$h = d_{HW} - d_{TW} = 12 \text{ ft} - 10 \text{ ft}$$
$$= 2 \text{ ft}$$

Use Eq. 3.50 to determine the flow capacity for the culvert.

$$Q = C_d A \sqrt{2gh}$$
$$= (1.0)(72 \text{ ft}^2)\sqrt{(2)\left(32.2 \frac{\text{ft}}{\text{sec}^2}\right)(2 \text{ ft})}$$
$$= 817 \text{ ft}^3/\text{sec}$$

13. Calculate the area from Eq. 3.14.

$$A = \pi r^2 = \pi\left(\frac{D}{2}\right)^2$$
$$= \frac{\pi\left(\frac{24 \text{ in}}{2}\right)^2}{\left(12 \frac{\text{in}}{\text{ft}}\right)^2}$$
$$= 3.14 \text{ ft}^2$$

Substitute the area and the given value of the velocity into the definition of flow.

$$Q = vA = \left(5 \frac{\text{ft}}{\text{sec}}\right)(3.14 \text{ ft}^2)$$
$$= 15.7 \text{ ft}^3/\text{sec} \quad (16 \text{ ft}^3/\text{sec})$$

The answer is (A).

14. For a rectangular channel,

$$A = bd = (10 \text{ ft})(5 \text{ ft})$$
$$= 50 \text{ ft}^2$$

From Eq. 3.16, the wetted perimeter is

$$WP = 2d + b = (2)(5 \text{ ft}) + 10 \text{ ft}$$
$$= 20 \text{ ft}$$

Using Eq. 3.8, the hydraulic radius is

$$R = \frac{A}{WP} = \frac{50 \text{ ft}^2}{20 \text{ ft}}$$
$$= 2.5 \text{ ft}$$

To find the velocity, use the Manning equation.

$$n = 0.013 \quad [\text{concrete, Table 3.1}]$$

$$v = \left(\frac{1.49}{n}\right)R^{2/3}\sqrt{S}$$
$$= \left(\frac{1.49}{0.013}\right)(2.5 \text{ ft})^{2/3}\sqrt{0.02 \frac{\text{ft}}{\text{ft}}}$$
$$= 29.9 \text{ ft/sec} \quad (30 \text{ ft/sec})$$

The answer is (C).

15. Most of the terms in the Manning equation can either be calculated from the given values or are given. Carry the unknown value of the diameter, D, through the calculation of the Manning flow.

The area of a circular pipe is

$$A = \pi r^2 = \pi\left(\frac{D}{2}\right)^2$$

The wetted perimeter of a circular pipe is calculated using Eq. 3.19.

$$WP = 2\pi r = \frac{2\pi D}{2} = \pi D$$

From Eq. 3.8, the hydraulic radius is

$$R = \frac{A}{WP} = \frac{\pi\left(\frac{D}{2}\right)^2}{\pi D}$$
$$= D/4$$

Substitute D into the Manning equation for flow, choosing a diameter of 48 in.

$$D = \frac{48 \text{ in}}{12 \frac{\text{in}}{\text{ft}}} = 4 \text{ ft}$$

$$Q = vA = \left(\frac{1.49}{n}\right) A R^{2/3} \sqrt{S}$$

$$= \left(\frac{1.49}{0.013}\right)\left(\pi\left(\frac{4 \text{ ft}}{2}\right)^2\right)\left(\frac{4 \text{ ft}}{4}\right)^{2/3}\sqrt{0.02\,\frac{\text{ft}}{\text{ft}}}$$

$$= 204 \text{ ft}^3/\text{sec} \quad (200 \text{ ft}^3/\text{sec}) \quad [\text{OK}]$$

A 48 in diameter pipe will convey 200 ft^3/sec when full.

The answer is (D).

16. The critical depth is 1.1 ft, and the channel width, b, is 10 ft. The specific discharge in a rectangular channel is defined by Eq. 3.22 as the discharge per foot of channel width.

$$q = \frac{Q}{b}$$

$$= Q/10 \text{ ft}$$

The gravitational acceleration is 32.2 ft/sec^2. Use Eq. 3.23 and solve for Q.

$$d_c = \frac{q^{2/3}}{g^{1/3}} = \frac{\left(\dfrac{Q}{10 \text{ ft}}\right)^{2/3}}{g^{1/3}}$$

$$1.1 \text{ ft} = \frac{\dfrac{Q^{2/3}}{(10 \text{ ft})^{2/3}}}{\left(32.2\,\dfrac{\text{ft}}{\text{sec}^2}\right)^{1/3}}$$

$$Q^{2/3} = 16.2$$

$$Q = 65 \text{ ft}^3/\text{sec}$$

The answer is (B).

17. Use the rectangular weir flow formula, Eq. 3.42.

$$Q = Cbd_H^{3/2} = (3.5)(10 \text{ ft})(1 \text{ ft})^{3/2}$$

$$= 35 \text{ ft}^3/\text{sec}$$

The answer is (C).

18. Using Eq. 3.50, the flow capacity of the culvert is

$$Q = C_d A \sqrt{2gh}$$

$$= (1.0)(60 \text{ ft}^2)\sqrt{(2)\left(32.2\,\frac{\text{ft}}{\text{sec}^2}\right)(2 \text{ ft})}$$

$$= 681 \text{ ft}^3/\text{sec} \quad (680 \text{ ft}^3/\text{sec})$$

The answer is (B).

19. From Table 3.7, the maximum velocity for a gravel lined channel is 4.0 ft/sec.

The answer is (C).

20. From Table 3.2, for a d/D of 3 ft/4 ft = 0.75,

$$\frac{A}{D^2} = 0.6318$$

Calculate the area of flow.

$$A = 0.6318D^2 = (0.6318)(4 \text{ ft})^2$$

$$= 10.1 \text{ ft}^2$$

From Table 3.2, for a d/D of 0.75, WP/D is 2.0944.

Calculate the wetted perimeter.

$$\text{WP} = 2.0944D = (2.0944)(4 \text{ ft})$$

$$= 8.38 \text{ ft}$$

Calculate the hydraulic radius from Eq. 3.8.

$$R = \frac{A}{\text{WP}} = \frac{10.1 \text{ ft}^2}{8.38 \text{ ft}}$$

$$= 1.21 \text{ ft}$$

The velocity is then calculated with the Manning equation, Eq. 3.7. Table 3.1 gives the Manning roughness coefficient for concrete as $n = 0.013$.

$$v = \left(\frac{1.49}{n}\right) R^{2/3}\sqrt{S}$$

$$= \left(\frac{1.49}{0.013}\right)(1.21 \text{ ft})^{2/3}\sqrt{0.02\,\frac{\text{ft}}{\text{ft}}}$$

$$= 18.4 \text{ ft/sec}$$

The flow rate through a 48 in pipe at 3 ft deep is

$$Q = vA = \left(18.4\,\frac{\text{ft}}{\text{sec}}\right)(10.1 \text{ ft}^2)$$

$$= 186 \text{ ft}^3/\text{sec}$$

Groundwater Engineering

Nomenclature

A	area	ft^2	m^2
b	aquifer thickness	ft	m
d	depth	ft	m
e	void ratio	–	–
h	head	ft	m
K	hydraulic conductivity or permeability	ft/day	m/d
L	length	ft	m
n	porosity	–	–
Q	flow rate	ft^3/day	m^3/d
P	pump power	ft-lbf/sec	N·m/s
r	radial distance	ft	m
s	drawdown of water table	ft	m
t	time of travel for groundwater from outer radius of wellhead to well	sec	s
T	aquifer transmissivity	ft^2/day	m^2/d
v_s	seepage velocity	ft/day	m/d
V	volume	ft^3	m^3
W	width	ft	m
z_1	elevation of water table at upstream section of gradient	ft	m
z_2	elevation of water table at downstream section of gradient	ft	m

Symbols

γ	specific weight of water	lbf/ft^3	N/m^3
η	pump efficiency	–	–

Subscripts

L	loss
p	pump
v	voids
t	total

1. INTRODUCTION

Groundwater engineering applies to the flow of water beneath the surface of the earth and refers to the infiltration term of the hydrologic cycle. Groundwater engineers design wells, retention basins, and infiltration trenches using the basic equations of well hydraulics, permeability, and recharge.

Water supply wells are drilled to withdraw water from the aquifer, as depicted in Fig. 4.1. A typical *aquifer profile* consists of the soil horizon, water table, aquifer, and the underlying bedrock or impermeable clay layer. The *soil horizon* extends approximately 4 ft to 5 ft (1.2 m to 1.5 m) from the ground surface and includes organic materials, air, and weathered rock in the forms of clay, silt, sand, and gravel. The *water table* begins below the soil horizon and marks the top of the aquifer. The *aquifer*, with saturated thickness b, is the water-bearing formation that yields water to wells. Bedrock or an impermeable clay layer usually underlies the aquifer.

Water supply yield from the aquifer can be estimated using several soil parameters: porosity, void ratio, permeability (hydraulic conductivity), transmissivity, and seepage.

Figure 4.1 Typical Aquifer Profile

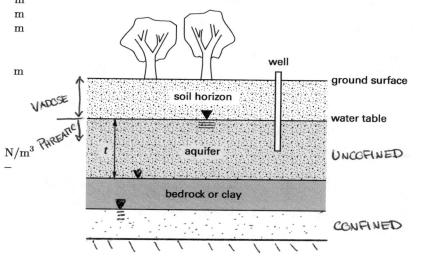

2. HYDRAULIC PROPERTIES OF AQUIFER SOILS

Porosity

The volume of water in an aquifer can be characterized by the *porosity*, n, or the volume of open spaces that contain the water between the soil particles. The porosity of soils in the aquifer is calculated using Eq. 4.1.

$$n = \frac{V_v}{V_t} = \frac{V_t - V_s}{V_t} \qquad 4.1$$

The volume of the open spaces, or voids, is V_v. The total volume is V_t.

For instance, the total volume, V_t, of the cube of soil illustrated in Fig. 4.2 is 1 ft^3. The volume of voids between the soil particles, V_v, is 0.476 ft^3. The porosity is

$$n = \frac{V_v}{V_t} = \frac{0.476 \text{ ft}^3}{1 \text{ ft}^3}$$
$$= 0.476$$

Figure 4.2 *Soil Porosity*

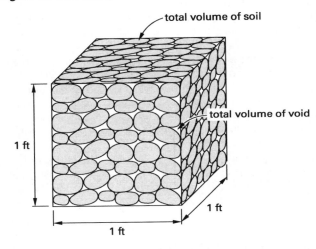

(not to scale)

Void Ratio

The *void ratio*, e, is the volume of the voids divided by the volume of the soil particles. The void ratio is related to the porosity, n, and is found from Eq. 4.2.

$$e = \frac{n}{1-n} = \frac{V_v}{V_s} = \frac{V_t - V_s}{V_s} \qquad 4.2$$

For the cube of soil depicted in Fig. 4.2, the void ratio is

$$e = \frac{n}{1-n}$$
$$= \frac{0.476}{1-0.476}$$
$$= 0.908$$

Permeability (Hydraulic Conductivity)

Permeability, K, also known as *hydraulic conductivity*, measures the velocity, or rate, that water moves through soil. Finely graded and more tightly packed soils, such as silt and clay, have low permeability. Openly graded and loosely packed soils, such as sand and gravel, have high permeability. Wells founded in aquifers with high permeability soils are usually capable of yielding greater water supply.

Seepage Analysis

Seepage is the movement of water through soils due to the difference in head from an upstream to a downstream point. Figure 4.3 shows a *seepage net* (or *flow net*), which illustrates the passage of flow from one point to another. The following steps may be used for drawing seepage nets.

step 1: Draw flow lines (stream lines) that originate upstream and end downstream. Flow lines must be approximately parallel to each other and are measured as ΔQ.

step 2: Draw equipotential lines perpendicular to the flow lines as contours of constant total head. The total head difference between adjacent equipotential lines should be the same (Δh).

step 3: Calculate the head at any point along the flow net using Eq. 4.3.

$$h = h_t - \sum \Delta h \qquad 4.3$$

Figure 4.3 *Seepage Net*

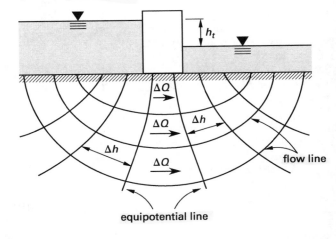

Transmissivity

The yield, or the amount of water that a well can provide, is defined by the *transmissivity*, T. The transmissivity is a function of the soil permeability and the aquifer thickness. Aquifers with higher soil permeability and greater thickness have higher water supply capacity. Equation 4.4 gives the transmissivity in terms of the permeability, K, and the saturated aquifer thickness, b.

$$T = Kb \qquad 4.4$$

Table 4.1 summarizes the hydraulic properties of aquifer soils.

Table 4.1 Hydraulic Properties of Aquifer Soils

soil	soil particle diameter (in)	porosity, n	permeability (hydraulic conductivity), K (ft/day)
clay	≤ 0.02	0.35–0.70	10^{-7}–0.001
silt	≤ 0.02	0.35–0.70	0.001–0.1
sand (fine to medium)	0.02–0.04	0.25–0.55	1–10
sand (coarse)	0.04–0.08	0.20–0.50	10–100
gravel (fine to medium)	0.08–0.6	0.25–0.40	100–1000
gravel (coarse)	0.6–2.5	0.25–0.40	> 1000

(Multiply ft/day by 0.3048 to obtain m/d.)
(Multiply in by 25.4 to obtain mm.)

Example 4.1

Using Table 4.1, calculate the transmissivity of a coarse sand aquifer with a saturated aquifer thickness of 50 ft.

Solution

From Table 4.1, the conductivity of coarse sand is

$$K = 10 \text{ ft/day}$$

The aquifer transmissivity is

$$T = Kb$$
$$= \left(10 \ \frac{\text{ft}}{\text{day}}\right)(50 \text{ ft})$$
$$= 500 \text{ ft}^2/\text{day}$$

3. GROUNDWATER CONTROL

Groundwater control is utilized especially during construction to intercept seepage, increase slope stability, and reduce loads on trenches. Uncontrolled groundwater can undermine the stability of foundations and excavations. Sites are dewatered by installing wells that are pumped to lower the groundwater table, which should be lowered to at least 3 ft (0.9 m) below the bottom of the excavation to ensure dry conditions. (See Fig. 4.4.)

The following *groundwater control methods* are employed by the U.S. Army Corps of Engineers.

- *Sump and ditch method:* Water is collected as it enters a ditch. Usually, water can only be lowered a few feet.

- *Well-point system method:* This is the most common dewatering method. The drawdown is limited to 15 ft (4.5 m) per well.

- *Deep well system method:* This system is commonly used for deep excavation and is suitable to dewater the perimeter of construction of tunnels and shafts.

- *Vertical sand drain method:* This method is usually effective for impermeable soils, such as clays, but less effective for permeable soils, such as sand and gravel.

- *Cut-off trench:* A slurry or clay filled wall is installed to deflect groundwater from entering the excavation.

Figure 4.4 Lowered Groundwater Table

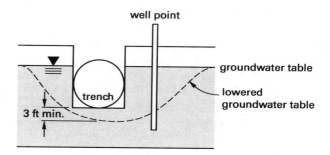

4. DARCY'S LAW

The flow rate of water that moves though soil is calculated using *Darcy's law* (derived by Henry Darcy in 1856 in Dijon, France) and measured in the laboratory using a soil permeameter. Darcy's law, Eq. 4.5, is the groundwater version of the continuity equation (see Eq. 2.3) used in open channel hydraulics.

$$Q = K\left(\frac{z_1 - z_2}{L}\right)A \qquad 4.5$$

The flow rate, Q, through the soil or the aquifer is given in unit volume per day. The permeability, or hydraulic conductivity, K, is given in unit length per day. The change in elevation of the water table at the upstream and downstream sections, $z_1 - z_2$, and the length of the section, L, are given in unit length. (See Fig. 4.5.) The quantity $(z_1 - z_2)/L$ is the *head loss gradient*. The cross-sectional area, A, of the water flow is given in unit area.

Figure 4.5 *Aquifer Cross Section Defining Darcy's Law Parameters*

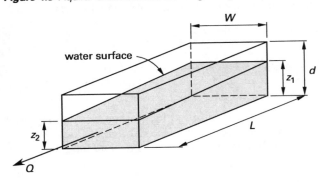

The *seepage velocity*, v_s, or the velocity of flow through voids of the soil, is calculated from Eq. 4.6.

$$v_s = K\left(\frac{z_1 - z_2}{Ln}\right) \qquad 4.6$$

n is the porosity of the soil, as defined in Eq. 4.1.

Example 4.2

Water flows through a 100 ft high × 100 ft wide fine sand aquifer section. The water table head is 100 ft at one end and 99 ft at the other end. The gradient length is 100 ft. Estimate the flow rate and the seepage velocity through the aquifer.

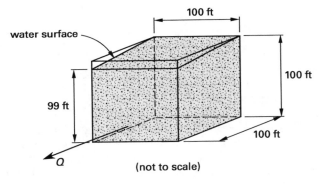

(not to scale)

Solution

From Table 4.1, the soil permeability for fine sand is

$$K = 1 \text{ ft/day}$$

From Table 4.1, the porosity for fine sand is

$$n = 0.25$$

The cross-sectional area of the aquifer section is

$$A = Ld = (100 \text{ ft})(100 \text{ ft})$$
$$= 10,000 \text{ ft}^2$$

Using Darcy's law, Eq. 4.5, the flow rate is

$$
\begin{aligned}
Q &= K\left(\frac{z_1 - z_2}{L}\right) A \\
&= \left(1 \ \frac{\text{ft}}{\text{day}}\right)\left(\frac{100 \text{ ft} - 99 \text{ ft}}{100 \text{ ft}}\right)(10,000 \text{ ft}^2)\left(7.48 \ \frac{\text{gal}}{\text{ft}^3}\right) \\
&= 748 \text{ gal/day}
\end{aligned}
$$

Using Eq. 4.6, the seepage velocity is

$$
\begin{aligned}
v_s &= K\left(\frac{z_1 - z_2}{Ln}\right) \\
&= \left(1 \ \frac{\text{ft}}{\text{day}}\right)\left(\frac{100 \text{ ft} - 99 \text{ ft}}{(100 \text{ ft})(0.25)}\right) \\
&= 0.04 \text{ ft/day}
\end{aligned}
$$

5. WELL HYDRAULICS

Wells are drilled into aquifers to withdraw groundwater, as depicted in Fig. 4.6. Water suppliers seek to drill high-yielding wells in aquifers that have high conductivity and greater aquifer thickness. Most single family dwellings require an individual well capacity of 5 gpm to 10 gpm (20 L/min to 40 L/min). Productive public water supply wells for utilities usually withdraw at least 100 gpm (380 L/min). High-yield wells can pump over 700 gpm (approximately 1 MGD (2650 L/min)).

Figure 4.6 *Well Hydraulics*

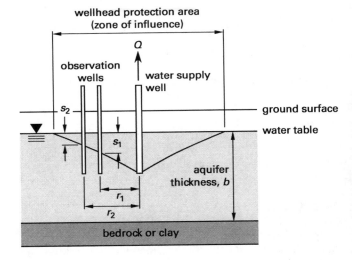

Well capacity, or yield, is related to the hydraulic conductivity (permeability) and the thickness of the aquifer. *Well yield* is estimated by installing at least two observation wells at various distances (radii) from the main water supply wells and conducting a long-term pump test. The water supply well is pumped and the *drawdown* (*cone of depression*) in the water table is measured at each of the observation wells.

Well capacity can be estimated from the pump test using the *Thiem equation* (originally derived in 1906 by George Thiem), Eq. 4.7.

$$Q = \frac{2\pi T(s_1 - s_2)}{\ln \dfrac{r_2}{r_1}} \qquad 4.7$$

The flow rate, or capacity of the well, Q, is calculated in unit volume per day. The radial distances from the water supply well of observation wells 1 and 2, r_1 and r_2, are given in unit length. The drawdowns from the water table at wells 1 and 2, s_1 and s_2, are given in unit length. The transmissivity, T, of the aquifer is given in unit area per day by Eq. 4.4.

Example 4.3

A pump test is performed. The drawdown from the water table is measured as 40 ft at observation well 1, drilled 100 ft from the water supply well. The drawdown is 30 ft at observation well 2, drilled 1000 ft from the main well. The aquifer is sampled to be coarse sand with a thickness of 50 ft. Calculate the capacity of the water supply well in gallons per minute.

Solution

From Table 4.1, the soil conductivity is $K = 30$ ft/day. Use Eq. 4.4 to find the transmissivity.

$$T = Kb = \left(30 \ \frac{\text{ft}}{\text{day}}\right)(50 \ \text{ft})$$
$$= 1500 \ \text{ft}^2/\text{day}$$

For the well capacity, use Eq. 4.7, and convert the well capacity units from cubic feet per day to gallons per minute.

$$Q = \frac{2\pi T(s_1 - s_2)}{\ln \dfrac{r_2}{r_1}}$$

$$= \frac{2\pi \left(1500 \ \dfrac{\text{ft}^2}{\text{day}}\right)(40 \ \text{ft} - 30 \ \text{ft})\left(7.48 \ \dfrac{\text{gal}}{\text{ft}^3}\right)}{\left(\ln \dfrac{1000 \ \text{ft}}{100 \ \text{ft}}\right)\left(24 \ \dfrac{\text{hr}}{\text{day}}\right)\left(60 \ \dfrac{\text{min}}{\text{hr}}\right)}$$

$$= 213 \ \text{gal/min} \quad (213 \ \text{gpm})$$

Example 4.4

A pump test is performed. The elevation of the water table after drawdown is measured at 220 ft above mean sea level (msl) at observation well 1, drilled at a distance of 200 ft from the water supply well. The drawdown level is 230 ft msl at observation well 2, drilled 400 ft from the main well. The initial water table is measured at elevation 240 ft msl. The well capacity during the pump test is 300 gpm. The aquifer thickness is measured to be 100 ft. Calculate the hydraulic conductivity and transmissivity of the aquifer, and classify the aquifer soil.

Solution

Subtract the water surface elevations after drawdown from the initial elevation of the water table.

$$s_1 = 240 \ \text{ft} - 220 \ \text{ft}$$
$$= 20 \ \text{ft}$$
$$s_2 = 240 \ \text{ft} - 230 \ \text{ft}$$
$$= 10 \ \text{ft}$$

Rearrange the Thiem equation, Eq. 4.7, for transmissivity.

$$T = \frac{Q\left(\ln \dfrac{r_2}{r_1}\right)}{2\pi(s_1 - s_2)}$$

$$= \frac{\left(300 \ \dfrac{\text{gal}}{\text{min}}\right)\left(60 \ \dfrac{\text{min}}{\text{hr}}\right)\left(24 \ \dfrac{\text{hr}}{\text{day}}\right)\left(\ln \dfrac{400 \ \text{ft}}{200 \ \text{ft}}\right)}{2\pi(20 \ \text{ft} - 10 \ \text{ft})\left(7.48 \ \dfrac{\text{gal}}{\text{ft}^3}\right)}$$

$$= 637 \ \text{ft}^2/\text{day}$$

Rearrange Eq. 4.4 for hydraulic conductivity.

$$K = \frac{T}{b}$$

$$= \frac{637 \ \dfrac{\text{ft}^2}{\text{day}}}{100 \ \text{ft}}$$

$$= 6.4 \ \text{ft/day}$$

Referring to Table 4.1, for a permeability, K, of 6.4 ft/day, the aquifer soil is likely to be fine to medium sand.

6. WATER QUALITY CONTAMINATION AND PREVENTION

Groundwater provides approximately 20% of the water supply withdrawals in the United States according to the U.S. Geological Survey. However, it is prone to contamination from urban, industrial, and agricultural sources. Sources of *groundwater contamination* include leaking septic systems, agricultural chemicals, and industrial dumping. The United States Environmental Protection Agency (EPA) gives contaminants standards, along with each contaminant's likely source and health effects. The federal government has taken steps to protect the groundwater quality through the following.

1. The *Safe Drinking Water Act* (SDWA) was originally passed by Congress in 1974 and was amended in 1986 and 1996. It requires many actions to protect

drinking water and its sources, such as rivers, lakes, reservoirs, springs, and groundwater wells.

2. *Wellhead Protection Programs* (WHPPs) were part of the 1986 amendments to the SDWA, which requires each state to develop and implement a state WHPP. A WHPP is a pollution prevention and management program used to protect underground sources of drinking water.

7. WELLHEAD PROTECTION

The Federal Safe Drinking Water Act Amendments of 1996 require that water suppliers and local governments protect the quality and quantity of drinking water wells through source water and wellhead protection programs. *Wellhead protection areas* (WHPA) are often delineated around the wells to limit land use activities, such as urban and suburban development, that might diminish the quantity and quality of recharge infiltrating into the aquifer.

The wellhead protection area is often defined as the *zone of influence*, or radius around the well, in which precipitation can recharge and travel through the groundwater to the well during pumping (see Fig. 4.7). The radius of a wellhead protection area can be defined by Eq. 4.8.

$$r = \sqrt{\frac{Qt}{\pi bn}} \qquad \qquad 4.8$$

Figure 4.7 *Wellhead Protection Area Radius*

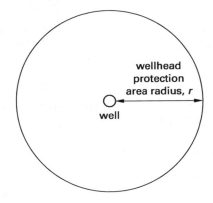

The flow rate or capacity of the well, Q, is given in unit volume per day. The travel time, t, for groundwater to flow from the outer radius of the wellhead area to the well is defined by many states, counties, and municipalities to be 5 years, given in units of seconds. The aquifer thickness, b, is given in unit length. The porosity of the soil is n.

Example 4.5

Define the radius, in miles, of a wellhead protection area for a groundwater travel time of 5 years, given that the well capacity is 1 MGD, the aquifer porosity is 0.4 (medium sand), and the aquifer thickness is 100 ft.

Solution

Convert the flow rate units from million gallons per day to cubic feet per second.

$$Q = \frac{10^6 \, \frac{\text{gal}}{\text{day}}}{\left(7.48 \, \frac{\text{gal}}{\text{ft}^3}\right)\left(24 \, \frac{\text{hr}}{\text{day}}\right)\left(60 \, \frac{\text{min}}{\text{hr}}\right)\left(60 \, \frac{\text{sec}}{\text{min}}\right)}$$
$$= 1.55 \, \text{ft}^3/\text{sec}$$

Convert the travel time units from years to seconds.

$$t = (5 \, \text{yr})\left(365 \, \frac{\text{day}}{\text{yr}}\right)\left(24 \, \frac{\text{hr}}{\text{day}}\right)\left(60 \, \frac{\text{min}}{\text{hr}}\right)\left(60 \, \frac{\text{sec}}{\text{min}}\right)$$
$$= 157{,}680{,}000 \, \text{sec}$$

Substituting into the wellhead protection area equation, Eq. 4.8, the radius of the wellhead protection area is

$$r = \sqrt{\frac{Qt}{\pi bn}}$$
$$= \frac{\sqrt{\dfrac{\left(1.55 \, \frac{\text{ft}^3}{\text{sec}}\right)(157{,}680{,}000 \, \text{sec})}{\pi(100 \, \text{ft})(0.4)}}}{5280 \, \frac{\text{ft}}{\text{mi}}}$$
$$= 0.26 \, \text{mi}$$

8. WELL PUMPS

Groundwater is usually pumped from the aquifer to the surface for distribution to the water supply system. For such pumps, *pump power* is calculated using Eq. 4.9.

$$P = \frac{\gamma Q h_p}{\eta} \qquad \qquad 4.9$$

The pump power, P, is given in units of foot-pounds force per second (newton·meters per second), the flow rate, Q, in unit volume per second, and the pump efficiency, η, is dimensionless. z_1 and z_2 are the upstream and downstream elevations, respectively. The pump head, h_p, in pipe systems is found from Eq. 4.10.

$$h_p = z_2 - z_1 + h_L \qquad \qquad 4.10$$

The head loss, h_L, is given in terms of the Darcy-Weisbach equation, Eq. 2.9; the Hazen-Williams equation, Eq. 2.11 and Eq. 2.12; and/or the sum of the minor losses (see Sec. 2.11) due to frictions at bends, tees, and valves, depending on the contributions to the losses in the system. (See Fig. 4.8.)

The *brake power*, P_n, is given by Eq. 4.11.

$$P_n = \frac{P}{\eta} \qquad 4.11$$

Figure 4.8 *Losses in a Well Pump System*

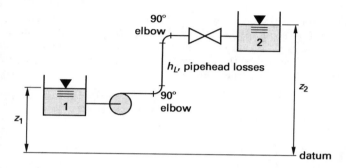

Example 4.6

A well pumps 1 MGD through a pipeline to a water tank, as shown in the following illustration. The head loss in the pipeline is 5 ft. The pump efficiency is 0.75. What is the necessary pump power in horsepower?

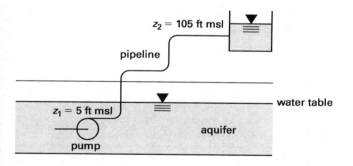

Solution

Convert Q into units of cubic feet per second.

$$Q = \frac{10^6 \ \frac{\text{gal}}{\text{day}}}{\left(7.48 \ \frac{\text{gal}}{\text{ft}^3}\right)\left(24 \ \frac{\text{hr}}{\text{day}}\right)\left(60 \ \frac{\text{min}}{\text{hr}}\right)\left(60 \ \frac{\text{sec}}{\text{min}}\right)}$$
$$= 1.55 \ \text{ft}^3/\text{sec}$$

The pump head, h_p, is given by Eq. 4.10.

$$h_p = z_2 - z_1 + h_L$$
$$= 105 \ \text{ft} - 5 \ \text{ft} + 5 \ \text{ft}$$
$$= 105 \ \text{ft}$$

Calculate the pump power using Eq. 4.9.

$$P = \frac{\gamma Q h_p}{\eta}$$
$$= \frac{\left(62.4 \ \frac{\text{lbf}}{\text{ft}^3}\right)\left(1.55 \ \frac{\text{ft}^3}{\text{sec}}\right)(105 \ \text{ft})}{(0.75)\left(550 \ \frac{\text{ft-lbf}}{\frac{\text{sec}}{\text{hp}}}\right)}$$
$$= 24.6 \ \text{hp}$$

9. INFILTRATION TRENCH DESIGN (SWALE, TRENCH)

Infiltration trenches are often required by wellhead protection ordinances to recharge aquifers during construction of new development. Infiltration trenches are usually filled with coarse gravel and perforated pipe (see Fig. 4.9). The optimal depth and width of the trench are usually the same, between 3 ft and 6 ft (0.91 m and 1.83 m) deep. The dimensions of the infiltration trench are related to the hydraulic conductivity or permeability of the soil by Eq. 4.12.

$$L = \frac{Q}{K(W + 2d)} \qquad 4.12$$

Figure 4.9 *Infiltration Trench*

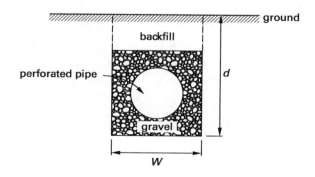

The length, L, width, W, and depth, d, of the infiltration trench are given in unit length. The hydraulic conductivity or permeability, K, is given in unit length per day. The flow rate, Q, is given in unit volume per day.

Example 4.7

A 5 ft deep × 5 ft wide infiltration trench is needed to recharge a 2 yr design flow of 2 ft³/sec. The soil permeability is 30 ft/day (medium sand). Determine the length of the trench.

Solution

Using Eq. 4.12, the length of the infiltration trench is

$$L = \frac{Q}{K(W+2d)}$$

$$= \frac{\left(2\ \dfrac{ft^3}{sec}\right)\left(60\ \dfrac{sec}{min}\right)\left(60\ \dfrac{min}{hr}\right)\left(24\ \dfrac{hr}{day}\right)}{\left(30\ \dfrac{ft}{day}\right)\left(5\ ft + (2)(5\ ft)\right)}$$

$$= 384\ ft$$

PRACTICE PROBLEMS

1. A well is drilled 200 ft deep where the water table is situated 20 ft below the ground surface. The well is designed to withdraw water from a 50 ft thick fine sand aquifer which overlays a 75 ft thick fine gravel aquifer. What is the transmissivity of the fine gravel aquifer?

  (A) 750 ft²/day

  (B) 2000 ft²/day

  (C) 5000 ft²/day

  (D) 7500 ft²/day

2. The flow rate through a 200 ft × 200 ft fine sand aquifer section is measured to be 2000 gal/day, and the change in water table elevation is measured to be 0.5 ft over a 100 ft long gradient. What is the estimated permeability of the aquifer?

  (A) 0.1 ft/day

  (B) 1.0 ft/day

  (C) 2.0 ft/day

  (D) 10 ft/day

3. A well capacity test is performed. The drawdown from the water table is measured as 50 ft at observation well 1, drilled 200 ft from the water supply well. The drawdown is 25 ft at observation well 2, drilled 1200 ft from the water supply well. The capacity of the water supply well is measured at 200 gpm. What is the estimated thickness of the saturated aquifer if the soil is classified as coarse sand?

  (A) 1.0 ft

  (B) 44 ft

  (C) 170 ft

  (D) 870 ft

4. What is the radius of a wellhead protection area for a groundwater travel time of 5 years, given that the well capacity is 2 MGD, the aquifer porosity is 0.3 (fine sand), and the aquifer thickness is 50 ft?

  (A) 0.50 mi

  (B) 0.55 mi

  (C) 0.60 mi

  (D) 0.80 mi

5. A 400 ft long infiltration trench is designed to capture the 2 yr flow of 1 ft³/sec from a 10 ac drainage area. Assuming a square trench, what are the cross-sectional dimensions of the infiltration trench if the permeability of the fine sand composing the trench is 10 ft/day?

(A) 5 ft × 5 ft

(B) 7 ft × 7 ft

(C) 15 ft × 15 ft

(D) 30 ft × 30 ft

SOLUTIONS

1. The permeability of the aquifer with fine gravel is

$$K = 100 \text{ ft/day} \quad \text{[fine gravel, Table 4.1]}$$

Using Eq. 4.4, the transmissivity of the aquifer is

$$T = Kb$$
$$= \left(100 \ \frac{\text{ft}}{\text{day}}\right)(75 \text{ ft})$$
$$= 7500 \text{ ft}^2/\text{day}$$

The answer is (D).

2. The head loss is

$$h_L = \frac{z_1 - z_2}{L} = \frac{0.5 \text{ ft}}{100 \text{ ft}}$$
$$= 0.005 \text{ ft/ft}$$

The aquifer cross-sectional area is

$$A = Ld = (200 \text{ ft})(200 \text{ ft})$$
$$= 40{,}000 \text{ ft}^2$$

The flow rate through the aquifer is

$$Q = \frac{2000 \ \frac{\text{gal}}{\text{day}}}{7.48 \ \frac{\text{gal}}{\text{ft}^3}}$$
$$= 267 \text{ ft}^3/\text{day}$$

Rearrange Eq. 4.5 to find the permeability of the fine sand aquifer.

$$K = \frac{QL}{(z_1 - z_2)A}$$
$$= \frac{\left(267 \ \frac{\text{ft}^3}{\text{day}}\right)(100 \text{ ft})}{(0.5 \text{ ft})(40{,}000 \text{ ft}^2)}$$
$$= 1.3 \text{ ft/day} \quad (1.0 \text{ ft/day})$$

The answer is (B).

3. From Table 4.1, the permeability of the coarse sand soil is

$$K = 10 \text{ ft/day}$$

The aquifer thickness is unknown, so from Eq. 4.4, the transmissivity is

$$T = Kb$$
$$= \left(10 \; \frac{\text{ft}}{\text{day}}\right) b$$

The water supply well capacity is given as 200 gpm. Convert to cubic feet per day.

$$Q = \frac{\left(200 \; \frac{\text{gal}}{\text{min}}\right)\left(60 \; \frac{\text{min}}{\text{hr}}\right)\left(24 \; \frac{\text{hr}}{\text{day}}\right)}{7.48 \; \frac{\text{gal}}{\text{ft}^3}}$$
$$= 38{,}500 \text{ ft}^3/\text{day}$$

Substitute into the Thiem equation, Eq. 4.7, to find the aquifer thickness.

$$Q = \frac{2\pi T(s_1 - s_2)}{\ln\frac{r_2}{r_1}} = \frac{2\pi\left(10 \; \frac{\text{ft}}{\text{day}}\right)b(s_1 - s_2)}{\ln\frac{r_2}{r_1}}$$

$$b = \frac{Q\ln\frac{r_2}{r_1}}{2\pi K(s_1 - s_2)}$$
$$= \frac{\left(38{,}500 \; \frac{\text{ft}^3}{\text{day}}\right)\ln\frac{1200 \text{ ft}}{200 \text{ ft}}}{2\pi\left(10 \; \frac{\text{ft}}{\text{day}}\right)(50 \text{ ft} - 25 \text{ ft})}$$
$$= 43.9 \text{ ft} \quad (44 \text{ ft})$$

The answer is (B).

4. Convert the well capacity, Q, from million gallons per day to cubic feet per second.

$$Q = \frac{2{,}000{,}000 \; \frac{\text{gal}}{\text{day}}}{\left(7.48 \; \frac{\text{gal}}{\text{ft}^3}\right)\left(24 \; \frac{\text{hr}}{\text{day}}\right)\left(60 \; \frac{\text{min}}{\text{hr}}\right)\left(60 \; \frac{\text{sec}}{\text{min}}\right)}$$
$$= 3.09 \text{ ft}^3/\text{sec}$$

Convert the groundwater travel time, t, from years to seconds.

$$t = (5 \text{ yr})\left(365 \; \frac{\text{day}}{\text{yr}}\right)\left(24 \; \frac{\text{hr}}{\text{day}}\right)\left(60 \; \frac{\text{min}}{\text{hr}}\right)\left(60 \; \frac{\text{sec}}{\text{min}}\right)$$
$$= 157{,}680{,}000 \text{ sec}$$

Use Eq. 4.8 to find the radius of the wellhead protection area.

$$r = \sqrt{\frac{Qt}{\pi b n}}$$
$$= \frac{\sqrt{\frac{\left(3.09 \; \frac{\text{ft}^3}{\text{sec}}\right)(157{,}680{,}000 \text{ sec})}{\pi(50 \text{ ft})(0.3)}}}{5280 \; \frac{\text{ft}}{\text{mi}}}$$
$$= 0.61 \text{ mi} \quad (0.60 \text{ mi})$$

The answer is (C).

5. Rearrange Eq. 4.12 to find the dimensions of the trench.

$$W + 2d = \frac{Q}{KL}$$
$$= \frac{\left(1 \; \frac{\text{ft}^3}{\text{sec}}\right)\left(60 \; \frac{\text{sec}}{\text{min}}\right)\left(60 \; \frac{\text{min}}{\text{hr}}\right)\left(24 \; \frac{\text{hr}}{\text{day}}\right)}{\left(10 \; \frac{\text{ft}}{\text{day}}\right)(400 \text{ ft})}$$
$$= 21.6 \text{ ft} \quad (21 \text{ ft})$$

Rounding to the nearest foot, the width, W, is 7 ft and the depth, d, is 7 ft.

The answer is (B).

5 Water Treatment

Nomenclature

A	area	ft^2	m^2
AW	atomic weight	amu	amu
d	depth	ft	m
D	diameter	ft	m
D	dose	mg/L	mg/L
G	mass fraction available active compound	–	–
G	mean velocity gradient	sec^{-1}	s^{-1}
h	height of water column	ft	m
LR	filter loading rate	gal/day-ft^2	m^3/d·m^2
N	number	–	–
p	pressure	lbf/in^2	kPa
P	power	ft-lbf/sec	W
P_y	purity of active compound	–	–
PDD	peak daily water demand	gal/day	m^3/d
PF	peaking factor	–	–
PWU	peak water use	gal/day	m^3/d
q_o	hydraulic loading rate	gal/day-ft^2	m^3/d·m^2
Q	flow rate	ft^3/sec	m^3/s
r	radius	ft	m
SG	particle specific gravity	–	–
t	storage capacity duration in reservoir	day	d
t_d	detention time	sec	s
t_o	flocculation mixing time	min	min
t_s	particle settling time in sedimentation tank	sec	s
v_s	particle settling velocity	ft/sec	m/s
V	volume	ft^3, gal	m^3, L
W	amount of chemical compound needed daily to treat water	lbm/day	kg/d

Symbols

γ	specific weight	lbf/ft^3	N/m^3
μ	absolute viscosity	lbf-sec/ft^2	Pa·s
ρ	density	lbm/ft^3	kg/m^3

Subscripts

ave	average
Cl	chloride
com	commercial
d	demand
du	dwelling unit
e	emergency
f	fire storage, floor, or flow-through
ind	industrial
O	oxygen
ofc	office
res	residential
t	tank

1. INTRODUCTION

Water treatment engineering includes the design of water supply treatment and pipeline distribution systems that can satisfy public demand and federal and state purity standards. It requires knowledge of the chemical and physical processes necessary to purify water to meet these standards. Water resources engineers must be able to proficiently estimate water demands and design closed conduit pipe networks (see Chap. 2).

The water treatment process removes impurities and pollutants from ground and surface water to make the water safe and healthy to drink in accordance with state and federal drinking water standards. Figure 5.1 shows the typical steps in the water treatment process.

2. APPLICABLE STANDARDS

Most water treatment standards in North America are derived from the *Recommended Standards for Water Works, Policies for the Review and Approval of Plans and Specifications for Public Water Supplies*, also known as the *Ten States Standards*. Links to the *Ten States Standards* are available at **www.ppi2pass.com/ CEwebrefs**. Excerpts from the standards are included in App. 5.A.

Figure 5.1 *Typical Water Treatment Process*

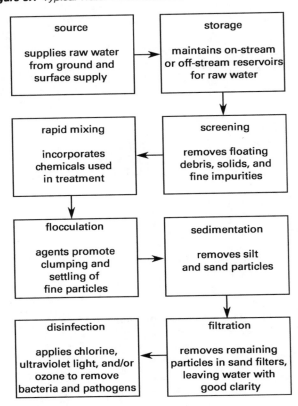

The *Source Water Protection Program* administered by the EPA under the Federal Safe Drinking Water Act classifies the following items as pollutants of concern.

- nutrients (e.g., nitrogen, phosphorus)
- pathogens (e.g., coliform, Cryptosporidium, Giardia)
- petroleum hydrocarbons (e.g., benzene, toluene)
- pesticides (e.g., alachlor, endrin, lindane)
- polychlorinated biphenyls (e.g., PCBs)
- organics (e.g., chloroform, PCE, TCE)
- metals (e.g., copper, iron, zinc)
- inorganics (e.g., chloride, fluoride, radon)

National Secondary Drinking Water Regulations (secondary standards) are non-enforceable guidelines regulating contaminants that may cause cosmetic effects (such as skin or tooth discoloration) in consumers, or aesthetic effects (such as taste, odor, or color) in drinking water. The EPA recommends secondary standards (given in App. 5.C) to water systems, but does not require systems to comply. However, states may choose to adopt them as enforceable standards.

The complete regulations regarding these contaminants are available from the *Code of Federal Regulations* website. For more information, read *Secondary Drinking Water Regulations: Guidance for Nuisance Chemicals.*

3. WATER DEMAND AND SUPPLY

Water Demand

Peak daily water demand, PDD, is estimated using Eq. 5.1.

$$\mathrm{PDD} = N_{\mathrm{du}}\mathrm{PWU}_{\mathrm{res}} + A_f\mathrm{PWU}_{\mathrm{com}}$$
$$+ A_f\mathrm{PWU}_{\mathrm{ofc}} + A_f\mathrm{PWU}_{\mathrm{ind}} \qquad 5.1$$

PWU is the *peak water use* from the residential, commercial, office, and industrial sectors, in gallons per day per dwelling unit, as given in Table 5.1. The number of *dwelling units* (du) is N_{du}. A_f is the floor area. When industrial uses have greater peak demand than the product of A_f and $\mathrm{PWU}_{\mathrm{ind}}$, substitute the actual demand into Eq. 5.1.

Table 5.1 summarizes peak daily demand usage (PWU) criteria, as well as minimum residual and service pressures, for residential, commercial, office, and industrial uses.

PDD can also be calculated on a per capita basis. *Normal residential water demands* are estimated between 75 gallons per capita per day (gpcd) and 100 gpcd (284 liters per capita per day and 378 liters per capita per day). Peak daily water demands range from

The *Ten States Standards* were assembled by the following Great Lakes member states and Canadian province.

[handwritten annotation:]
Demand [geometric method / constant %]

$\ln P_2 = \ln P_1 + K_P(t_2 - t_1)$

P = population
K_P = geometric growth rate
t = time

- Illinois
- Indiana
- Iowa
- Michigan
- Minnesota
- Missouri
- New York
- Ohio
- Ontario
- Pennsylvania
- Wisconsin

The United States Environmental Protection Agency (EPA) sets *drinking water standards* (given in App. 5.B) in accordance with the Federal Safe Drinking Water Act of 1996, with amendments. Water treatment systems are designed according to these standards to remove impurities in drinking water and to ensure that water is safe to drink.

Table 5.1 Daily Water Demand Criteria

	PWU_{res}	PWU_{com}	PWU_{ofc}	PWU_{ind}
fire fighting flows (gpm) over 5 hr	1000	1000	1000	1500
minimum service pressure (psi)	35	35	35	35
lots ≤ 1 ac (gpd)	400 per detached du, 250 per attached du	0.5 per ft^2	0.3 per ft^2	0.5 per ft^2 or actual, whichever is more
lots ≥ 1 ac (gpd)	500 per du			

(Multiply ft^2 by 0.0929 to obtain m^2.)
(Multiply gal by 0.00379 to obtain m^3.)
(Multiply ac by 0.405 to obtain ha.)

Adapted from online forms created by University of Delaware, Water Resources Agency.

150 gpcd to 200 gpcd (568 liters per capita per day to 757 liters per capita per day). The ratio of peak daily to average daily water demands is defined as the *peaking factor* (PF). The peaking factor can range from 1.5 for older neighborhoods with smaller lots to 2.0 for new neighborhoods with larger lots that have higher outdoor water needs.

Example 5.1

Estimate the peak daily water demand for a new subdivision with 2000 detached residential dwelling units with $^1/_4$ ac lots and 500,000 ft^2 of new office space.

Solution

The customer data is given in terms of dwelling units and office space. Therefore, use Eq. 5.1 and the peak daily demand criteria from Table 5.1 for a number, N_{du}, of 2000 detached residential dwelling units having lots less than 1 ac, and an area, A_f, of 500,000 ft^2 of office space.

$$\begin{aligned}
\text{PDD} &= N_{du}\text{PWU}_{res} + A_f\text{PWU}_{com} \\
&\quad + A_f\text{PWU}_{ofc} + A_f\text{PWU}_{ind} \\
&= (2000 \text{ du})\left(400 \ \frac{\text{gal}}{\text{day-du}}\right) + 0 \ \frac{\text{gal}}{\text{day}} \\
&\quad + (500,000 \text{ ft}^2)\left(0.3 \ \frac{\text{gal}}{\text{day-ft}^2}\right) + 0 \ \frac{\text{gal}}{\text{day}} \\
&= 950,000 \text{ gal/day}
\end{aligned}$$

Example 5.2

In a city with a population of 50,000, the majority of housing is over 50 years old and on $^1/_4$ ac lots. If the normal residential demand is 100 gpcd, what is the normal and peak daily water demand for the city?

Solution

Given the population data and the normal residential demand of 100 gpcd, the city's normal demand is

$$\begin{aligned}
\text{normal demand} &= (50,000 \text{ people})(100 \text{ gpcd}) \\
&= 5,000,000 \text{ gal/day} \quad (5.0 \text{ MGD})
\end{aligned}$$

For older neighborhoods with smaller lots, the peaking factor is 1.5, so the peak daily demand is

$$(5.0 \text{ MGD})(1.5) = 7.5 \text{ MGD}$$

Raw Water Storage

Raw (untreated) water storage is usually provided by reservoirs, either on-stream or off-stream. *On-stream reservoirs* are situated along rivers or streams and are filled by gravity flow. The volume and yield of an on-stream reservoir depends on the size of the drainage area that flows into it.

Off-stream reservoirs are also known as pumped storage facilities. At such facilities, water is pumped from a waterway via pipeline. The reservoir is usually situated at a high elevation. The water is released from the reservoir back into the water supply system, usually by means of gravity flow through a pipeline.

Duration of Reservoir Storage Capacity

The *storage capacity duration*, t, of a reservoir can be estimated from Eq. 5.2 by dividing the reservoir's volume, V, by the flow rate of water released by the reservoir, Q.

$$t = \frac{V}{Q} \hspace{2cm} \text{5.2}$$

Example 5.3

Calculate the storage capacity duration for a 300 MG reservoir that has a water supply yield of 3 MGD.

Solution

Use Eq. 5.2.

$$\begin{aligned}
t &= \frac{V}{Q} \\
&= \frac{300 \text{ MG}}{3 \text{ MGD}} \\
&= 100 \text{ days}
\end{aligned}$$

Volume of Reservoir Storage Capacity

The volume of a reservoir's storage capacity is found using the average end area (see Sec. 3.15). The average end area calculates the volume by averaging the two end areas and multiplying by unit length. Thus, the average end area can also be used to determine the relationship between a reservoir's elevation, surface area, and volume.

Example 5.4

Calculate the volume of a 4 ft deep reservoir. The elevation-surface area relationship is given in the following table.

column 1, depth (ft)	column 2, incremental depth (ft)	column 3, area, A (ft^2)
0	0	100,000
1	1	200,000
2	1	300,000
3	1	400,000
4	1	500,000

Solution

Use the average end area method (see Sec. 3.15), expanding by three additional columns: average area (column 4), incremental volume (column 5), and cumulative volume (column 6).

step 1: Estimate the average area (column 4) in the ith row by summing the areas, then divide by 2.

$$A_{ave} = \frac{A_1 + A_2}{2}$$

step 2: Calculate the incremental volume (column 5) by multiplying the average area (column 4) by the incremental depth in feet (column 2).

step 3: Cumulative volume (column 6) is calculated by adding the incremental volume at the preceding depth to the incremental volume at the current depth. At the depth of 1 ft, the cumulative volume is

$$0 \text{ ft}^3 + 150,000 \text{ ft}^3 = 150,000 \text{ ft}^3$$

At the depth of 4 ft, the cumulative volume is

$$750,000 \text{ ft}^3 + 450,000 \text{ ft}^3 = 1,200,000 \text{ ft}^3$$

step 4: Convert the total volume of the reservoir to acre-feet.

$$\frac{1,200,000 \text{ ft}^3}{43,560 \frac{\text{ft}^3}{\text{ac-ft}}} = 27.55 \text{ ac-ft}$$

The completed table is

depth (ft)	incremental depth (ft)	area, A (ft^2)	average area, A_{ave} (ft^2)	incremental volume (ft^3)	cumulative volume (ft^3)
0	0	100,000	0	0	0
1	1	200,000	150,000	150,000	150,000
2	1	300,000	250,000	250,000	400,000
3	1	400,000	350,000	350,000	750,000
4	1	500,000	450,000	450,000	1,200,000

Treated Water Storage and Water Pressure

Treated water storage is provided by covered water tanks situated at the highest elevations in a water supply system. The treated water tank volume is sized for one or two days of reserve supply. Water is released from the elevated water tank back into the system by means of gravity to generate sufficient water pressure.

According to *Ten States Standards* for water supply systems, the *minimum water pressure* at a service connection must be 35 psi (242 kPa) Normal working pressures in the water distribution network should be between 60 psi and 80 psi (413 kPa and 552 kPa).

Water pressure provided by an elevated water tank is given by Eq. 5.3. γ is the specific weight of water, which is dependent on the water's temperature. Water is most commonly assumed to be 32°F (0°C) with a corresponding specific weight of 62.4 lbf/ft^3 (1000 kg/m^3). h is the height of the water column, found as the change in elevation between the supply tank water level and the discharge point.

$$p = \gamma h \qquad \text{5.3}$$

Example 5.5

What is the water pressure at a service connection to a home at elevation 100 ft msl, if the water level in the storage tank is 32°F at elevation 200 ft msl?

Solution

Use Eq. 5.3.

$$p = \gamma h$$
$$= \frac{\left(62.4 \, \frac{\text{lbf}}{\text{ft}^3}\right)(200 \text{ ft} - 100 \text{ ft})}{\left(12 \, \frac{\text{in}}{\text{ft}}\right)^2}$$
$$= 43.3 \text{ lbf/in}^2$$

Water Tank Storage Volume

The volume, V_t, of a treated *water storage tank* is sized according to Eq. 5.4.

$$V_t = V_d + V_f + V_e \qquad \text{5.4}$$

The volume of demand in excess of the maximum daily demand, V_d, is usually taken as 25% of the maximum daily demand. Therefore, V_d is given in Eq. 5.5 as the product of 0.25 times the peaking factor (PF), the population, and the normal per capita water demand.

$$V_d = 0.25(\text{PF})(\text{population})$$
$$\times (\text{normal per capita water demand}) \qquad \text{5.5}$$

The *fire storage volume*, V_f, is the fire fighting flow rate multiplied by the duration of fire (usually 5 hr). The emergency storage for one day at an average daily demand, V_e, is the product of the population and the normal per capita water use.

Example 5.6

A town of 2800 people has a normal per capita water use of 150 gpcd and a fire fighting flow requirement of 5000 gpm over a 5 hr period. The peaking factor is 2.0. Calculate the water tank storage volume needed for the town.

Solution

Use Eq. 5.5 to find the volume in excess of the maximum daily demand, V_d.

$$V_d = (0.25)(\text{PF})(\text{population})$$
$$\times (\text{normal per capita water demand})$$
$$= (0.25)(2.0)(2800 \text{ people})\left(150 \frac{\text{gal}}{\text{person-day}}\right)(1 \text{ day})$$
$$= 210{,}000 \text{ gal}$$

Use Eq. 5.4 to find the volume of the tank, V_t.

$$V_t = V_d + V_f + V_e$$
$$= 210{,}000 \text{ gal} + \left(5000 \frac{\text{gal}}{\text{min}}\right)\left(60 \frac{\text{min}}{\text{hr}}\right)(5 \text{ hr})$$
$$+ (2800 \text{ people})\left(150 \frac{\text{gal}}{\text{person-day}}\right)(1 \text{ day})$$
$$= 2{,}130{,}000 \text{ gal} \quad (2.1 \text{ MG})$$

4. HYDRAULIC LOADING RATES AND DETENTION TIMES

Hydraulic loading rates, sometimes called HLR, are estimated to design various stages (e.g., mixing, flocculation, sedimentation, and filtration) in both the water treatment and waste water treatment processes (see Chap. 7 and Chap. 8). The *hydraulic loading rate* (also known as *surface loading rate*, *surface overflow rate*, and *overflow rate*), q_o, is the flow per unit area and is calculated using Eq. 5.6.

$$q_o = \frac{Q}{A} = \frac{Q}{\pi\left(\dfrac{D}{2}\right)^2} \qquad \textbf{5.6}$$

Water treatment processes also require calculation of *detention time*, t_d, in hours. Detention time is the time it takes for water to flow through a storage tank (i.e., the volume per flow rate) and is found from Eq. 5.7.

$$t_d = \frac{V}{Q} = \frac{Ad}{Q} = \frac{\pi\left(\dfrac{D}{2}\right)^2 d}{Q} \qquad \textbf{5.7}$$

Volume, V, when given in cubic feet, is the cross-sectional area, A, of the tank multiplied by its depth, d. Flow rate, Q, is given in gallons per minute, so its units must be converted to cubic feet for calculating t_d.

Example 5.7

Calculate the hydraulic loading rate for the sand filter shown.

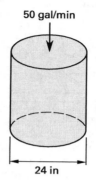

50 gal/min

24 in

Solution

Use the hydraulic loading rate formula, Eq. 5.6.

$$q_o = \frac{Q}{A} = \frac{Q}{\pi\left(\dfrac{D}{2}\right)^2}$$
$$= \frac{\left(50 \dfrac{\text{gal}}{\text{min}}\right)\left(60 \dfrac{\text{min}}{\text{hr}}\right)\left(24 \dfrac{\text{hr}}{\text{day}}\right)}{\pi\left(\dfrac{24 \text{ in}}{(2)\left(12 \dfrac{\text{in}}{\text{ft}}\right)}\right)^2}$$
$$= 22{,}918 \text{ gal/day-ft}^2$$

Example 5.8

Calculate the detention time of a flocculation tank that is 20 ft in diameter and 10 ft deep with a flow rate of 50 gpm.

Solution

Using Eq. 5.7, the detention time is

$$t_d = \frac{V}{Q} = \frac{Ad}{Q} = \frac{\pi\left(\dfrac{D}{2}\right)^2 d}{Q}$$
$$= \frac{\pi\left(\dfrac{20 \text{ ft}}{2}\right)^2 (10 \text{ ft})\left(7.48 \dfrac{\text{gal}}{\text{ft}^3}\right)}{\left(50 \dfrac{\text{gal}}{\text{min}}\right)\left(60 \dfrac{\text{min}}{\text{hr}}\right)}$$
$$= 7.83 \text{ hr}$$

5. RAPID MIXING

Water treatment chemicals are blended with raw water through a process of *rapid mixing*, in which the detention time is low, usually between 10 sec and 30 sec. To promote rapid mixing, the tank volume is small and square in cross section, usually no greater than 300 ft^3. A vertical shaft impeller is fitted into the square tank.

Tank volume and mixing time can be estimated by the rapid mixing equation, Eq. 5.8.

$$P = G^2 \times \mu$$

$$Gt_o = \frac{1}{Q}\sqrt{\frac{PV}{\mu}} \qquad \textbf{5.8}$$

In Eq. 5.8, the mixing time, t_o, is given in seconds, and the power requirement, P, is given in unit length-mass per second. The volume, V, of the rapid mixer is given in cubic length and the flow rate, Q, in cubic length per second. The absolute (dynamic) viscosity of 60°F water, μ, is approximately 0.000023 lbf-sec/ft^2 (0.0011 Pa·s). The mean velocity gradient, G, is usually between 900 sec^{-1} and 1000 sec^{-1} for a mixing time of 10 sec to 30 sec.

Example 5.9

Calculate the volume and dimensions of a rapid mixing tank with a flow rate of 5 ft^3/sec, a mixing time of 10 sec, a steep mean velocity gradient of 1000 sec^{-1}, and a power requirement of 200 ft-lbf/sec. The water is 60°F.

Solution

Use Eq. 5.8 to solve for V.

$$Gt_o = \frac{1}{Q}\sqrt{\frac{PV}{\mu}}$$

$$(1000\ \text{sec}^{-1})(10\ \text{sec}) = \frac{1}{5\ \frac{\text{ft}^3}{\text{sec}}}\sqrt{\frac{\left(200\ \frac{\text{ft-lbf}}{\text{sec}}\right)V}{0.000023\ \frac{\text{lbf-sec}}{\text{ft}^2}}}$$

$$50{,}000\ \frac{\text{ft}^3}{\text{sec}} = \sqrt{\left(8{,}695{,}652\ \frac{\text{ft}^3}{\text{sec}^2}\right)V}$$

$$V = \frac{\left(50{,}000\ \frac{\text{ft}^3}{\text{sec}}\right)^2}{8{,}695{,}652\ \frac{\text{ft}^3}{\text{sec}^2}}$$

$$= 288\ \text{ft}^3$$

Since the mixing tank is usually square in cross section, design a 6 ft deep × 7 ft wide × 7 ft long square concrete tank to provide a volume of 294 ft^3.

6. FLOCCULATION

Coagulating chemicals are added to turbid raw water to remove particulates through *flocculation*. After rapid mixing, the particles are gently agitated for 20 min to 30 min. During this period, the particles combine to form larger particles, also known as *floc*. Because the larger particles have greater size and density, they can be removed more easily through settling. The settling of floc is a form of sedimentation. Slowly rotating paddles are used to achieve flocculation (see Fig. 5.2).

Figure 5.2 *Circular Flocculation/Sedimentation Basin*

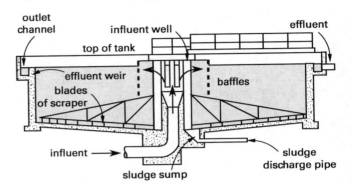

Adapted from *Unified Facilities Criteria, Water Supply: Water Treatment.* UFC 3-230-80A. January 2004. United States Department of Defense.

Typical dimensions for sedimentation and flocculation tanks are given in Table 5.2.

Table 5.2 *Water Treatment Sedimentation/Flocculation Tank Dimensions*

	parameter	value
rectangular tanks	tank depth	10–15 ft
	tank width	30 ft
	tank length	100–200 ft
	tank aspect ratio	3:1, length:width
	tank material	concrete
circular tanks	tank depth	9–14 ft
	diameter	25–200 ft

(Multiply ft by 0.3048 to obtain m.)

Table 5.3 gives values for the *mean velocity gradient, G,* and the mean velocity gradient times the *flocculation mixing time, t_o.*

Table 5.3 *Flocculation Criteria*

turbidity	G (sec^{-1})	Gt_o (sec/sec)
low	20–70	60,000–200,000
high	100–150	90,000–180,000

Adapted from the *Recommended Standards for Water Works, Policies for the Review and Approval of Plans and Specifications for Public Water Supplies.* Water Supply Committee of the Great Lakes—Upper Mississippi River Board of State and Provincial Public Health and Environmental Managers, 2007.

Example 5.10

Determine the volume and diameter of a circular flocculation tank that is 10 ft deep and needed for highly turbid water that has an average flow of 1000 gpm, a velocity gradient of 100 sec^{-1}, and a Gt_o of 120,000 sec/sec.

Solution

Determine t_o from the given values of G and Gt_o.

$$t_o = \frac{Gt_o}{G} = \frac{120{,}000 \ \frac{\text{sec}}{\text{sec}}}{(100 \ \text{sec}^{-1})\left(60 \ \frac{\text{sec}}{\text{min}}\right)}$$

$$= 20 \ \text{min}$$

Use t_o for the detention time, t_d, in Eq. 5.7. The volume of the basin is

$$V = Qt_o = \frac{\left(1000 \ \frac{\text{gal}}{\text{min}}\right)(20 \ \text{min})}{7.48 \ \frac{\text{gal}}{\text{ft}^3}}$$

$$= 2674 \ \text{ft}^3$$

The depth of the tank is 10 ft. The surface area is

$$A = \frac{2674 \ \text{ft}^3}{10 \ \text{ft}}$$

$$= 267 \ \text{ft}^2$$

The radius, r, and diameter, D, of the tank are found from the circular surface area.

$$A = \pi r^2$$
$$267 \ \text{ft}^2 = \pi r^2$$
$$r = \sqrt{\frac{267 \ \text{ft}^2}{\pi}}$$
$$= 9.22 \ \text{ft}$$

$$D = 2r = (2)(9.22 \ \text{ft})$$
$$= 18.4 \ \text{ft}$$

The dimensions of the tank are 18.4 ft diameter × 10 ft deep.

7. SEDIMENTATION

Sedimentation is used to remove particles from sediment-laden water. The settling rate is dictated by the water density and viscosity, and by the shape, dimensions, and specific gravity of the sediment. In water treatment plants, sedimentation is used after chemical coagulants have been added to produce particles that can settle out. In wastewater treatment, sedimentation is a process of sludge thickening (discussed in Chap. 7 and Chap. 8). The typical sedimentation parameters for water treatment are given in Table 5.4.

Sedimentation tanks are circular or rectangular in shape. Table 5.2 gives typical dimensions. Circular tanks are used in small to medium applications, or when ease of

Table 5.4 Water Treatment Sedimentation Parameters*

parameter	value
particle specific gravity, SG	1.0 (mud particles)
	1.4 (organic matter)
	2.65 (sand)
particle settling velocity, v_s	see Fig. 5.3 (dependent on particle diameter and specific gravity)
detention time, t_d	1–10 hr
hydraulic loading rate, q_o	600–1200 gal/day-ft^2
flow-through velocity, v_f	1 ft/min, maximum

(Multiply ft by 0.3048 to obtain m.)
(Multiply gal/day-ft^2 by 0.04075 to obtain m^3/d·m^2.)
*Sediment will settle more slowly in cold water than in warm water.

sludge removal is needed. Rectangular tanks are used in large capacity municipal or industrial applications.

The minimum settling time, t_s, is given in seconds by Eq. 5.9 from the tank depth, d, and the *particle settling velocity*, v_s.

$$t_s = \frac{d}{v_s} \qquad \text{5.9}$$

The particle settling velocity, v_s, in feet per second, is given in Fig. 5.3.

Figure 5.3 Settling Velocity of Particles

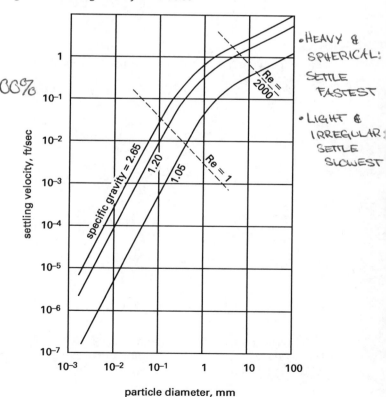

Reprinted with permission from Michael R. Lindeburg, PE, *Civil Engineering Reference Manual for the PE Exam.* 12th ed., © 2011. Professional Publications, Inc.

The *detention time, t_d*, also known as *hydraulic residence time*, is calculated from Eq. 5.10 and is given in seconds by the tank volume, V, and the flow rate, Q.

$$t_d = \frac{V}{Q} \qquad 5.10$$

The surface area, A, is the ratio of the flow rate, Q, to the *hydraulic loading rate, q_o*, as shown in Eq. 5.11.

$$A = \frac{Q}{q_o} \qquad 5.11$$

Example 5.11

Assuming a sand particle diameter of 0.1 mm, a flow rate of 2,000,000 gal/day, and a hydraulic loading rate of 600 gal/day-ft^2, determine the deepest rectangular tank dimensions for sedimentation in water treatment. Check for settling time.

Solution

To determine the tank dimensions, first find its area.

$$A = \frac{Q}{q_o}$$

$$= \frac{2,000,000 \ \dfrac{\text{gal}}{\text{day}}}{600 \ \dfrac{\text{gal}}{\text{day-ft}^2}}$$

$$= 3333 \ \text{ft}^2$$

From Table 5.2, a rectangular tank has a length:width aspect ratio of 3:1.

$$A = lw = 3w^2$$
$$w^2 = \frac{A}{3}$$
$$= \frac{3333 \ \text{ft}^2}{3}$$
$$= 1111 \ \text{ft}^2$$
$$w = 33.3 \ \text{ft}^2 \quad (33 \ \text{ft}^2)$$

$$l = 3w$$
$$= (3)(33.3 \ \text{ft})$$
$$= 99.9 \ \text{ft} \quad (100 \ \text{ft})$$

Select the maximum tank depth, d, of 15 ft from the range given in Table 5.2. The tank dimensions are 33 ft wide, 100 ft long, and 15 ft deep. The tank volume is

$$V = dA = (15 \ \text{ft})(3333 \ \text{ft}^2)$$
$$= 49,995 \ \text{ft}^3$$

Check detention time using Eq. 5.7. The detention time should fall within the range of 1 hr to 10 hr given in Table 5.4.

$$t_d = \frac{V}{Q}$$

$$= \frac{(49,995 \ \text{ft}^3)\left(7.48 \ \dfrac{\text{gal}}{\text{ft}^3}\right)\left(24 \ \dfrac{\text{hr}}{\text{day}}\right)}{2,000,000 \ \dfrac{\text{gal}}{\text{day}}}$$

$$= 4.5 \ \text{hr}$$

8. FILTRATION

Filtration removes algae, particles, floc, and other materials remaining in the water after sedimentation. In water treatment, filters are square in cross section. Table 5.5 lists the different filter types and characteristics.

Table 5.5 Filter Types and Characteristics

filter type	bed materials	media thickness (in)	loading rate, LR (gpm/ft^2)
slow sand filter	sand and gravel	24–30	1–2
rapid sand filter	sand and gravel	24–30	2–3
dual-media filter	sand and coal	12 (sand) 18 (coal)	4–6
multi-media filter	sand, coal, and granular activated carbon	12 (sand) 18 (coal) 12 (granulated active carbon)	5–10

(Multiply in by 2.54 to obtain mm.)
(Multiply gpm/ft^2 by 40.75 to obtain L·min/m^2.)

Water treatment plants usually have three to six filters, so at least one filter can be taken out of service to be backwashed and cleaned while the remaining filters continue to function. (Section 8.8 discusses the mechanisms of filtration and filter cleaning.)

In water treatment, the *filter loading rate*, LR, is calculated from Eq. 5.12 as the ratio of the flow, Q, to the area, A, of the filter.

$$\text{LR} = \frac{Q}{A} \qquad 5.12$$

Example 5.12

Calculate the total cross-sectional area of six filters that are needed for a rapid sand filter having a flow rate of 1000 gpm and a loading rate of 3 gpm/ft^2.

Solution

Rearrange Eq. 5.12 and solve for area.

$$A = \frac{Q}{LR} = \frac{1000 \ \dfrac{gal}{min}}{3 \ \dfrac{gal}{min\text{-}ft^2}}$$

$$= 333 \ ft^2$$

There are six filters. Each filter's cross-sectional area is

$$\frac{A}{6} = \frac{333 \ ft^2}{6}$$

$$= 56 \ ft^2$$

Square filters with an area of 56 ft² will be 7.5 ft × 7.5 ft. Specify six filters, each 7.5 ft × 7.5 ft.

9. DISINFECTION

Disinfection is necessary to remove bacteria and other pathogens during the last step in the water treatment process. Disinfection can be achieved through either physical or chemical means. *Physical disinfecting processes* use energy sources, such as the following, to destroy or inactivate microorganisms.

- ultraviolet (UV) light
- electronic radiation
- gamma radiation
- sound
- heat

Water treatment *chemical disinfecting processes* include

- chlorine in various forms, including Cl_2 gas, chlorine dioxide, ClO_2, and hypochlorite, ClO^-
- ozone, O_3
- halogens, including bromine, Br_2, iodine, I, and bromine chloride, BrCl
- metals, including copper, Cu^{2+}, and silver, Ag^+
- kaliumpermanganate ($KMnO_4$)
- phenols
- alcohols
- soaps and detergents

The most common form of chemical disinfection is chlorination, an early 20th century advancement that has virtually eliminated cholera and other waterborne diseases in many parts of the developed world. Chlorine can be added as a solid, in liquid solutions, or as a gas. A free chlorine residual (the amount of chlorine remaining in drinking water when it reaches the tap)

of 0.2 mg/L to 0.5 mg/L within 10 minutes after chlorination is needed to achieve disinfection.

When chlorine gas is added to water, the chlorine is hydrolyzed to form hypochlorous acid, hydrochloric acid, and a proton.

$$Cl_2 + H_2O \rightarrow Cl^- + H^+ + HOCl$$

This equation is more commonly written as

$$Cl_2 + H_2O \rightarrow HCl + HOCl$$

The hypochlorous acid, HOCl, ionizes and dissociates into a hypochlorite ion plus a proton.

$$HOCl \rightleftharpoons H^+ + ClO^-$$

When solid calcium hypochlorite is added to water, it dissociates to form hypochlorite ions, ClO^-, and calcium ions, Ca^{+2}.

$$Ca(ClO)_2 \rightarrow Ca^{+2} + 2ClO^-$$

Table 5.6 summarizes the atomic weights, AW, of the elements involved in chlorination.

Table 5.6 *Atomic Weight of the Elements Involved in Chlorination*

element	atomic weight (amu)
hydrogen	1.0
calcium	40.1
oxygen	16.0
chloride	35.5

To calculate the amount, W, in units mass per day of a chemical compound needed to treat water, use the *standard dose equation*, Eq. 5.13.

$$W = \frac{DQ}{P_y G} \tag{5.13}$$

D is the desired *dose* in milligrams per liter of the active substance, in this case, hypochlorite. The flow rate, Q, is given in million gallons per day. The *fractional purity* of the compound is P_y, which measures the proportion of a compound in a pure state.

The *fraction of available hypochlorite*, G, is calculated from Eq. 5.14.

$$G = \frac{2AW_O + 2AW_{Cl}}{AW_{total}} \tag{5.14}$$

Example 5.13

How many pounds of calcium hypochlorite are needed to treat water if the flow through the treatment plant is 5 MGD, the hypochlorite ion dose is 25 mg/L, and the purity of the calcium hypochlorite is 98%?

Solution

Calculate the molecular weight of calcium hypochlorite $Ca(ClO)_2$ from the atomic mass units of its constituents (see Table 5.6).

$$(1)(40.1) = 40.1 \quad \text{[calcium]}$$
$$(2)(16.0) = 32.0 \quad \text{[oxygen]}$$
$$(2)(35.5) = 71.0 \quad \text{[chloride]}$$
$$\text{total} \quad \overline{143.1}$$

The fraction of available hypochlorite, G, is

$$G = \frac{2AW_O + 2AW_{Cl}}{AW_{total}}$$
$$= \frac{32.0 + 70.0}{143.1}$$
$$= 0.713$$

The amount, W, of calcium hypochlorite needed to treat the water is determined from the dose equation, Eq. 5.13.

$$W = \frac{DQ}{P_y G}$$
$$= \frac{\left(25 \ \frac{mg}{L}\right)(5 \ MGD)\left(8.345 \ \frac{lbm\text{-}L}{mg\text{-}MG}\right)}{(0.98)(0.713)}$$
$$= 1492 \ lbm/day$$

10. WATER SOFTENING

Water softening removes the hardness in treated water caused by high levels of calcium and magnesium ions. *Hardness*, as measured by $CaCO_3$, should not exceed 500 mg/L in public drinking water. A desirable level for public water is 100 mg/L to 150 mg/L. Water hardness can be reduced using a *lime-soda process*. In this process, lime $(Ca(OH)_2)$ and soda ash (Na_2CO_3) are added to hard water to cause precipitation of excess sediments, such as calcium carbonate $(CaCO_3)$ and magnesium hydroxide $(Mg(OH)_2)$. Water hardness removal limits are 30 mg/L of $CaCO_3$ and 10 mg/L of $Mg(OH)_2$.

Chemicals used in water softening include calcium hydroxide, $Ca(OH)^2$ (lime), and sodium carbonate, Na_2CO_3 (soda ash). The following reactions occur with lime-soda ash softening.

When lime is added,

$$hardness + lime \ \rightarrow \ precipitate$$
$$CO_2 + Ca(OH)_2 \ \rightarrow \ CaCO_3 + H_2O$$
$$Ca(HCO_3)_2 + Ca(OH)_2 \ \rightarrow \ 2CaCO_3 + 2H_2O$$
$$Mg(HCO)_2 + Ca(OH)_2 \ \rightarrow \ CaCO_3 + MgCO_3 + H_2O$$
$$MgCO_3 + Ca(OH)_2 \ \rightarrow \ CaCO_3 + Mg(OH)_2$$

When $Ca(OH)_2$ (lime) and Na_2CO_3 (soda ash) are added, $CaCO_3$ (precipitate) is formed.

$$MgSO_4 + Ca(OH)_2 \ \rightarrow \ Mg(OH)_2 + CaSO_4$$
$$CaSO_4 + Na_2CO_3 \ \rightarrow \ CaCO_3 + Na_2SO_4$$

Example 5.14

Determine the theoretical quantity of softening chemicals (lime and soda ash) needed to reduce hardness to zero for water with the following chemistry.

$$CO_2 = 10 \ mg/L$$
$$Ca^{+2} = 100 \ mg/L$$
$$Mg^{+2} = 10 \ mg/L$$
$$Na^+ = 5 \ mg/L$$
$$CaCO_3 \ (alkalinity) = 100 \ mg/L$$
$$SO_4^{-2} = 100 \ mg/L$$
$$Cl^- = 10 \ mg/L$$

Solution

Calculate the equivalent weight and milliequivalent per liter for each parameter.

parameter	concentration (mg/L)	equivalent weight	meq/L
CO_2	10	22	0.45
Ca^{+2}	100	20	5.0
Mg^{+2}	10	12	0.83
Na^+	5	23	0.22
$CaCO_3$	100	50	2.0
SO_4^{-2}	100	48	2.1
Cl^-	10	35	0.28

Calculate the meq/L of lime and soda ash needed.

parameter	meq/L	lime	soda ash
CO_2	0.45	0.45	0
$Ca(HCO_3)_2$	2.0	2.3	0
$CaSO_4$	1.0	0	1.0
$MgSO_4$	0.83	0.83	0.83
total		3.58	1.83

The lime required is

$$lime \ required = quantity \ of \ stoichiometry$$
$$+ \ excess \ lime$$
$$= \left(3.58 \ \frac{meq}{L}\right)\left(28 \ \frac{mg}{meq}\right)$$
$$= 100 \ mg/L$$

The soda ash required is

$$\text{soda ash required} = \left(1.83 \; \frac{\text{meq}}{\text{L}}\right)\left(53 \; \frac{\text{mg}}{\text{meq}}\right)$$
$$= 97 \; \text{mg/L}$$

11. ADVANCED WATER TREATMENT

Advanced water treatment processes are employed to remove organic compounds, filter out viruses and cysts, and remove salt from water. Three common types of water treatment processes are

- membrane treatment
- activated carbon
- desalination

Membrane treatment removes contaminants and impurities by forcing water through very small pores. Membranes may also be employed to reduce turbidity, remove organics, and desalinate water. Required membrane pore sizes depend on the type of contaminate being treated and are given in Table 5.7.

Table 5.7 *Membrane Pore Sizes*

impurity	membrane pore size (μm)
ions	0.001
viruses	0.01–0.1
bacteria	0.1–10.0
Cryptosporidium	2.0–5.0
Giardia	5.0–15.0

Activated carbon treatment removes contaminants by absorption into the pores of carbon materials. Activated carbon is mixed into water as a powder. The normal dosage of activated carbon is 5 mg/L, with a contact time of 15 minutes. Activated carbon is especially suited to remove organic compounds, such as pesticides, *volatile organic chemicals* (VOCs), and trihalomethanes.

Desalination removes salt from water by reverse osmosis and distillation. Reverse osmosis removes salt by flushing saline water through a membrane that is more permeable to fresh water than to salt ions. Therefore, when the saline water is pumped at high pressure through the membrane, the treated water gets separated into two tanks—fresh water and saline water.

12. WATER DISTRIBUTION SYSTEMS

Water distribution systems are networks of looped pipes under pressure that are designed to deliver treated drinking water to counties. Components of water distribution systems, shown in Fig. 5.4, include

- supply, including surface water (streams, rivers, and lakes) and groundwater (wells)
- reservoirs
- pumps
- treatment
- storage tanks
- piped networks, including *transmission mains* (> 48 in (121 cm)), *feeder mains* (18–48 in (45–121 cm)), and *distribution mains* (< 18 in (45 cm))
- customer service connections/providers

Figure 5.4 *Water Distribution System*

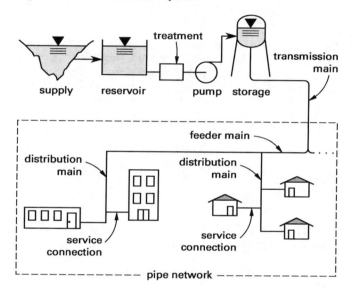

In a water distribution system, water is withdrawn from the supply and treated, then pumped to water-storage tanks. The treated water is released from the tanks using gravity (tanks are located at a system's highest topographical elevation). Water is then delivered under pressure via a looped network of transmission, feeder, and water mains to the customer service connection. The standard working pressures in the distribution system are

- minimum pressure: 35 psi (240 kPa)
- normal pressure: 60 psi to 65 psi (410 kPa to 450 kPa)
- maximum pressure: 90 psi (620 kPa)

Water Resources

PRACTICE PROBLEMS

1. Calculate the hydraulic loading rate for a sand filter with a flow rate of 100 gpm and a diameter of 18 in.

2. Calculate the detention time of a circular 24 ft diameter, 12 ft deep flocculation tank in which the flow rate is 100 gpm.

3. A reservoir is 80 ft deep and covers 40 ac. The volume of the reservoir is 500 MG, and the outlet pipeline has a capacity of 9 cfs. How many days of storage does the reservoir have?

4. Survey design drawings of a proposed reservoir indicate the cross-sectional areas at elevations 100 ft, 101 ft, 102 ft, and 103 ft are 50,000 ft^2, 100,000 ft^2, 150,000 ft^2, and 200,000 ft^2, respectively. What is the volume of the reservoir at elevation 103 ft?

5. The top of a water tank is at elevation 1360 ft msl and the base is at 1300 ft msl. The tank water level is 50 ft deep. The water tank provides pressure to a water distribution system serving customers at elevations between 1290 ft msl and 1110 ft msl. What is the water pressure from the tank at a service connection to a home at elevation 1200 ft msl?

6. A city of 5000 people has a normal per capita water use of 150 gpcd and a fire fighting flow requirement of 5000 gpm over 5 hr. Using a conservative peaking factor of 2.0, calculate the water tank storage volume needed for the city.

7. Calculate the volume of a rapid mixing tank with a flow rate of 6 cfs, water temperature of 60°F, a mixing time of 15 sec, a mean velocity gradient of 1000 sec^{-1}, and a power requirement of 300 ft-lbf/sec.

8. A 10 ft deep circular flocculation tank is to be filled with low turbidity water with an average flow of 2000 gpm. The mean velocity gradient is 50 sec^{-1}, and the Gt_o value is 150,000 sec/sec. Determine the volume and diameter required for the tank to accommodate the flow.

9. A 10 ft deep rectangular sedimentation tank has a sand particle diameter of 0.1 mm and a flow rate of 1,000,000 gal/day. Choose a conservative surface overflow rate, and check for settling time. Determine the length and width required for the tank to accommodate the flow.

10. Calculate the filter cross-sectional area for each of the 10 rapid sand filters with a flow rate of 10,000 gpm. Choose a rapid loading rate for sand.

11. Determine the number of pounds of calcium hypochlorite needed to treat water, given that the flow through the treatment plant is 1 MGD, the hypochlorite ion dose is 25 mg/L, and the purity of the calcium hypochlorite is 98%.

12. Which of the following is not a step in a water treatment process?

(A) aerobic digester

(B) sedimentation

(C) screening

(D) filtration

13. What is the peak water demand for a city of 5000 people? Use a peaking factor of 1.5.

(A) 0.3 MGD

(B) 0.75 MGD

(C) 5.0 MGD

(D) 6.5 MGD

14. A house at 1100 ft above msl is connected to a water tank with a water level of 1200 ft msl. What is the water pressure at the connection to the house?

(A) 43 psi

(B) 86 psi

(C) 6200 psi

(D) 12,000 psi

15. What is the hydraulic loading rate for a filter with a flow of 100 gpm and a diameter of 24 in?

(A) 7.9 gal/day-ft^2

(B) 32 gal/day-ft^2

(C) 12,000 gal/day-ft^2

(D) 46,000 gal/day-ft^2

16. Calculate the detention time of a flocculation tank (20 ft diameter × 10 ft deep) with a flow rate of 100 gpm.

(A) 3.1 hr

(B) 3.9 hr

(C) 7.8 hr

(D) 130 hr

17. What is the volume of a 15 ft deep sedimentation tank needed for a flow rate of 1 MGD, assuming a sand particle diameter of 0.1 mm (0.004 in) and a hydraulic loading rate of 600 gal/day-ft^2?

(A) 2500 ft^3

(B) 25,000 ft^3

(C) 50,000 ft^3

(D) 75,000 ft^3

18. What is the loading rate for a rapid sand filter?

(A) 1 gpm/ft^2

(B) 3 gpm/ft^2

(C) 6 gpm/ft^2

(D) 8 gpm/ft^2

19. How many pounds of calcium hypochlorite are needed to treat water, given that the flow is 1 MGD, the hypochlorite ion dose is 25 mg/L, and the purity of the hypochlorite is 98%?

(A) 300 lbm/day

(B) 400 lbm/day

(C) 500 lbm/day

(D) 3000 lbm/day

20. What is the recommended *Ten States Standards* minimum water pressure in a water supply distribution system?

(A) 20 psi

(B) 25 psi

(C) 35 psi

(D) 45 psi

SOLUTIONS

1. Use the hydraulic loading rate formula (Eq. 5.6).

$$q_o = \frac{Q}{A} = \frac{Q}{\pi\left(\frac{D}{2}\right)^2}$$

$$= \frac{\left(100\ \frac{\text{gal}}{\text{min}}\right)\left(60\ \frac{\text{min}}{\text{hr}}\right)\left(24\ \frac{\text{hr}}{\text{day}}\right)}{\pi\left(\frac{1.5\ \text{ft}}{2}\right)^2}$$

$$= 81{,}500\ \text{gal/day-ft}^2$$

2. Use Eq. 5.7.

$$t_d = \frac{V}{Q} = \frac{Ad}{Q} = \frac{\pi r^2 d}{Q}$$

$$= \frac{\pi\left(\frac{24\ \text{ft}}{2}\right)^2(12\ \text{ft})\left(7.48\ \frac{\text{gal}}{\text{ft}^3}\right)}{\left(100\ \frac{\text{gal}}{\text{min}}\right)\left(60\ \frac{\text{min}}{\text{hr}}\right)}$$

$$= 6.8\ \text{hr}$$

3. Use Eq. 5.2.

$$t_d = \frac{V}{Q}$$

$$= \frac{500{,}000{,}000\ \text{gal}}{\left(9\ \frac{\text{ft}^3}{\text{sec}}\right)\left(7.48\ \frac{\text{gal}}{\text{ft}^3}\right)\left(60\ \frac{\text{sec}}{\text{min}}\right)}$$

$$\times\left(60\ \frac{\text{min}}{\text{hr}}\right)\left(24\ \frac{\text{hr}}{\text{day}}\right)$$

$$= 86\ \text{days}$$

4. Set up the average end area calculation as shown in the following table. Columns 1, 2, and 3 tabulate the given elevation, incremental elevation, and area terms.

Calculate the incremental volume (column 5) by multiplying the average area (column 4) by the incremental depth in feet (column 2).

Cumulative volume (column 6) is calculated by adding the incremental volume at the preceding depth to the incremental volume at the current depth.

elevation (ft)	incr. elevation (ft)	area (ft^2)	ave. area volume (ft^2)	incr. volume (ft^3)	cum. volume (ft^3)
100	0	50,000	0	0	0
101	1	100,000	75,000	75,000	75,000
102	1	150,000	125,000	125,000	200,000
103	1	200,000	175,000	175,000	375,000

The total volume of the reservoir at elevation 103 ft is 375,000 ft^3.

5. Calculate the height.

$$h = 1300 \text{ ft} + 50 \text{ ft} - 1200 \text{ ft} = 150 \text{ ft}$$

Use Eq. 5.3.

$$p = \gamma h = \frac{\left(62.4 \ \frac{\text{lbf}}{\text{ft}^3}\right)(150 \text{ ft})}{\left(12 \ \frac{\text{in}}{\text{ft}}\right)^2}$$

$$= 65 \text{ lbf/in}^2$$

6. Use Eq. 5.5 for the volume in excess of the maximum daily demand.

$$V_d = (0.25)(\text{PF})(\text{population})$$
$$\times (\text{normal per capita water demand})$$
$$= (0.25)(2.0)(5000 \text{ people})(150 \text{ gpcd})$$
$$= 375{,}000 \text{ gal}$$

The volumes of treated water storage tanks are sized according to Eq. 5.4.

$$V_t = V_d + V_f + V_e$$
$$= 375{,}000 \text{ gal} + \left(5000 \ \frac{\text{gal}}{\text{min}}\right)\left(60 \ \frac{\text{min}}{\text{hr}}\right)(5 \text{ hr})$$
$$+ (5000 \text{ people})(150 \text{ gpcd})$$
$$= 2{,}625{,}000 \text{ gal} \quad (2.6 \text{ MG})$$

7. Use Eq. 5.8. The absolute viscosity of water at 60°F is 0.000023 lbf-sec/ft^2.

$$Gt_h = \frac{1}{Q}\sqrt{\frac{PV}{\mu}}$$

$$(1000 \text{ sec}^{-1})(15 \text{ sec}) = \left(\frac{1}{6 \ \frac{\text{ft}^3}{\text{sec}}}\right)\sqrt{\frac{\left(300 \ \frac{\text{ft-lbf}}{\text{sec}}\right)V}{0.000023 \ \frac{\text{lbf-sec}}{\text{ft}^2}}}$$

$$90{,}000 \ \frac{\text{ft}^3}{\text{sec}} = \sqrt{\left(13{,}043{,}478 \ \frac{\text{ft}^3}{\text{sec}^2}\right)V}$$

$$V = \frac{\left(90{,}000 \ \frac{\text{ft}^3}{\text{sec}}\right)^2}{13{,}043{,}478 \ \frac{\text{ft}^3}{\text{sec}^2}}$$

$$= 621 \text{ ft}^3$$

8. Determine t_o.

$$t_o = \frac{Gt_o}{G}$$

$$= \frac{150{,}000 \ \frac{\text{sec}}{\text{sec}}}{(50 \text{ sec}^{-1})\left(60 \ \frac{\text{sec}}{\text{min}}\right)}$$

$$= 50 \text{ min}$$

Rearrange Eq. 5.7 to determine the volume.

$$V = Qt_o$$

$$= \frac{\left(2000 \ \frac{\text{gal}}{\text{min}}\right)(50 \text{ min})}{7.48 \ \frac{\text{gal}}{\text{ft}^3}}$$

$$= 13{,}400 \text{ ft}^3$$

The surface area, A, is

$$A = \frac{V}{d} = \frac{13{,}400 \text{ ft}^3}{10 \text{ ft}}$$

$$= 1340 \text{ ft}^2$$

$$A = \pi r^2$$

$$r = \sqrt{\frac{1340 \text{ ft}^2}{\pi}}$$

$$= 20.7 \text{ ft}$$

The diameter, D, is

$$D = 2r = (2)(20.7 \text{ ft})$$

$$= 41.4 \text{ ft}$$

The dimensions of the tank are 41.4 ft × 10 ft.

9. From Table 5.4, use a surface overflow rate of 600 gal/ft^2-day. The surface area of the tank is given by the ratio of the flow rate, Q, to the hydraulic loading rate, q_o.

$$A = \frac{Q}{q_o} = \frac{1{,}000{,}000 \ \frac{\text{gal}}{\text{day}}}{600 \ \frac{\text{gal}}{\text{ft}^2\text{-day}}}$$

$$= 1667 \text{ ft}^2$$

For a rectangular tank with a length to width aspect ratio of 3:1,

$$A = lw = 3w^2$$

$$= 1667 \text{ ft}^2$$

$$l = 3w = (3)(23.6 \text{ ft})$$

$$= 70.8 \text{ ft}$$

The tank volume is

$$V = dA = (10 \text{ ft})(1667 \text{ ft}^2)$$
$$= 16,670 \text{ ft}^3$$

Check detention time using Eq. 5.7.

$$t_d = \frac{V}{Q}$$

$$= \frac{(16,670 \text{ ft}^3)\left(7.48 \frac{\text{gal}}{\text{ft}^3}\right)\left(24 \frac{\text{hr}}{\text{day}}\right)}{1,000,000 \frac{\text{gal}}{\text{day}}}$$

$$= 3.0 \text{ hr} \quad [\text{within detention time of 1–10 hr, OK}]$$

10. Table 5.5 gives a loading rate of 3 gpm/ft^3 for a rapid sand filter. Rearrange Eq. 5.12 for the filter loading rate and substitute the given flow rate, Q, of 10,000 gpm and the selected LR.

$$A = \frac{Q}{\text{LR}} = \frac{10,000 \frac{\text{gal}}{\text{min}}}{3 \frac{\text{gal}}{\text{min-ft}^2}}$$

$$= 3333 \text{ ft}^2$$

Given that 10 filters are needed, each filter's cross-sectional area is

$$A_{\text{each}} = \frac{3333 \text{ ft}^2}{10}$$

$$= 333.3 \text{ ft}^2$$

Given that filters are square in cross section, specify 10 filters, each 18.26 ft × 18.26 ft, for a total of 3334 ft^2 of filter cross-sectional area.

11. To find G, the fraction of available chlorine, first calculate the molecular weight of calcium hypochlorite, $Ca(ClO)_2$, from the atomic weights of its constituent atoms.

$$(1)(40.1) = 40.1 \quad [\text{calcium}]$$
$$(2)(16.0) = 32.0 \quad [\text{oxygen}]$$
$$(2)(35.5) = 71.0 \quad [\text{chloride}]$$
$$\text{total} \quad \overline{143.1}$$

From Eq. 5.14, the fraction of available chlorine, G, from hypochlorite ion is

$$G = \frac{2\text{AW}_O + 2\text{AW}_{Cl}}{\text{AW}_{\text{total}}} = \frac{32.0 + 71.0}{143.1}$$

$$= 0.720$$

The factor of 8.345 lbm-L/mg-MG in the standard dosage equation converts the units to pounds per day.

$$W = \frac{DQ}{P_y G}$$

$$= \frac{\left(25 \frac{\text{mg}}{\text{L}}\right)(1 \text{ MGD})\left(8.345 \frac{\text{lbm-L}}{\text{mg-MG}}\right)}{(0.98)(0.720)}$$

$$= 296 \text{ lbm/day}$$

12. The aerobic digester is part of a wastewater treatment process, not a water treatment process.

The answer is (A).

13. A population of 5000 with a per capita water use of 100 gpcd and a peaking factor of 1.5 is given.

The normal water demand is the product of the population and the gallons per capita day.

$$(5000 \text{ people})(100 \text{ gpcd}) = 500,000 \text{ gal/day}$$

The peak daily water demand, PDD, is the product of the normal water use and the peaking factor.

$$\text{PDD} = (1.5)\left(500,000 \frac{\text{gal}}{\text{day}}\right)$$

$$= 750,000 \text{ gal/day} \quad (0.75 \text{ MGD})$$

The answer is (B).

14. Use Eq. 5.3.

$$p = \gamma h = \frac{\left(62.4 \frac{\text{lbf}}{\text{ft}^3}\right)(1200 \text{ ft} - 1100 \text{ ft})}{\left(12 \frac{\text{in}}{\text{ft}}\right)^2}$$

$$= 43.3 \text{ lbf/in}^2 \quad (43 \text{ psi})$$

The answer is (A).

15. The hydraulic loading rate is given by Eq. 5.6.

$$q_o = \frac{Q}{A} = \frac{Q}{\pi\left(\frac{D}{2}\right)^2}$$

$$= \frac{\left(100 \frac{\text{gal}}{\text{min}}\right)\left(60 \frac{\text{min}}{\text{hr}}\right)\left(24 \frac{\text{hr}}{\text{day}}\right)\left(12 \frac{\text{in}}{\text{ft}}\right)^2}{\pi\left(\frac{24 \text{ in}}{2}\right)^2}$$

$$= 45,800 \text{ gal/day-ft}^2 \quad (46,000 \text{ gal/day-ft}^2)$$

The answer is (D).

16. The detention time is calculated using Eq. 5.7.

$$t_d = \frac{Ad}{Q} = \frac{\pi r^2 d}{Q}$$

$$= \frac{\pi \left(\dfrac{20 \text{ ft}}{2}\right)^2 (10 \text{ ft}) \left(7.48 \dfrac{\text{gal}}{\text{ft}^3}\right)}{\left(100 \dfrac{\text{gal}}{\text{min}}\right)\left(60 \dfrac{\text{min}}{\text{hr}}\right)}$$

$$= 3.9 \text{ hr}$$

The answer is (B).

17. The surface overflow rate is given as 600 gal/day-ft^2.

The surface area of the tank is given by the ratio of the given flow rate, Q, of 1 MGD, to the hydraulic loading rate, q_o.

$$A = \frac{Q}{q_o} = \frac{1{,}000{,}000 \dfrac{\text{gal}}{\text{day}}}{600 \dfrac{\text{gal}}{\text{day-ft}^2}}$$

$$= 1667 \text{ ft}^2$$

The tank volume is

$$V = Ad = (1667 \text{ ft}^2)(15 \text{ ft})$$

$$= 25{,}000 \text{ ft}^3$$

Check detention time.

$$t_d = \frac{V}{Q}$$

$$= \frac{(25{,}000 \text{ ft}^3)\left(7.48 \dfrac{\text{gal}}{\text{ft}^3}\right)\left(24 \dfrac{\text{hr}}{\text{day}}\right)}{1{,}000{,}000 \dfrac{\text{gal}}{\text{day}}}$$

$$= 4.5 \text{ hr} \quad \text{[within detention time of 1–10 hr, OK]}$$

The answer is (B).

18. The loading rate is 2–3 gpm/ft^2 according to Table 5.5.

The answer is (B).

19. Calculate the molecular weight of calcium hypochlorite, $Ca(ClO)_2$.

$$(1)(40.1) = 40.1 \qquad \text{[calcium]}$$
$$(2)(16.0) = 32.0 \qquad \text{[oxygen]}$$
$$(2)(35.5) = 71.0 \qquad \text{[chloride]}$$
$$\text{total} \quad \overline{143.1}$$

Using Eq. 5.14, the available hypochlorite ion, G, from the calcium hypochlorite is

$$G = \frac{2AW_O + 2AW_{Cl}}{AW_{total}} = \frac{32.0 + 71.0}{143.1}$$

$$= 0.720$$

The amount, W, of calcium hypochlorite needed to treat the water is determined by the standard dosage equation, Eq. 5.13.

$$W = \frac{DQ}{P_y G}$$

$$= \frac{\left(25 \dfrac{\text{mg}}{\text{L}}\right)(1 \text{ MGD})\left(8.345 \dfrac{\text{lbm-L}}{\text{mg-MG}}\right)}{(0.98)(0.720)}$$

$$= 296 \text{ lbm/day} \quad (300 \text{ lbm/day})$$

The answer is (A).

20. According to *Ten States Standards*, the normal working pressure in the distribution system should not be less than 35 psi.

The answer is (C).

Topic II: Wastewater

Chapter

example

Min. Flow ($ft^3/s/mi^2$)

0.5   summer

1.0   fall/winter

4.0   spring

Total Max. Daily Loads (TMDL)

$MOS = \dfrac{MOS}{Q_{total}}$

$$TMDL = \Sigma WLA + \Sigma LA + MOS$$

— margin of safety

adopt WQ standards → monitor waters → assess waters → list threat. waters → develop TMDLs

→ control point sources via NPDES permits

↘ manage nonpoint sources via other programs

$$MCR = TCR \sqrt{1 + CVCR^2}$$

$$M(MASS) = MCR(MVR)(62.45 \times 10^{-6})$$

EMC: event mean conc.
MCR: mean EMC for site (mg/L)
TCR: site median pollutant conc. (mg/L)
CVCR: coef. of variation EMCs
M(MASS): mean pollutant mass loading (lb/event)
MCR: mean runoff conc. (mg/L)
MVR: mean storm event runoff tc (ft^3)

Inorganic Contaminants (IOC's)
- non-carbon based
- arsenic, lead, copper

Organic Contaminants
- volatile (VOCs): low boiling points (benzene)
- synthetic (SOCs)

Max. Contaminant Level Goals (MCLGs)
- unenforceable; no adverse impact

Max. Contaminant Levels (MCLs)
- enforceable; federal min. standards

SDW Act:
- disinfectants
- IOCs
- OCs
- radionuclides

see App. 5B

Macronutrients
C:N:P ratio
100:17:5 — municipal WW
↗ 106:16:1 — bio. tmt. in WWTPs
↳ 100:5:1 phytoplankton
exceed ratio? = pollutants

Micronutrients
trace (calcium, magnesium)

6 Water and Wastewater Composition and Chemistry

Nomenclature

A	volume of odor-causing sample	mL	mL
B	volume of odor-free sample	mL	mL
BOD	biochemical oxygen demand	mg/L	mg/L
C^*	equilibrium oxygen concentration at standard pressure	mg/L	mg/L
C_p	equilibrium oxygen concentration at nonstandard pressure	mg/L	mg/L
COD	chemical oxygen demand	mg/L	mg/L
D	oxygen deficit	mg/L	mg/L
DO	dissolved oxygen	mg/L	mg/L
F	dilution factor for BOD testing	–	–
g_j	number of positive tubes in jth dilution	–	–
K'	reaction rate coefficient	day^{-1}	d^{-1}
K'	rate constant	day^{-1}	d^{-1}
m_j	amount of original sample in each tube in jth dilution	mg	mg
n	number of dilutions	–	–
p	pressure	atm	atm
p_{wv}	partial pressure of water vapor	atm	atm
Q	influent flow rate	gal/day	L/d
r	rate	mg/L-day	mg/L·d
t	incubation time	day	d
t_c	sag time	sec	s
t_j	number of tubes in jth dilution	–	–
T	temperature	°F	K
ThOD	theoretical oxygen demand	mg/L	mg/L
TON	threshold odor number	–	–
v	velocity	ft/sec	m/s
x_c	critical sag distance	ft	m

Symbols

θ	temperature variation constant	–	–
θ_T	temperature multiplier for oxygen calculation	°F	°C
λ	concentration of a target microorganism	mg/L	mg/L
Φ	van't Hoff-Arrhenius oxygen temperature coefficient	–	–

Subscripts

0	at $t=0$
5	5 day
c	critical
d	deoxygenation
eff	effluent
f	final
i	initial
r	reoxygenation or reference temperature
s	sample
sat	saturated
t	time or total
T	temperature
u	ultimate

1. MICROORGANISMS

Microorganisms (microscopic organisms including bacteria, protozoans, yeast, viruses, and algae) are found virtually everywhere in nature, including hostile environments. Microorganisms that live in extreme conditions are called *extremeophiles*.

Microorganisms have an important place in all ecosystems, and also in the lives of most multicellular organisms. For mankind, they are important because they participate in the earth's element cycles (such as the carbon cycle), and because they are used in the creation of certain types of food, medicines, and biological weapons. Microorganisms are also useful in the treatment of wastewater.

Microorganisms may be grouped according to the environmental conditions, specifically the temperature conditions, in which they exist. There are four groups based on optimal growth temperatures.

- *Psychrophiles* grow best at low temperatures. Typically, the optimum temperature is less than 68°F (20°C).

- *Mesophiles* grow best at midrange temperatures. Typically, the optimum temperature is between 68°F and 104°F (20°C and 40°C). Most water treatment systems fall in this range.

- *Thermophiles* grow best at moderately hot temperatures. Typically, the optimum temperature is between 104°F and 158°F (40°C and 70°C). Few wastewater treatment systems utilize thermophiles.

- *Hyperthermophiles* grow best at hot temperatures. Typical habitats are hot springs and deep-sea hydrothermal vents. Optimal temperatures exceed 158°F (70°C) and may go well over 212°F (100°C). Hyperthermophiles have not been extensively exploited because their habitats are difficult environments to work in.

2. PATHOGENS

A *pathogen* is any biological agent that causes disease or illness to its host. The term is most often used for agents that disrupt the normal physiology of a multicellular animal or plant. However, pathogens can also infect unicellular organisms. Pathogens may be introduced into water streams through animal and human fecal contamination. Pathogenic organisms include bacteria, protozoa, viruses, and worms.

In water treatment, the two pathogens of particular concern are *Giardia lamblia* and *Cryptosporidium parvum*. Since the late 1980s, Giardia and Cryptosporidium have been recognized as two of the most common causes of waterborne disease in humans in the United States.

Cryptosporidium is a protozoan pathogen that causes a gastrointestinal illness called *cryptosporidiosis*. Both the disease and the parasite are commonly known as "crypto." These organisms are commonly found in lakes and rivers, especially when the water is contaminated with sewage and animal wastes. Cryptosporidium is capable of completing its life cycle within a single host, resulting in cyst stages that are excreted in feces and are capable of transmission through ingestion (via water) to a new host. In the cyst stage, the parasite is encased by a protective outer shell, making it highly resistant to disinfection processes.

Giardia lamblia is a flagellated protozoan parasite that infects the gastrointestinal tract and causes *giardiasis*. Like Cryptosporidium, Giardia are capable of completing their life cycle within a single host and then exiting the host as cysts. Infection occurs through ingestion of dormant cysts via contaminated water or by the fecal-oral route. The Giardia cyst can survive for weeks to months in cold water (e.g., contaminated wells and water systems, mountain streams, and reservoirs). Giardia is found in soil, food, or water, or on surfaces that have been contaminated with feces from infected humans or animals. Source water protection and optimization of filtration processes have been identified as the best methods for ensuring the elimination of Giardia and Cryptosporidium parasites from water streams.

3. VIRUSES

Viruses are microscopic parasites that infect cells in biological organisms. Viruses are obligate intracellular parasites. They can reproduce only by invading and controlling other cells, because they lack the cellular machinery for reproduction. Viruses recruit the intracellular reproductive machinery of the cells they infect to reproduce themselves. The term "virus" usually refers to those parasites that infect *eukaryotes* (organisms with nuclei and other membranous organelles, which may be single-celled or multicellular). The term *bacteriophage* or *phage* is used to describe viruses that infect *prokaryotes* (bacteria and bacteria-like organisms lacking a nucleus and other membranous organelles). Typically, virus and phage particles carry a small amount of nucleic acid (either DNA or RNA) surrounded by a protective coat consisting of proteins, and sometimes lipids and/or glycoproteins.

4. BACTERIA

Bacteria are microscopic, unicellular prokaryotes with relatively simple cell structures lacking cell nuclei and organelles (e.g., mitochondria and chloroplasts). Bacteria are the most abundant of all organisms and are found in air, soil, and water, and as symbionts in other organisms. Bacteria range in size from 0.5 μm to 5.0 μm. Bacteria generally have cell walls, like plant and fungal cells, but with a different composition (peptidoglycans). Many move around using flagella, which are different in structure from the flagella of other groups.

5. FUNGI

Fungus or *fungi* are eukaryotic organisms that digest their food externally and then absorb the nutrients. Fungi are the primary decomposers of dead plant and animal matter in many ecosystems. Fungi are heterotrophs (i.e., they do not fix their own carbon through photosynthesis, but use the carbon fixed by other organisms) and therefore are not true plants. Fungi cells have cell walls surrounding them. Fungi are a monophyletic group, meaning all varieties of fungi come from a common ancestor.

6. ALGAE

Algae comprise several different groups of organisms that capture light energy and convert carbon dioxide and water into simple sugars through photosynthesis. All algae have photosynthetic machinery and produce oxygen as a by-product of photosynthesis. Algae range from simple single-celled organisms to multicellular ones, some with fairly complex differentiated forms (e.g., seaweed). All lack leaves, roots, flowers, and other organ structures characteristic of higher plants.

Algae are traditionally regarded as simple plants. However, some are closely related to more complex plants. Other algae are classified in different protist groups alongside protozoans that are considered to be more animal-like. They are distinguished from other protozoans in that they are usually photoautotrophic. Photoautotrophic organisms can produce organic materials that can be used for cellular functions from carbon sources in the presence of light. This is not a hard and fast distinction, because some groups of algae contain species that are mixotrophic, deriving energy both from photosynthesis and the uptake of organic carbon by osmosis or consumption of other microorganisms. Some unicellular species rely entirely on external energy sources and have a reduced or absent photosynthetic apparatus. Microscopic forms that live suspended in the water column, called *phytoplankton*, provide the food base for most marine food chains.

In very high densities (called *algal blooms*), algae may discolor the water and out-compete or poison other life forms. Algal blooms in surface waters may lead to *eutrophication* of that water (see Sec. 6.14). In eutrophied water, the dissolved oxygen concentration reaches an extremely low level as the algae are degraded by various organisms that consume oxygen in the process.

7. PROTOZOA

Protozoa are single-celled eukaryotes that have some characteristics usually associated with animals, most notably mobility and heterotrophy. Protozoa are often grouped in the *Protista* kingdom with algae. In some newer schemes, however, most algae are classified in the kingdoms *Plantae* and *Chromista*, and in such cases, the remaining forms may be classified as a kingdom Protozoa. Protozoa are divided on the basis of locomotion and include *Flagellates*, *Amoeboids*, *Sporozoans*, and *Ciliates*. Protozoans are between 0.01 μm and 0.05 μm in their largest dimension. Protozoans are ubiquitous throughout aqueous environments and the soil. Protozoans such as Cryptosporidium and Giardia cause disease in humans and must be removed from water prior to ingestion.

8. INDICATOR ORGANISMS

Indicator organisms are used to assess the quality and disinfection efficiency of treated effluent waters. Indicator organisms are chosen because they often accompany pathogens, although they do not necessarily cause disease in humans. Individual pathogens are difficult to detect in large volumes of water. In contrast, indicator organisms are more easily identified, and they are present in higher concentrations when the pathogens are present, making them useful for water quality analysis. Examples of common indicator organisms used in water treatment are total coliform and the enterovirus.

The most common indicator organism for the evaluation of drinking water or treated wastewater is *total coliform* (TC), unless there is a reason to focus on a specific pathogen. *Coliform* is a group of bacteria that is readily found in soil, decaying vegetation, and animal feces, but is not normally present in treated surface water or deep groundwater. Therefore, its presence in water samples indicates a potential health risk.

The most common indicator organisms for wastewater evaluation are *fecal coliform* (FC) and *fecal streptococcus* (FS). Fecal coliform is a harmless bacteria commonly found in human or animal intestine. The ratio of fecal coliform to fecal streptococcus (*FC:FS*) can be used to indicate whether the observed bacteria levels are caused by human activities (e.g., wastewater discharges or septic tank malfunctions) or stormwater runoff (e.g., from animal wastes). An FC:FS ratio between 0.7 and 1.0 indicates a predominance of animal wastes. In a mixed pollution, an FC:FS ratio between 1.0 and 2.0 is an ambiguous range. An FC:FS ratio greater than 2.0 indicates human wastes. *Escherichia coli* (*E. coli*) is being used more frequently as an indicator organism. E. coli is a bacterium commonly found in the human intestine and feces, making it a good indicator of whether water is contaminated by human waste.

The *Colilert Presence/Absence method* is used to determine the presence of coliform in water samples. The test involves a two-part examination in which a 100 mL sample of water is combined with a growth medium and incubated for 24 hr at 95°F (35°C). A change in the appearance of the sample after that time indicates the presence of total coliform (TC). The second part of the examination involves an assessment using varying types of biological identification tests to determine whether the coliform is derived from fecal material (fecal coliform). Three results may be yielded from this type of analysis: TC absent, TC present, and FC present.

- *TC absent* indicates that no coliform was detected in the sample, and the water from which the sample was collected may be considered free of bacterial and protozoan contamination.

- *TC present* indicates that coliform is present in the sample, and the water from which the sample was collected is not safe to drink. TC present results in the effluent water being highly chlorinated and/or an advisory being issued to the receiving public.

- *Fecal coliform present* indicates that fecal coliform is present in the sample, and the water is not safe to drink. This finding results in the effluent water being highly chlorinated and/or an advisory being issued to the receiving public.

9. METABOLISM AND METABOLIC PATHWAYS

Metabolic processes are chemical processes necessary for the maintenance of life, which occur within living organisms. In metabolism, some substances are broken down to yield energy, while other substances that are necessary for life are synthesized. A *metabolic pathway* is a series of chemical reactions catalyzed by enzymes, resulting in either the formation of a metabolic product to be used or stored by the cell, or in the initiation of another metabolic pathway. Many pathways are elaborate and involve a step-by-step modification of the initial substance to shape it into the product with the exact desired chemical structure. These products are called *metabolites.*

There are two general metabolic pathways: anabolism and catabolism. *Anabolism* is biosynthesis. Anabolic pathways produce metabolites including pyruvate, acetyl-CoA, and citric acid cycle intermediates. These metabolites serve as the starting materials for biological production. *Catabolism* is a set of metabolic pathways in which long-chained molecules[1] (polymers) are broken down into smaller units (monomers) in order to produce energy and various building blocks for cellular material.

There are four characteristics of metabolic pathways.

1. Metabolic pathways are irreversible.

2. Every metabolic pathway has a first committed step.

3. All metabolic pathways are regulated.

4. Metabolic pathways in eukaryotic cells occur in specific cellular locations.

For the environmental engineer, the catabolic process of cellular respiration is of special interest. Respiring cells that feed on waste decompose the waste in a series of oxidative steps that pass electrons from one chemical to another. Carbon is commonly the electron donor.

10. DECOMPOSITION OF WASTE

Waste materials may be broken down in a number of different ways that are generally classified according to the terminal electron acceptor (e.g., oxygen) in the respiratory processes that result in the decomposition. The various decomposition routes are carried out by

[1]Examples of long-chained molecules typically encountered in wastewater engineering include polysaccharides, fatty acids, nucleic acids, and proteins.

different types of organisms and typically produce different end products. Environmental conditions (e.g., temperature, pH, and oxygen content) determine the decomposition routes and the various organisms involved.

Aerobic Decomposition

Aerobic decomposition is the breakdown of organic materials by aerobic organisms (aerobes) using oxygen as the terminal electron acceptor. *Aerobes* are organisms that use oxygen to oxidize substrates to obtain energy (cellular respiration). The by-products of aerobic oxidation are carbon dioxide and water.

There are three classes of aerobes.

- *Obligate aerobes* need oxygen for cellular respiration.

- *Facultative aerobes* can use oxygen but also have anaerobic methods for producing energy.

- *Microaerophiles* can use oxygen, but only at low micromolar concentrations. The growth of these organisms is inhibited at normal atmospheric oxygen concentrations (approximately 21%). Microaerophiles are also capable of carrying out anaerobic respiration.

Anoxic Decomposition

Anoxic decomposition is the decomposition of a material in the absence of oxygen. It is accomplished by several different types of microbes: hydrolytic bacteria, primary fermenters, fatty acid fermenters (syntrophs), methanogens and acetogens, and sulfate-reducing bacteria.

Anaerobic Decomposition

Anaerobic decomposition is the degradation of a waste by anaerobic organisms in low oxygen conditions. *Anaerobic organisms* do not require oxygen for growth and energy production. Anaerobic organisms can use nitrogen, phosphorus, iron species, sulfate, and carbon dioxide as terminal electron acceptors, depending on the availability of these nutrients. The by-products from anaerobic degradation are dependent on the species of the terminal electron donor. Nitrate is reduced to nitrogen gas; ferric iron (Fe^{3+}) is reduced to ferrous iron (Fe^{2+}); sulfates are reduced to hydrogen sulfide; and carbon dioxide is reduced to methane.

There are three types of anaerobic organisms.

- *Aerotolerant anaerobes* do not use oxygen as the terminal electron acceptor, but are capable of living in the presence of oxygen, and may therefore play a role in the aerobic degradation of a waste.

- *Obligate anaerobes* die when exposed to atmospheric levels of oxygen.

- *Facultative anaerobes* may use oxygen when it is present, but are otherwise anaerobic.

11. MOST PROBABLE NUMBER METHOD

Serial dilution tests measure the concentration of a target microorganism in a sample with an estimate called the *most probable number* (MPN). This method provides a statistically accurate determination of the concentration of viable microorganisms in a water sample and is particularly useful for dilute water samples. This type of test is required in water analyses because microorganisms are rarely evenly distributed throughout a sample. Therefore, microorganisms are assumed to be randomly distributed throughout the sample. The goal of the MPN method is to dilute the sample into test tubes of growth medium to such a degree that the growth medium in the tubes will sometimes, but not always, contain viable organisms. Every sample that contains even one viable organism will produce detectable growth or change when added to a tube or plate of growth medium. The individual tubes of the sample are independent.

To obtain estimates over a broad range of possible concentrations, serial dilutions are used. The number of tubes, and the number of tubes with growth at each dilution, provides an estimate of the original, undiluted concentration of microorganisms in the sample. Standard MPN procedures use a minimum of three dilutions, with three tubes, five tubes, or ten tubes per dilution. The statistical variability of bacterial distribution is best estimated by using as many tubes as is practical. After incubation, the pattern of positive and negative tubes is noted and the MPN is determined. The MPN can be determined using the following three methods.

1. *Reference to standardized tables:* When using standardized tables, the serial dilutions must be done in a succession of 1:10 dilutions in the appropriate growth medium. The tubes chosen for comparison with the table must be in consecutive order of tenfold dilutions.

2. *Iteration* of the following equation (in cases where tables cannot be used) and solving for the concentration in the undiluted sample, λ.

$$\sum_{j=1}^{n} \frac{g_j m_j}{1 - e^{-\lambda m_j}} = \sum_{j=1}^{n} t_j m_j \qquad 6.1$$

Equation 6.1 is based on a Poisson distribution. n is the number of dilutions, g_j is the number of positive tubes in the jth dilution, m_j is the amount of the original sample put in each tube in the jth dilution, and t_j is the number of tubes in the jth dilution.

3. The *Thomas equation*, Eq. 6.2, was developed by H. A. Thomas and is used to estimate coliform MPN in water samples.

$$\frac{\text{MPN}}{100 \text{ mL}} = \frac{(\text{number of positive tubes})(100)}{\sqrt{\begin{array}{c}(\text{mL of sample in negative tubes}) \\ \times (\text{mL of sample in all tubes})\end{array}}} \qquad 6.2$$

12. FOOD CHAIN

A *food chain* describes a single pathway that energy and nutrients may follow in an ecosystem. Put simply, a food chain outlines the route by which organisms get their food. There is one organism per trophic level in a food chain, and trophic levels[2] are therefore easily defined. Food chains start with a *primary producer* (plant) and end with a *top predator* (animal). Plants are producers because they are capable of using light energy to convert carbon dioxide and water into biomass (food). Animals cannot do this and therefore must consume plants and/or other animals, hence the term consumers.

Consumers are divided into three categories: *herbivores*, *carnivores*, and *omnivores*. Herbivores and carnivores consume only plants or only animals, respectively, while omnivores consume both. Bacteria and fungi are classified as decomposers, as they help break down organic materials and release mineral salts back into the food chain for absorption by plants as nutrients. Progressing from the bottom to the top of a food chain, there is a decrease in the amount of biomass at each level. With each transfer, some energy is lost to the environment. For the food chain to be sustainable, there must be more producers than consumers.

Nutrients are defined as elements that are essential to the growth and reproduction of living organisms. There are two classes of nutrients: *macronutrients* and *micronutrients*. The macronutrients are carbon[3], nitrogen, and phosphorus. For macronutrients, a minimum concentration is required to support organic life. These minimums are represented by *C:N:P* (carbon:nitrogen: phosphorus) *ratios*. If any of the macronutrients are present in excess based on the C:N:P ratio[4], they are considered to be pollutants.

Micronutrients are trace elements that are essential for the existence of organic life. Examples of micronutrients are sodium (Na^+), calcium (Ca^{2+}), magnesium (Mg^{2+}), potassium (K^+), copper (Cu^{2+}, Cu^{3+}), iron (Fe^{2+}, Fe^{3+}), and manganese (Mn^{2+}). At high concentrations, micronutrients are considered to be toxins.

13. BIOACCUMULATION

Bioaccumulation is the accumulation of contaminants in an organism as a result of exposure to and/or ingestion of the contaminant. Specific contaminants of concern for bioaccumulation include pesticides, such as DDT, PCBs, and PAHs, and heavy metals, such as mercury,

---

[2]The *trophic level* is an organism's position in a food chain. Trophic levels describe the sequence of consumption and energy transfer through the environment.

[3]The carbon concentration for determining the C:N:P ratio is generally measured as the BOD_5 in a wastewater sample.

[4]Municipal wastewater typically has a C:N:P ratio of around 100:17:5. The C:N:P ratio is somewhat higher for settled sludge in primary clarification processes at around 100:19:6. The optimum C:N:P ratio for biological treatment processes in wastewater treatment plants is 100:5:1.

Wastewater

copper, cadmium, chromium, lead, nickel, zinc, and tin. Bioaccumulation results in the organism having a higher concentration of the contaminant than the concentration in the surrounding environment. The level at which a given substance is bioaccumulated depends on the rate of uptake, the mode of uptake, the rate at which the contaminant is removed from the organism, transformation of the substance by metabolic processes, the fat content of the organism, the hydrophobicity[5] of the substance, and other biological, environmental, and physical factors. In general, more-hydrophobic compounds are more likely to bioaccumulate in organisms such as fish.

14. EUTROPHICATION

Eutrophication is a process in which water bodies, such as lakes, estuaries, or slow-moving streams, receive excess nutrients that stimulate excessive growth of plants such as algae, periphyton-attached algae, nuisance plants, and weeds. Eutrophication is considered a form of pollution despite being a natural process, because it promotes excessive plant growth and decay, favors certain weedy species over others, and causes severe water quality problems. Enhanced plant growth, often in the form of an algal bloom, reduces the amount of dissolved oxygen in the water as the plant material decays. When the dissolved oxygen concentration in a water body reaches a significantly low level (approximately 1% to 30% of the dissolved oxygen saturation concentration, or less than 3 mg/L), the water body is characterized as hypoxic. Healthy aquatic systems are typically characterized by a dissolved oxygen concentration greater than 80% of the saturated value.

Once the dissolved oxygen concentration drops to a low level, aquatic organisms such as fish become oxygen starved and die. Nitrogen, phosphorus, and carbon are the nutrients primarily responsible for eutrophication. Sources of these nutrients include fertilizers, deposition of nitrogen from the atmosphere, erosion of soil containing nutrients, agricultural and urban runoff, and sewage treatment plant discharges. Nitrogen is of particular concern as it is generally the limiting nutrient for plants. The different forms of nitrogen encountered in wastewater treatment, which must be controlled to prevent eutrophication, include the following.

- ammonia, NH_3
- nitrate, NO_3^-
- nitrite, NO_2^-
- organic nitrogen, N

Total Kjeldahl nitrogen (TKN) is defined as the sum of ammonia and organic nitrogen forms.

---

[5]*Hydrophobicity* is a term used to describe the interaction between a compound and water. Hydrophobic substances are water hating and therefore will not easily dissolve in water. Hydrophobicity is often quantified in terms of an octanol-water partition coefficient.

Nitrogen in the form of ammonia is toxic to many aquatic organisms. In receiving waters, ammonia is oxidized to nitrate, which creates an oxygen demand (see Sec. 6.17), and lowers the dissolved oxygen concentrations in the receiving water. Organic and inorganic forms of nitrogen may cause eutrophication to occur in nitrogen-limited waters.

Humans can accelerate the eutrophication process with runoff from agriculture and development, pollution from septic systems and sewers, and other human-related activities that increase the flux of inorganic nutrients and organic substances into water bodies. When detergents contained large amounts of phosphates, municipal wastewater was an injurious source of phosphorus to lakes. Formulation changes to detergents have decreased the amount of phosphates in many municipal wastewaters.

15. TASTE AND ODOR ISSUES

Taste and odor are two primary *aesthetic* properties of water. These concerns are termed aesthetic because they describe the perceptions and not the safety of the water. In general, taste and odor problems are caused by algae, dissolved organic matter, petroleum compounds, and various inorganic materials, such as chloramines, ammonia, and iron. The "rotten egg smell" is one of the most common odor problems. It is caused by hydrogen sulfide, which can be produced by sulfate-reducing bacteria. Taste and odor issues may be treated using oxidation and carbon adsorption processes.

In water quality analysis, taste and odor are typically assessed through subjective techniques, though analytical methods using mechanical sensors (sniffers) are in development. For both taste and odor issues, a wheel diagram linking observations to likely causes is often used. For odor problems, a threshold odor test can be used. Odor can be described or measured in terms of a *threshold odor number* (TON) using Eq. 6.3.

$$\text{TON} = \frac{A + B}{A} \qquad 6.3$$

A is the volume of the odor-causing sample in milliliters, and B is the volume of the odor-free sample in milliliters. The odor strength is determined by a positive detection of odor in successive dilutions of a water sample into odor-free water. In each sample, the total sample volume to be used is kept constant at 200 mL. The TON for a given water sample is the value (dilution ratio) at which an odor is detected.

Example 6.1

A wastewater treatment plant is discharging effluent into a local pristine river. To avoid causing odor complaints, the TON of the effluent must be determined. Various sample volumes are diluted to 200 mL using doubly deionized water, which is perfectly odorless. A panel of 10 experts is charged with smelling each diluted sample, from the strongest to the most dilute. The number of experts that first detected no odor at a given sample volume is shown in the following table. Determine the TON of the effluent.

sample no.	sample volume, A (mL)	number of experts first detecting no odor
1	200	0
2	185	0
3	170	0
4	155	0
5	140	0
6	125	0
7	110	0
8	95	0
9	90	1
10	80	0
11	70	0
12	60	1
13	50	0
14	40	7
15	30	1
16	20	0
17	10	0
18	1	0

Solution

Equation 6.3 gives the TON in terms of the sample volume, A, and the diluent volume, B. The TONs have been calculated using Eq. 6.3 and entered in the table. The TON of the sample (sample 14) at which odor is most frequently nondetectable is 5.0.

$$\text{TON} = \frac{A + B}{A}$$

sample no.	sample volume, A (mL)	diluent volume, B (mL)	number of experts first detecting no odor	TON
1	200	0	0	1.0
2	185	15	0	1.1
3	170	30	0	1.2
4	155	45	0	1.3
5	140	60	0	1.4
6	125	75	0	1.6
7	110	90	0	1.8
8	95	105	0	2.1
9	90	110	1	2.2
10	80	120	0	2.5
11	70	130	0	2.9
12	60	140	1	3.3
13	50	150	0	4.0
14	40	160	7	5.0
15	30	170	1	6.7
16	20	180	0	10
17	10	190	0	20
18	1	199	0	200

16. DISSOLVED OXYGEN

Dissolved oxygen (DO) is the amount of molecular oxygen dissolved in water (see Table 6.1). Oxygen dissolves in water by diffusion from the surrounding atmosphere, by aeration (turbulence), and by release from plants as a by-product of photosynthesis. The concentration of dissolved oxygen in a body of water is proportional to the concentration of oxygen in the atmosphere contacting the water, according to Henry's law of partial pressures. Dissolved oxygen concentrations are also dependent on environmental conditions, such as temperature, salinity, and altitude, as well as factors such as the types of microbes present and the organic composition of the water. When temperature and/or salinity increase, the dissolved oxygen concentration decreases.

As dissolved oxygen levels in water fall below 5.0 mg/L, aquatic life is put under stress. This value is generally used as a minimum acceptable dissolved oxygen concentration for supporting a large fish population, though it varies with temperature. Dissolved oxygen levels that remain below 2 mg/L for a few hours can result in large fish kills. High dissolved oxygen concentration improves the taste of water, though it also speeds up pipe corrosion. In water and wastewater analyses, dissolved oxygen concentrations are most commonly measured using electronic dissolved oxygen meters and DO probes.

The *saturated dissolved oxygen concentration*, C_p, is the maximum concentration possible for the water at the given atmospheric pressure and temperature (see Fig. 6.1). If the dissolved oxygen concentration is substantially less than C_p, it is an indication that the water may be polluted. This is because the organisms in the water are using the oxygen to break down organic matter. The *percent saturation value* for dissolved oxygen at a given temperature can be determined using a saturation chart for samples at sea level and average barometric pressure. A line is drawn from the known dissolved oxygen concentration, in parts per million, to the centigrade temperature of the water. The intercept point on the scale is the percent saturation value.

The saturated dissolved oxygen concentration for samples not at sea level or average barometric pressure (1 atm) may also be calculated. The equilibrium oxygen concentration at nonstandard pressure is calculated using Eq. 6.4 from the American Public Health Association (APHA).

$$C_p = C^* p \left(\frac{\left(1 - \frac{p_{wv}}{p}\right)(1 - \theta_T p)}{(1 - p_{wv})(1 - \theta_T)} \right) \qquad 6.4$$

C_p is the equilibrium oxygen concentration at nonstandard pressure in milligrams per liter. C^* is the equilibrium oxygen concentration, in milligrams per liter, at standard pressure of 1 atm. p is the nonstandard pressure. p_{wv} is the partial pressure of water vapor in

Table 6.1 Oxygen Solubility in Fresh Water Exposed to Water-Saturated Air at Standard Pressure (1 atm)

temperature (°C)	oxygen solubility (mg/L)
4.0	13.107
5.0	12.770
6.0	12.447
7.0	12.139
8.0	11.843
9.0	11.559
10.0	11.288
11.0	11.027
12.0	10.777
13.0	10.537
14.0	10.306
15.0	10.084
16.0	9.870
17.0	9.665
18.0	9.467
19.0	9.276
20.0	9.092
21.0	8.915
22.0	8.743
23.0	8.578
24.0	8.418
25.0	8.263
26.0	8.113
27.0	7.968
28.0	7.827
29.0	7.691
30.0	7.559
31.0	7.430
32.0	7.305
33.0	7.183
34.0	7.065
35.0	6.950
36.0	6.837
37.0	6.727
38.0	6.620

Reprinted from L. S. Clescerl, A. E. Greenberg, and A. D. Eaton, eds. *Standard Methods for the Examination of Water and Wastewater*, 16th ed., the American Public Health Association, the American Water Works Association, and the Water Environment Federation.

atmospheres, based on Eq. 6.5. T is the absolute temperature in kelvins.

$$\ln p_{wv} = 11.8571 - \frac{3840.70}{T} - \frac{216{,}961}{T^2} \qquad 6.5$$

The factor θ_T is calculated from Eq. 6.6, where $T_{°C}$ is in degrees Celsius.

$$\theta_T = 9.75 \times 10^{-4} - (1.426 \times 10^{-5} T_{°C})$$
$$+ (6.436 \times 10^{-8} T_{°C}^2) \qquad 6.6$$

Figure 6.1 Saturation Alignment Chart for Determining the Percent Saturation for a Water Sample of Known Temperature and Dissolved Oxygen Concentration

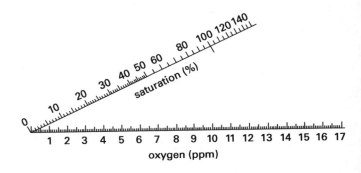

The percent saturation value is then determined from the measured dissolved oxygen concentration, DO, as shown in Eq. 6.7.

$$\% \text{ saturation} = \frac{\text{DO}}{C_p} \times 100\% \qquad 6.7$$

The difference between the saturated oxygen concentration, DO_{sat}, and the actual dissolved oxygen concentration, DO, is the *oxygen deficit*, D, calculated using Eq. 6.8.

$$D = \text{DO}_{sat} - \text{DO} \qquad 6.8$$

17. OXYGEN DEMAND

The strength and ultimate treatability of a wastewater is determined by a number of different parameters. Of particular interest in water treatment is the *oxygen demand* of a wastewater. The oxygen demand is the amount of oxygen that must be supplied for the oxidation of the constituent wastes. This may be determined theoretically through balancing chemical equations for a given system. The theoretically determined value is known as *theoretical oxygen demand* (ThOD).

The two most common measurable parameters are the *biochemical oxygen demand* (BOD) and the *chemical oxygen demand* (COD). These two values measure different quantities, and thus different characteristics, of the wastewater. The biological oxygen demand is a measurement of the amount of dissolved oxygen required by microorganisms for the biochemical oxidation of organic matter in the water. The magnitude of this value will tell the engineer the amount of dissolved oxygen that must be supplied to the water for the organic matter to be sufficiently oxidized by the microorganisms. COD is a measure of the oxygen requirement for the chemical oxidation of the waste in a water sample. The oxygen demand (theoretical, biochemical, and chemical) is expressed in units of milligrams per liter, which indicates the mass of oxygen consumed per liter of solution.

Theoretical Oxygen Demand (ThOD)

The *theoretical oxygen demand* (ThOD) is a measure of the amount of oxygen that is required to break down a given compound or set of compounds into the basic constituents of carbon dioxide and water. The theoretical oxygen demand of a wastewater is determined by summing the amount of dissolved oxygen required to oxidize the compound(s) of interest in the wastewater via a balanced chemical equation. For the oxidation of methane, four oxygen atoms (two molecules of dissolved oxygen) oxidize one molecule of methane to carbon dioxide and water, as shown in Eq. 6.9.

$$CH_4 + 2O_2 \rightarrow CO_2 + 2H_2O \qquad \textit{6.9}$$

If the theoretical oxygen demand is high and the biological oxygen demand is low, then the wastewater is not a candidate for biological treatment. Chemical treatment (oxidation) is required to degrade the organic materials in wastewater.

Biochemical Oxygen Demand (BOD)

Biochemical oxygen demand, BOD, is a measure of the quantity of oxygen required by aerobic microorganisms for the biological oxidation of organic matter. The biochemical oxygen demand value is a measure of the concentration of biodegradable organic matter present in a water sample and is an indicator of the general quality of the water. High biochemical oxygen demand levels (greater than several milligrams per liter) generally indicate high levels of organic pollution.

Some general statements may be made. Up to a BOD of approximately 5000 mg/L, aeration processes may be used to degrade the waste (e.g., aerobic reactors/degradation). However, for stronger wastes (BOD greater than 5000 mg/L), anaerobic treatment will be necessary. Although exceptions may be made to these general ranges, process selection will commonly follow this line.

BOD is measured at a fixed temperature over a given period of time. Microorganisms are taken from an activated sludge system (see Chap. 8 and Chap. 9) either at the given utility or at a similar one, then used as seed for the test sample. The source of the sample is important because the microorganisms may vary considerably from one wastewater source to the next. The seed is typically generated by diluting activated sludge with deionized water. All samples being tested in any one batch are inoculated with an equal volume of seed, including the sample control of deionized water saturated with oxygen. The BOD test is done by diluting the wastewater sample with deionized water saturated with oxygen,

then sealing the sample and placing it in the dark to prevent photosynthesis. This sample is kept at 68°F (20°C) and the dissolved oxygen (DO) is measured after 5 days. The apparent BOD for the control is subtracted from the test result to provide the corrected value. The loss of dissolved oxygen in the sample, once corrections have been made for the degree of dilution, is called the BOD_5. In other words, it is the BOD following 5 days of incubation.

In the *ultimate BOD* (BOD_u) test, DO is repeatedly measured using a dissolved-oxygen meter until an equilibrium dissolved oxygen concentration has been reached. The BOD at time t, BOD_t, may be calculated from the BOD_u according to Eq. 6.10.

$$BOD_t = BOD_u(1 - e^{-K't}) \qquad \textit{6.10}$$

K' is the *reaction rate coefficient* (per day), and t is the incubation time, in days, at which the BOD value is calculated. The reaction rate coefficient is a function of the solution temperature and must therefore be corrected for the specific temperature of the test solution. This can be done using the van't Hoff-Arrhenius equation, Eq. 6.11.

$$K'_T = K'_r \Phi^{T_1 - T_2} \qquad \textit{6.11}$$

K'_T is the reaction rate coefficient at the desired temperature, K'_r is the reaction rate coefficient at the reference temperature, and Φ is the dimensionless temperature coefficient[6] (see Table 6.2). T_2 is always taken to be 68°F (20°C), because this is the standard temperature at which the reaction rate coefficient, K'_r, is measured.

Table 6.2 *Temperature Coefficients for Various Biological Processes*

	Φ value	
process	range	typical
activated sludge	1.00–1.04	1.02
aerated lagoons	1.06–1.12	1.08
trickling filters	1.02–1.14	1.08

Adapted from G. Tchobanoglous, F. L. Burton, and H. D. Stensel, and the staff of Metcalf & Eddy, Inc. *Wastewater Engineering: Treatment, Disposal, Reuse,* 3rd ed., the McGraw-Hill Book Companies.

The biochemical oxygen demand for a given water sample is a function of time and generally follows the function shown in Fig. 6.2. The biochemical oxygen demand increases until a maximum value, or ultimate biochemical oxygen demand (BOD_u), is reached. The ultimate biochemical oxygen demand is independent of temperature. However, the biochemical oxygen demand measured at time t will change with temperature (i.e.,

---

[6]Although the value of Φ is assumed to be constant, it can vary substantially with temperature. Therefore, care must be taken to select appropriate values for the dimensionless temperature coefficient, Φ, for different temperature ranges and processes.

Figure 6.2 *Evolution of the Biochemical Oxygen Demand (BOD) for a Wastewater Sample with Time*

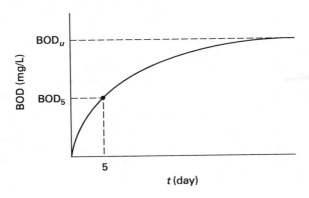

decrease with decreasing temperature) because the biological activity will change.

The dissolved oxygen concentration in a sample is measured at time zero and at a time of 5 days (typical value) following incubation at 68°F (20°C). The biochemical oxygen demand at time t is then calculated according to Eq. 6.12 and Eq. 6.13.

$$BOD_t = \frac{DO_i - DO_f}{F} \qquad 6.12$$

$$F = \frac{V_s}{V_t} \qquad 6.13$$

The sample volume is V_s. V_t represents the total volume of the diluted sample and is always equal to 300 mL. t is the number of days of incubation, and F is the dilution factor. If the dissolved oxygen concentration is less than 2 mg/L, the sample should be discarded because anaerobic conditions exist, and aerobic bacteria can no longer degrade the waste. If the dissolved oxygen concentration is greater than 7 mg/L, the sample should also be discarded because it has likely been exposed to the atmosphere, making the results erroneous.

Example 6.2

Calculate the BOD_u (ultimate BOD) and the 3 day BOD for a wastewater sample having a BOD_5 of 160 mg/L at 20°C and a reaction rate coefficient of $K' = 0.21$ day^{-1}.

Solution

Rearrange Eq. 6.10.

$$BOD_u = \frac{BOD_t}{1 - e^{-K't}} = \frac{BOD_5}{1 - e^{-K't}}$$

$$= \frac{160 \frac{mg}{L}}{1 - e^{-(0.21\,day^{-1})(5\,days)}}$$

$$= 246 \text{ mg/L}$$

The value of the reaction rate coefficient, K', will depend strongly on the strength of the wastewater sample. The reaction rate coefficient for temperature does not need to

be corrected because the given temperature, 20°C, is the same as the standard temperature for the reaction rate coefficient.

Calculate the BOD from Eq. 6.10.

$$BOD_{3\,day} = BOD_u(1 - e^{-K't})$$

$$= \left(246 \frac{mg}{L}\right)\left(1 - e^{-(0.21\,day^{-1})(3\,day)}\right)$$

$$= 115 \text{ mg/L}$$

Example 6.3

A wastewater sample was taken from an activated sludge reactor in order to measure the BOD_5. The reaction rate coefficient, K', was determined to be 0.23 day^{-1} at a temperature of 20°C (standard test conditions). The activated sludge reactor operates at a temperature of 23°C. Φ equals 1.02, a typical value for activated sludge systems. Determine the reaction rate coefficient, K', in the reactor.

Solution

Solve for the temperature-corrected reaction rate coefficient using Eq. 6.11.

$$K'_T = K'_r \Phi^{T_1 - T_2}$$

$$= (0.23 \text{ day}^{-1})(1.02)^{23°C - 20°C}$$

$$= 0.24 \text{ day}^{-1}$$

Chemical Oxygen Demand (COD)

Chemical oxygen demand, COD, is a measure of the oxygen requirement for the chemical oxidation of the waste in a water sample. The COD of a wastewater is most often higher than the BOD because more compounds can be chemically oxidized than can be biologically oxidized. COD is measured through oxidation of a sample by strong chemical oxidants. Potassium permanganate and potassium dichromate are the two most widely used oxidants in COD tests. Chemical oxidation of most organic compounds is 95–100% of the theoretical value.

One limitation of the measurement of COD is that it cannot differentiate between levels of biologically active organic substances and biologically inactive substances. An advantage of the COD test, however, is that it can be completed in a fraction of the time required by a BOD. For instance, while BOD takes 5–7 days to determine, the COD test can be completed in just three hours.

The type of treatment—chemical or biological—required for a given water sample may be determined by comparing the BOD to the COD measured for that water. If the ratio of BOD to COD is 0.1 or greater, the wastewater is biodegradable; if not, chemical oxidation is most likely required.

18. TOTAL ORGANIC CARBON

The *total organic carbon* (TOC) concentration in a wastewater is a measure of the amount of organic material in a water source. The TOC value represents the amount of carbon covalently bonded in organic molecules. Total organic carbon is measured using a carbon analyzer, which is an infrared detection system. Total organic carbon is a distinctly different quantity from biochemical oxygen demand and chemical oxygen demand because it does not describe how much material may be either biologically or chemically oxidized. It simply describes how much organic material is in a water source. Generally, if the total organic carbon is high, both the biochemical oxygen demand and chemical oxygen demand are likely to be high, though this is not always the case.

19. REOXYGENATION

Reoxygenation is the process by which molecular oxygen is added to a body of water. Reoxygenation is a function of the characteristics of the water system, such as flow volume, flow velocity, and inflow. Environmental factors such as temperature and altitude also affect the rate of reoxygenation. The *rate of reoxygenation*, r_r, is determined by Eq. 6.14.

$$r_r = K_r'(\text{DO}_{\text{sat}} - \text{DO}) = K_r'D \qquad 6.14$$

K_r' is the reoxygenation rate constant. Values for the reoxygenation rate constant typically range from 0 d^{-1} to 100 d^{-1}, depending on the characteristics of the flowing water body. Values of reoxygenation rate constants for different water bodies are summarized in Table 6.3.

Table 6.3 Reoxygenation Constants for Different Bodies of Water

water body	K_r' at 20°C, base e
small ponds and backwaters	0.1–0.23
sluggish streams and large lakes	0.23–0.35
large streams of low velocity	0.35–0.46
large streams of normal velocity	0.46–0.69
swift streams	0.69–1.15
rapids and waterfalls	>1.15

lower / *higher*

Reprinted from H. S. Peavy, D. Rowe, and G. Tchobanoglous. *Environmental Engineering.* 1985. McGraw-Hill, NY.

Example 6.4

A wastewater treatment plant discharged 5 MGD of treated effluent with a BOD$_5$ of 6.5 mg/L. The BOD reoxygenation rate coefficient, K_r', is 0.18 day^{-1}. What is the loading in pounds of ultimate BOD per day?

Solution

Calculate the BOD$_u$ for the effluent using Eq. 6.10.

$$\text{BOD}_u = \frac{\text{BOD}_t}{1 - e^{-K't}} = \frac{6.5 \; \dfrac{\text{mg}}{\text{L}}}{1 - e^{-(0.18 \, \text{day}^{-1})(5 \, \text{days})}}$$

$$= 10.95 \text{ mg/L}$$

Multiply the effluent BOD$_u$ by the effluent flow rate using the appropriate unit conversion factor.

$$\text{loading} = Q_{\text{eff}}\text{BOD}_u$$

$$= \left(5 \; \frac{\text{MG}}{\text{day}}\right)\left(10.95 \; \frac{\text{mg}}{\text{L}}\right)\left(8.345 \; \frac{\text{lbm-L}}{\text{mg-MG}}\right)$$

$$= 457 \text{ lbm/day}$$

20. DEOXYGENATION

Deoxygenation is the process by which dissolved oxygen is removed from a body of water. For environmental applications, the principal concern is deoxygenation resulting from the introduction of pollutants and wastewater into a water system. Deoxygenation is dependent on a number of factors, including temperature, the types of microbes present in the water, and the amount and strength (BOD) of the influent wastewater. The *rate of deoxygenation*, $r_{d,t}$, can be determined according to Eq. 6.15.

$$r_{d,t} = -K_d'\text{DO} \qquad 6.15$$

K_d' is the deoxygenation rate constant (per day). For wastewater treatment plant effluents, the deoxygenation rate constant value ranges from 0.05 d^{-1} to 0.1 d^{-1}.

The deoxygenation rate constant value is corrected for changes in temperature using the temperature variation constant, θ, which ranges from 1.135 for temperatures between 39°F and 68°F (4°C and 20°C) to 1.056 for temperatures between 68°F and 86°F (20°C and 30°C). T_2 is taken to be equal to 68°F (20°C). The temperature corrected deoxygenation rate constant ($K_{d,T}'$) is determined from Eq. 6.16.

temp. correction

$$K_{d,T_1}' = K_{d,T_2}'\theta_d^{T_1-T_2} \qquad 6.16$$

21. OXYGEN SAG CURVE

Oxygen sag curves for flowing water systems, also known as *dissolved oxygen* (DO) *sag curves*, are plots of dissolved oxygen concentration as a function of time or distance from a point of discharge of pollution or wastewater (see Fig. 6.3). These curves are used to show the evolution of the dissolved oxygen concentration in a water system after a wastewater or other pollution source has been added to it. When organic matter is introduced into water, the dissolved oxygen concentration decreases to a minimum before gradually recovering to the saturation level as a result of deoxygenation and reoxygenation processes.

Figure 6.3 *Dissolved Oxygen (DO) Sag Curve*

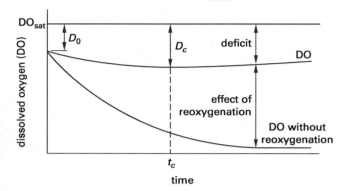

Deoxygenation of the water occurs as a result of degradation of organic materials by bacteria, fungi, and invertebrates. The amount of deoxygenation that occurs is, therefore, dependent on the strength, or the biochemical oxygen demand, of the influent wastewater. The time or distance to which the water must travel until the dissolved oxygen concentration recovers its previous value is determined by the influent wastewater strength and volume, and by the characteristics of the water system (photosynthetic plants present, velocity, and volume).

The *Streeter-Phelps equation*, Eq. 6.17, which accounts for both deoxygenation and reoxygenation processes, may be used to model the change in the dissolved oxygen deficit, D_t, in a flowing water body after wastewater is discharged into it. Therefore, it can be used to develop theoretical oxygen sag curves based on the characteristics of a water system and wastewater source.

time is only variable

$$D_t = \frac{K'_d \text{BOD}_u (e^{-K'_d t} - e^{-K'_r t})}{K'_r - K'_d} + D_0 e^{-K'_r t} \qquad 6.17$$

D_0 is the oxygen deficit when the pollutant first enters the stream at time zero, K'_r is the reoxygenation rate constant (see Table 6.3), and BOD_u is the ultimate BOD of the influent wastewater.

The minimum of the oxygen sag curve is known as the *sag time*, t_c. Sag time is the time that it takes the water to flow until the oxygen deficit is greatest and aquatic organisms experience the greatest stress. The critical sag time is calculated according to Eq. 6.18.

$$t_c = \frac{1}{K'_r - K'_d} \ln\left(\left(\frac{K'_d \text{BOD}_u - K'_r D_0 + K'_d D_0}{K'_d \text{BOD}_u}\right)\left(\frac{K'_r}{K'_d}\right)\right) \qquad 6.18$$

In Eq. 6.18, the ratio K'_r/K'_d is known as the *self-purification constant*. The greater the reoxygenation rate constant, K'_r, the greater the self-purification constant, and the greater the ability of the water body to recover quickly from oxygen deficits. Values of the reoxygenation rate constant are given in Table 6.3.

The *critical sag distance*, x_c, is calculated from the velocity of the flowing water and the sag time using Eq. 6.19.

x_c location of lowest Do concentration

$$x_c = v t_c \qquad 6.19$$

→ dead zone

The critical oxygen deficit, D_c, is found using Eq. 6.20.

$$D_c = \left(\frac{K'_d \text{BOD}_u}{K'_r}\right) e^{-K'_d t_c} \qquad 6.20$$

Example 6.5

A large, slowly moving river receives a wastewater with a BOD_u of 300 mg/L. At 20°C, the reoxygenation rate constant is estimated to be 0.7 day^{-1}, and the deoxygenation rate constant is 0.3 day^{-1}. The oxygen deficit at the discharge point is 1.6 mg/L. The temperature of the mixed wastewater and river water is 17°C, and the BOD_u is 35 mg/L. Using temperature coefficients of 1.130 for deoxygenation and 1.024 for reoxygenation, calculate the critical oxygen deficit.

Solution

Correct the reoxygenation rate constant, K'_r, and the deoxygenation rate constant, K'_d, for temperature using Eq. 6.16.

$$K'_{r,17°C} = K'_r \theta^{T_1 - T_2} = (0.7 \text{ day}^{-1})(1.024^{17°C-20°C})$$
$$= 0.65 \text{ day}^{-1}$$
$$K'_{d,17°C} = K'_d \theta_d^{T_1 - T_2} = (0.3 \text{ day}^{-1})(1.130^{17°C-20°C})$$
$$= 0.21 \text{ day}^{-1}$$

Calculate the minimum of the oxygen sag curve. Refer to Table 6.3 to select the reoxygenation rate constant, K_r. Because the river is large and slow moving, select a K_r value of 0.4 days^{-1} base e (this translates to a value of 1.49 days^{-1}).

$$t_c = \frac{1}{K'_r - K'_d} \ln\left(\left(\frac{K'_d \text{BOD}_u - K'_r D_0 + K'_d D_0}{K'_d \text{BOD}_u}\right)\left(\frac{K'_r}{K'_d}\right)\right)$$

$$= \left(\frac{1}{0.65 \text{ day}^{-1} - 0.21 \text{ day}^{-1}}\right)$$

$$\times \ln\left(\left(\frac{\begin{array}{c}(0.21 \text{ day}^{-1})\left(35 \frac{\text{mg}}{\text{L}}\right)\\ - (0.65 \text{ day}^{-1})\left(1.6 \frac{\text{mg}}{\text{L}}\right)\\ + (0.21 \text{ day}^{-1})\left(1.6 \frac{\text{mg}}{\text{L}}\right)\end{array}}{(0.21 \text{ day}^{-1})\left(35 \frac{\text{mg}}{\text{L}}\right)}\right)\left(\frac{0.65 \text{ day}^{-1}}{0.21 \text{ day}^{-1}}\right)\right)$$

$$= 2.34 \text{ days}$$

It is the BOD_u of the mixed wastewater and river water that is used in the calculation.

Calculate the critical oxygen deficit using Eq. 6.20.

$$D_c = \left(\frac{K_d' \text{BOD}_u}{K_r}\right) e^{-K_d' t_c}$$

$$= \left(\frac{(0.21 \text{ day}^{-1})\left(35 \dfrac{\text{mg}}{\text{L}}\right)}{0.65 \text{ day}^{-1}}\right)\left(e^{-(0.21 \text{ day}^{-1})(2.34 \text{ days})}\right)$$

$$= 6.92 \text{ mg/L}$$

22. TOXICITY

Toxicity is a measure of the degree to which something is toxic or poisonous. The three types of toxic entities are chemical, biological, and physical. Toxicity can refer to the effect on a whole organism or the effect on a component of an organism. Agents that affect only specific types of cells or organs are called *cytotoxins*. Toxicity is affected by many different factors, such as the pathway of administration (applied to the skin, ingested, inhaled, or injected), the exposure time (acute or chronic), the number of exposures, the physical form of the toxin (solid, liquid, or gas), the genetic makeup of the target individual, and the individual's overall health. *Acute exposure* is a single exposure, in a time of less than one day, to a toxic substance that may result in severe biological harm or death. *Chronic exposure* is continuous exposure to a toxin over an extended period of time, such as months to years. The toxicity of a substance is generally reported in terms of the probability that the substance is lethal to a given percentage of the population. For example, toxicity is often reported in terms of an LD_{50} value, which is the *lethal dose* or concentration at which 50% of an exposed population will die.

23. DISINFECTION

Disinfection, as described in Chap. 5, is the removal, deactivation, or killing of pathogenic microorganisms. The exact mechanism by which a disinfectant, such as chlorine, kills a microorganism is not fully understood. It is believed that cell death results from cell wall corrosion (in plants, fungi, and bacteria), changes in cell permeability, changes in enzyme activity (caused by structural change in the enzymes), and damage to the cellular DNA. These disturbances interfere with a microorganism's ability to reproduce, leading to the population of microorganisms dying out. Oxidizing disinfectants also react with organic matter in the water, reducing its nutritional value.

24. STREAM DEGRADATION

Surface water systems, particularly rivers and streams, may become impaired or degraded when exposed to pollution sources. Degradation of water quality may also result from riparian habitat loss and excessive erosion with the loss of surface cover. The water quality is degraded in terms of clarity, temperature, pH, oxygen content, and solid content (TS, TSS, TDS). Organic pollutants, as measured by BOD and COD levels, can result in low DO levels as microorganisms degrade the waste and, in the process, consume oxygen (see Sec. 6.21). Excessive nutrient (nitrogen and phosphorus) concentrations from agricultural and urban runoff can cause algal blooms, which cause oxygen depletion. A physical degradation process is the diversion of excessive amounts of water from a stream so that the stream can no longer support a minimum or base flow. Diversions may be for agricultural, industrial, or municipal purposes. Surface water degradation impacts aquatic species and raises the level of treatment required to produce drinking water.

25. WATER SAMPLING AND MONITORING

It is sometimes necessary to measure and subsequently analyze any number of water quality (WQ) parameters prior to the design of a wastewater treatment system or during the operation of an existing system. Common WQ parameters in wastewater treatment include pH, DO, BOD, temperature, conductivity, turbidity, fecal coliform, indicator organisms, and volumetric flow measurements. Collected data must be both accurate and representative of the given system being sampled to be useful. This is accomplished through the development and execution of *quality assurance* (QA) and *quality control* (QC) procedures, commonly referred to as *QA/QC procedures*. QA specifies the measures used to produce data of a known precision and bias and includes documented procedures, work instructions, staff training, and record keeping. QC procedures are a set of technical activities that ensure that information collected is accurate and precise, and is recorded and reported in an approved manner. Examples of QC measures include use of blank samples to detect contaminants, field replicates, sample splits, equipment blanks, ambient condition blanks, trip blanks, surrogates, matrix spikes, audits of processes and results, and cleaning procedures. QA/QC procedures apply to the total water quality monitoring process, including sample collection and storage, measurement procedures, and data analysis, interpretation, and management.

Another critical tool for ensuring the accuracy of WQ samples and data is the *chain of custody protocol and report*. The chain of custody is a protocol developed to provide a legal record of the persons having contact with a sample from the moment of collection to final disposal. Chain of custody documentation may include:

- sample label requirements
- container seals to prevent tampering
- sample analysis request sheets
- a chain of custody log
- field notes and logbooks related to sample collection
- laboratory logbooks and analysis logbooks

PRACTICE PROBLEMS

1. Which of the following is a commonly used indicator organism?

    (A) coliform

    (B) Cryptosporidium

    (C) Giardia

    (D) algae

2. If a wastewater sample has a fecal coliform to fecal streptococcus ratio of 0.9, what is a likely influent source?

    (A) a sports complex

    (B) an industrial development

    (C) a housing complex

    (D) a stockyard

3. During a BOD test, a wastewater sample is found to have a DO concentration in excess of 7 mg/L. What does this result indicate?

    (A) Anaerobic conditions are present.

    (B) Anoxic conditions are present.

    (C) Biological organisms are absent in the sample.

    (D) The sample has been exposed to the atmosphere for an extended period of time.

4. Most nearly, what percentage of the theoretical oxygen demand (ThOD) is the chemical oxygen demand (COD)?

    (A) 60%

    (B) 70%

    (C) 80%

    (D) 90%

5. On what time scale is the BOD for a given wastewater sample generally reported?

    (A) 0.5 day

    (B) 2 days

    (C) 3 days

    (D) 5 days

6. BOD is a measure of what wastewater characteristic?

    (A) biodegradability

    (B) toxicity

    (C) color

    (D) odor

SOLUTIONS

1. Section 6.8 discusses the use of indicator organisms to assess the quality of treated wastewater. Coliform bacteria often accompany pathogenic organisms and are, therefore, an indicator of the presence of pathogens in a water sample.

The answer is (A).

2. From Sec. 6.8, different wastewaters will have different proportions of microorganisms, particularly coliform and *Streptococcus*. For this reason, the FC:FS ratio may be used to characterize the origin of a given wastewater. An FC:FS ratio between 0.7 and 1.0 indicates a large amount of animal waste in the wastewater. Therefore, a stockyard is the most likely influent source.

The answer is (D).

3. Section 6.17 discusses the methodology for measuring and reporting the BOD of wastewater samples. When interpreting the results from the BOD test, it is important to know what samples may have been contaminated or are obviously erroneous. DO concentrations less than 2 mg/L indicate anaerobic conditions, while concentrations greater than 7 mg/L indicate exposure to the atmosphere.

The answer is (D).

4. From Sec. 6.17, COD is a measure of the amount of material in a wastewater sample that can be chemically oxidized. In general, more than 95% of the materials in a wastewater sample are able to be chemically oxidized.

The answer is (D).

5. Section 6.17 discusses the methodology for measuring and reporting the BOD of wastewater samples. Because most biological processes in wastewater treatment plants have a hydraulic retention time on the order of 5 days, it is important to know how much oxygen is required at this time scale to degrade the materials in a wastewater.

The answer is (D).

6. Section 6.17 describes the role of the BOD value in characterizing the strength of a wastewater. Quantitatively, it details the amount of oxygen required to biochemically oxidize organics in a given sample. Qualitatively, it determines if the wastewater can be treated through biological or physicochemical processes. In essence, BOD measures the biodegradability of the wastewater.

The answer is (A).

7 Wastewater

Nomenclature

A	area	ft^2	m^2
BOD	biochemical oxygen demand	mg/L	mg/L
C	characteristic of flow	gal/day	ML/d
d	depth	ft	m
D	diameter	ft	m
l	length	ft	m
n	Manning roughness coefficient	–	–
OLR	organic loading rate	lbm/person-day	g/person·d
P_e	population equivalent	1000s of people	1000s of people
Q	flow rate	gal/day	ML/d
R	hydraulic radius	ft	m
S	longitudinal slope	ft/ft	m/m
TDS	total dissolved solids	mg/L	mg/L
TS	total solids	mg/L	mg/L
TSS	total suspended solids	mg/L	mg/L
v	flow velocity	ft/sec	m/s
v_o	flow velocity of full pipe	ft/sec	m/s
V	volume	ft^3	m^3
VSS	volatile suspended solids	mg/L	mg/L
w	width	ft	m
WP	wetted perimeter	ft	m

Subscripts

5	5 day
f	final (mixture)
t	total
u	ultimate

1. DOMESTIC WASTEWATER

Domestic wastewater is wastewater discharged from residences, commercial buildings, and other institutions. It contains sewage only, not rain runoff. Domestic wastewater is typically characterized by strengths of approximately 250 mg/L BOD or 250 mg/L TSS. TSS are the *total suspended solids* in the wastewater. The amount of domestic wastewater produced per person is between 50 gal/capita-day and 250 gal/capita-day (190 L/capita·d and 950 L/capita·d). The production rate is dependent on the types of appliances used to dispose of the wastewater and on personal practices.

2. INDUSTRIAL WASTEWATER

Industrial wastewater is wastewater generated by manufacturing or other industrial systems that discharge into a domestic sewer system. These waste streams must meet criteria defined by the local municipality before they are discharged into the sewer system. In most cases, the effluent is of moderate to weak strength and does not contain toxic agents.

3. MUNICIPAL WASTEWATER

Municipal wastewater is the total fluid flow collected in a sanitary sewer system. It may be a combination of domestic and industrial wastewater and stormwater runoff. Combined sewers convey both wastewater and rain runoff, while separated sewers convey either wastewater or stormwater runoff only. Most older sewer systems are combined, while newer systems are separated in order to reduce treatment costs. Municipal wastewater is, therefore, a complex mixture of human waste, suspended solids, runoff, debris, and a variety of chemicals that come from residential, commercial, and industrial activities. Substances such as pharmaceuticals, therapeutic products, and endocrine disrupting compounds are also commonly found in municipal wastewater and are a particular concern. Wastewater characteristics vary from location to location, depending upon the population and industrial sector served, land uses, groundwater levels, and degree of separation between stormwater and sanitary wastes.

4. MUNICIPAL WASTEWATER COMPOSITION

Municipal wastewater is usually characterized by a gray color, musty odor, and a solids content of approximately 0.1%. Raw influent BOD_5 values for municipal wastewater in the United States are approximately 150 mg/L. The organic components may consist of carbohydrates, proteins, fats and greases, surfactants, oils, pesticides, and phenols. Inorganic components may consist of heavy metals, nitrogen, phosphorus, sulfur, chlorides, and toxic compounds. Though the concentrations of these constituents vary with location, generally they are all present to some degree in municipal wastewaters.

Wastewater contains a higher portion of dissolved solids than suspended. Approximately 85–90% of the inorganic component and 55–60% of the organic component is dissolved. Gases commonly dissolved in municipal wastewater are hydrogen sulfide, methane, and ammonia (all of which result from the decomposition of organic matter present in the wastewater), and oxygen, carbon dioxide, and nitrogen.

The biological composition of wastewater is complex. It includes a wide variety of protista, plants, animals, and pathogenic microorganisms. The pathogenic organisms originate mostly from humans who are infected or who are carriers of a disease.

5. WASTEWATER CHARACTERISTICS

Wastewater sources are characterized by a number of different physical characteristics, including density, color, turbidity, and organic content (measured by the BOD_5 value; see Sec. 6.17). Each of these different characteristics is significant in determining which treatment processes are appropriate for treating the water to an acceptable level.

Density is temperature dependent and varies slightly with the concentration of particulate matter and ionic makeup. Typical units for density are pounds per cubic foot (kilograms per cubic meter).

Color is a relatively good indicator of the wastewater's age, with respect to the length of time it has been in treatment. In the later stages of treatment, when the sulfide produced under anaerobic conditions reacts with the metals in the waste, wastewater tends to take on a gray to black color characteristic of metallic sulfides.

A relatively good indicator of the amount of particulate matter in a water sample is *turbidity*, which is a measure of the light scattering properties of a suspension. Turbidity is one of the most common measures of water quality and is most often reported in terms of *nephelometric turbidity units* (NTU).

6. SOLIDS IN WASTEWATER

Materials in a water source, wastewater or otherwise, are commonly divided into several different categories depending on their physical state in aqueous media. In each case, such materials are noted as solids. The sum of all solid materials in a water sample is taken as the *total solids* (TS), which is composed of settleable solids, total suspended solids (TSS), and total dissolved solids (TDS). The total solids concentration is determined by evaporating a well-mixed sample volume in a weighed dish, drying it to constant weight in an oven at 217.4–221.0°F (103–105°C), subtracting the weight of the empty dish, and dividing by the original sample volume.

Settleable solids are materials that readily settle out of suspension in a set amount of time. The settleable solids concentration is determined using an *Imhoff cone*, which allows the solids to settle from suspension over a period of 45 min. The settleable solids are then reported as the amount of solid material at the bottom of the cone.

Suspended solids, generally measured and referred to as total suspended solids (TSS), are all solid materials (inorganic or organic) that are retained on a 0.45 μm filter. In water treatment, TSS are responsible for the turbidity or cloudiness of a water sample and create both health and aesthetic concerns. Currently, the EPA discharge limit for TSS in treated municipal wastewater discharged into a receiving water body (river) is 30 mg/L.

TSS may be further fractionated into volatile (VSS) and nonvolatile (NVSS) suspended solids. *Volatile solids* are those solids lost on ignition at 1022°F (550°C). VSS may contribute to any of the solid fractions that make up the total solids concentration. However, volatile solids are primarily found only in the suspended and settleable solids fractions. Assessment of VSS is useful to the wastewater treatment plant operator because it gives a rough approximation of the amount of organic matter present in the solid fraction of wastewater, activated sludge, and industrial wastes. *Nonvolatile suspended solids* are those solids not lost on ignition.

The *total dissolved solids* (TDS) are the portion of solids in water that can pass through a filter with a pore size of 2 μm. The more minerals dissolved in the water, the higher the total dissolved solids. *Dissolved solids* refers to any minerals, metals, or other salts (cations or anions) dissolved in water. TDS is mostly composed of various ionic species, such as calcium, magnesium, sodium, iron, manganese, bicarbonate, chloride, sulfate, nitrate, and carbonate.

7. BIOCHEMICAL AND CHEMICAL OXYGEN DEMAND (BOD AND COD)

Chemical oxygen demand (COD) is a measure of the oxygen requirement for the chemical oxidation of the waste in a water sample. *Biochemical oxygen demand* (BOD) is a measure of the quantity of oxygen required by aerobic microorganisms for the biological oxidation of organic matter (see Sec. 6.17). The BOD value is a measure of the concentration of biodegradable organic matter present in a water sample and is an indicator of the general quality of the water. High BOD_5 levels (> 2 mg/L) are generally indicative of a water body having been polluted by organic wastes. Pristine rivers will have a BOD_5 of less than 1 mg/L. Moderately polluted rivers may have BODs from 2 mg/L to 8 mg/L. Tertiary treated municipal sewage has a BOD_5 value of about 20 mg/L. Untreated sewage has a BOD_5 of 200 mg/L or more. This value is highly dependent on the amount of groundwater or surface water infiltration into the sewer system. High BOD values can be caused by high levels of organic pollution and/or high nitrate levels, which trigger high plant growth.

8. SOLUBLE BOD

After secondary wastewater treatment, soluble organic matter that is resistant to biological degradation remains in the clarifier effluent (see Sec. 8.4). This organic matter is defined as soluble BOD. *Soluble BOD* is measured as the BOD remaining in a water sample that has been filtered in the standard suspended solids test. Soluble BOD is also referred to as *refractory organics*. These materials are generally removed through adsorption processes, such as powdered or activated carbon, which are highly efficient at removing soluble organic materials from water streams.

9. SEWER SYSTEMS

Sewers are hydraulic conveyance structures that carry wastewater to a treatment plant or other authorized point of discharge. Sewers are commonly referenced by the type of wastewater they transport. For example, storm sewers carry stormwater, industrial sewers carry industrial wastes, and sanitary sewers carry both domestic sewage and industrial wastes. Another type of sewer, known as a combined sewer, is prevalent in older communities, but is no longer constructed. Combined sewers carry domestic sewage, industrial waste, and stormwater. According to the EPA, approximately 20% of the population is served by combined systems, and 46% of the population is served by separate systems.

There are two general types of conveyance used in sewer systems to transport wastewater: *gravity* and *force*

collection. In gravity collection conveyance systems, the wastewater is transported by gravity along a downward sloping pipe gradient. These sewers are designed so that the slope and size of the pipe are adequate to maintain fluid flow toward the discharge point without surcharging manholes or pressurizing the pipe. Conventional gravity sewers are typically used in urban areas with consistently sloping ground because excessively hilly or flat areas result in deep excavations and drive up construction costs. Conventional gravity sewers remain the most common conveyance for collecting and transporting domestic wastewater. Minimum sewer pipe slopes are reported in Table 7.1.[1]

Table 7.1 *Minimum Pipe Slope for Various Sewer Pipe Lengths*

diameter		slope (ft/100 ft)	
(in)	(mm)	pipe length ≤ 5 ft	pipe length > 5 ft
8	200	0.47	0.42
10	250	0.34	0.31
12	310	0.26	0.24
14	360	0.23	0.22
24	610	0.08	0.088
30	760	0.07	0.07

Adapted from *Collection Systems Technology Fact Sheet—Sewers, Conventional Gravity*, USEPA, 2002.

Forced collection conveyance systems are used primarily in rural and semi-rural areas where excavation costs and low population densities make gravity systems less practical. These systems use pumping stations and watertight pipes (i.e., no manholes) to transport the wastewater from collection points to *waste and wastewater treatment plants* (WWTP). Because the piping network is watertight and under pressure, no extraneous flows may enter the collection system, which reduces the flow rate entering the WWTP. The pressurized system also means that smaller pipes may be used for conveying wastewater. A variety of designs may be used, but the two major types of forced collection systems are the septic tank effluent pump system and the grinder pump system. Both systems contain a septic tank (typically a buried concrete box having an inlet, a pump vault, and an effluent line) and a pumping system for transporting the wastewater from the tank to the forced conveyance system. The difference between the two systems is that the pump in the grinder pump system grinds any solids that are present in the water prior to discharging the flow into the pressurized conveyance system. There is no minimum average daily flow velocity for forced collection systems, but grinder pump systems must reach a flow velocity of 3 ft/sec to 5 ft/sec (0.9 m/s to 1.5 m/s)

[1]According to the *Recommended Standards for Wastewater Facilities* (*Ten States Standards*), all sewers, when flowing full, shall be designed and constructed to give mean velocities not less than 2.0 ft/sec (0.61 m/s), based on Manning's formula using a roughness coefficient value, *n*, of 0.013.

at least once daily to flush the system. Approximately 50% of the BOD, 75% of the TSS, 100% of the grit, and 90% of greases are removed in the septic tank prior to discharge into the conveyance system.

10. SEWER PIPE MATERIALS

Sewer lines and the associated fixtures and fittings may be made of a wide variety of materials, including ductile iron, cast iron, vitrified clay, concrete, asbestos cement, polyvinyl chloride (PVC), acrylonitrile-butadiene-styrene plastic (ABS), and reinforced plastic mortar pipe. Selection of the most appropriate pipe material depends on the site conditions (soil and waste characteristics) and the respective characteristics of the piping material (e.g., physical strength, friction coefficient, and corrosion resistance). The characteristics of the soil in which a pipe is placed affect the rates of corrosion, with the most corrosive soils being those having poor aeration, high acidity, and high moisture content.

Ductile iron pipe is used for sewers requiring a high resistance to external loading or a high degree of toughness and ductility. It is well-suited for most sanitary sewers, including river crossings, piping at wastewater treatment facilities, pipe located in unstable soils, highway and rail crossings, water line crossings, depressed sewers, and above-ground piping. The use of ductile iron is limited by its susceptibility to corrosion from wastewaters containing acids and from aggressive soils (acidic or alkaline soils). For this reason, it is usually cement lined or has an interior coating.

Cast iron pipe is normally only used for building connections, such as interior drainage and waste and vent piping. Cast iron pipe is resistant to internal and external corrosion when provided with a bituminous coating and is not subject to abrasion from grit, sand, or gravel.

Vitrified clay pipe is manufactured from clay and shale products fired at 2000°F (1100°C) to form an ideal material for sewer use. This type of piping has a high resistance to corrosion from acids and alkalis, as well as to scouring and erosion. This resistance provides a distinct advantage for industrial waste sewers or sanitary sewers subject to hydrogen sulfide generation. However, vitrified clay is rather brittle, which may inhibit its application under certain conditions.

Concrete sewer pipe is appropriate for applications requiring large diameters or high strength. Care should be taken when specifying concrete pipe to assure that it is suitable for the environment in which it will be installed. Unlined concrete pipe is subject to scouring by wastewaters carrying grit and sand at high velocities. Reinforced concrete pipe is used where high external loadings are anticipated and large diameters or tight joints are required. The disadvantage of reinforced concrete is that it is susceptible to corrosion, a critical disadvantage where hydrogen sulfide is generated in substantial quantities. However, special plastic or clay liner plates and coatings of coal-tar, coal-tar epoxy,

vinyl, or epoxy mortar can be applied to the pipe for corrosion protection.

Asbestos cement pipe is made from a mixture of asbestos fibers and Portland cement. Asbestos cement pipes are susceptible to corrosion in environments containing hydrogen sulfide, acid wastes, or aggressive soils. Plastic linings may be used with these pipes to protect against corrosion.

Polyvinyl chloride (PVC) is chemically inert to most acidic and alkaline wastes, and is totally resistant to biological attack. As it is nonconductive, PVC pipe is immune to nearly all types of underground corrosion caused by galvanic or electrochemical reactions, in addition to corrosion from aggressive soils. Its durability, light weight, high strength-to-weight ratio, long laying lengths, watertight joints, and smooth interior surfaces make PVC pipe an attractive alternative for use in sewer systems. The disadvantages of PVC include chemical instability caused by long-term exposure to sunlight, excessive pipe deflection under trench loadings when installed improperly or subjected to high-temperature wastes, and brittleness when exposed to very cold temperatures.

Acrylonitrile-butadiene-styrene (ABS) *composite pipe* consists of two concentric thermoplastic tubes integrally connected across the annulus by a truss-like bracing. The annular void space is filled with Portland cement concrete or other suitable material to form a bond between the inner and outer tubes. It is termed a "semi-rigid" pipe because it resists deflection better than most other plastics. The pipe is light in weight and resists attack by acids, alkalis, and biological growths.

Reinforced plastic mortar pipe is composed of a siliceous sand aggregate reinforced with glass fibers and embedded in a thermosetting polyester resin. It is ideally suited for large diameter applications and performs extremely well in resisting pipe wall deflection and internal/external corrosion. The unique fiberglass/resin construction provides optimum protection against attack from a wide range of chemically aggressive environments, including hydrogen sulfide and other sewer gases, most natural soils, salt, and brackish water, and galvanic or electrolytic reactions. Other advantages include its light weight and smooth, glass-like interior surface.

11. SANITARY SEWERS

The two main functions of a sanitary sewer are to carry the peak discharge for which it is designed and to transport suspended solids so that deposits in the sewer are kept to a minimum. The sewer must have adequate capacity for the peak flow and be able to operate properly at minimum flows. The *peak flow* is the maximum flow through the sewer, which determines the hydraulic capacity of sewers, pump stations, and treatment plants. Minimum flows must be considered in sewer design to ensure that material does not collect in the sewer piping.

12. POPULATION ESTIMATES

Sewage flows are estimated based on population data for a given area. Forecasts of commercial and industrial flows are also helpful. Sewage flow is generally reported in terms of gallons per capita-day (gpcd). This per capita flow varies from 50 gpcd to 140 gpcd and may be as high as 160 gpcd where industrial flows are included. Typically, a value of 125 gpcd is used to convert population to average sewage flow, including commercial and industrial flow. Average design flows for different types of developments are reported in Table 7.2.

Table 7.2 *Average Sewer Design Flows for Various Types of Developments*

type of development	design flow (gpd)
residential	
general	100/capita
single family	370/capita
townhouse unit	300/unit
apartment unit	300/unit
commercial	
general	2000/ac
motel	130/unit
office	20/employee
industrial (varies with type of industry)	
general	10,000/ac
warehouse	600/ac
school site (general)	16/student

(Multiply ac by 4046.87 to obtain m^2.)

13. PEAK FACTORS

A *peak factor* is used to convert the average flow to peak flow. For sanitary sewage flow, peak factors are the ratio of the highest dry weather peak of the year to the average daily dry weather flow. Sewage flow peak factors vary from 1.3 to 3.5. As average flows increase, peak factors decrease. Whenever possible, the peak factors should be adjusted to flow studies in the local area.

14. INFILTRATION AND INFLOW

In the context of sanitary sewer flows, *infiltration and inflow* (I/I) are terms that describe the groundwater and stormwater entering sanitary sewers. *Infiltration* is groundwater that enters sanitary sewers through leaks in pipes. *Inflow* is stormwater that is directed to the sanitary sewers through connections such as roof downspouts, driveway drains, and groundwater sump pumps. Other sources may include seepage from the water table through leaky pipes, leakage from manholes, or illegal draining from household downspouts. Two sources of infiltration are poorly constructed manholes and/or connections, and improperly laid sewer laterals.[2] Sewer

---

[2]A *sewer lateral* is the length of sewer pipe that runs from a building or residence to the main sewer.

laterals frequently have a total length greater than the collecting system and may contribute as much as 90% of infiltration. House connections should receive the same specifications, construction, and inspection as public sewers.

15. EXTRANEOUS FLOWS

Sanitary sewer design quantities should include consideration of the various non-sewage components that inevitably become a part of the total flow. The cost of transporting, pumping, and treating sewage obviously increases as the quantity of flow delivered to the pumps or treatment facility increases. Therefore, extraneous flow should be kept within economically justifiable limits by proper design and construction practices and adequately enforced connection regulations.

16. SEWER DESIGN REQUIREMENTS

Sewer system design is based on the hydraulic requirements of the system layout. The fluid flow velocity must be sufficiently high to prevent the deposition of solids in the pipe, but not high enough to induce excessive turbulence and wear of pipe materials. The minimum acceptable scouring velocity is 2 ft/sec (0.6 m/s) and the maximum recommended value is 20 ft/sec (6.1 m/s). Changes in the direction, flow quantity, or pipe size must be accompanied by an appropriate change in the sewer elevation to maintain an acceptable head loss. Pipe elevations and slopes are calculated once the peak flows, which incorporate future increases, are determined. Sewer pipes with a diameter up to 1.3 ft (0.38 m) are designed to operate half full at peak flow, while larger ones operate at three-quarters full at peak flow.

Sewer system piping, like the piping described in Sec. 3.5, is sized using the empirical *Manning equation*, Eq. 7.1.

GRAVITY

$$v = \left(\frac{1.49}{n}\right) R^{2/3} \sqrt{S} \quad \text{[U.S. only]} \qquad 7.1$$

FORCE: 89.2-2

In customary U.S. units, v is the flow velocity in feet per second, n is the *Manning roughness coefficient* (see Table 3.1), R is the hydraulic radius in feet, and S is the slope of the channel. The hydraulic radius, R, is defined by Eq. 7.2 as the area in flow, A, in square feet, divided by the pipe's wetted perimeter, WP, in feet.

$$R = \frac{A}{\text{WP}} \qquad 7.2$$

The Manning equation is used to describe sewer system performance by evaluating flow monitoring data on a scatter graph displaying flow depth and velocity data. *Scatter graphs* have characteristic patterns that provide information about the conditions within the sewer line and the impact of these conditions on the sewer

Wastewater

Wastewater

capacity. A calculated pipe curve, based on the dependence of velocity, v, on $R^{2/3}$ embodied in the Manning equation, may then be compared to measured data in a scatter graph.

In Fig. 7.1 (known as a *hydraulic elements graph*), the flow velocity, v, and flow depth, d, are normalized to those conditions that will occur when the pipe is full (v_o and D, respectively). These *hydraulic element ratios* (v/v_o and d/D) allow for the easy determination of conditions under any set of pipe flow conditions.

Figure 7.1 *Relationships Between Flow Velocity and Depth in a Sewer Pipe as Described by the Manning Equation*

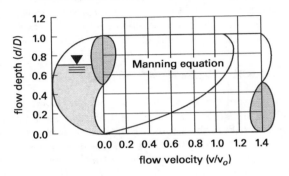

The value of the roughness coefficient, n, is normally assumed to be constant throughout the depth of flow. The n values for smooth bore pipe are actually affected by depth of flow, velocity of flow, and quality of construction. It is recommended that a conservative value of n be used in design to accommodate variable flow conditions, the deposition of debris and other foreign materials, the buildup of slime and grease on pipe surfaces, the loss of hydraulic capacity of flexible pipe due to ring deflection, and pipe misalignment caused during construction or settlement. These variations all affect the n value reported for the clean pipe and must be accounted for in design.

17. SEWER PIPE CORROSION

Corrosion of piping and other conveyance materials by hydrogen sulfide is a major concern in sewer design. Production of hydrogen sulfide in sewer lines is a function of several sewage characteristics, including temperature, strength, flow velocity, age, pH, and sulfate concentration. Sulfides are produced in the biofilm that forms at the pipe-sewage interface. A *biofilm* is a complex arrangement of microorganisms marked by the excretion of protective and adhesive polymeric substances. Here, bacteria convert sulfates to sulfides, which form *hydrogen sulfide* gas. When the hydrogen sulfide gas contacts the wet surface of the upper part of the sewer pipe, it is rapidly oxidized by bacteria, resulting in sulfuric acid. Unless neutralized, the sulfuric acid can corrode the piping material and other sewer components.

Because sulfate is present in sufficient quantities in nearly all domestic wastewater sources, hydrogen sulfide generation is a concern for most utilities. The factor that determines whether sulfide buildup occurs in a stream of sewage is whether or not oxygen is absorbed at the surface of the stream fast enough to oxidize the hydrogen sulfide diffusing out of the biofilm. The oxygen demand is dependent on the characteristics of the wastewater, while the oxygen absorption rate is dependent principally upon flow velocity. A high flow velocity may reduce sulfide buildup depending upon the wastewater BOD and the temperature. However, high flow velocities may also be damaging if any hydrogen sulfide is present in a stream of sewage, because the rate of sulfide release increases with increased flow rate. Turbulence caused by junctions, changes in pipe size, drops, and so on, will cause a relatively rapid release of hydrogen sulfide gas. Therefore, proper pipe material selection, flow velocity, and sewer layout are critical for preventing damage from hydrogen sulfide corrosion.

Sewer piping is subject to corrosion from acidic or alkaline soils, electrolytic decomposition attack, and temperature-induced damage. Different pipe materials display various levels of resistance to these factors. Therefore, prior to pipe material selection, determination of the soil and site characteristics is necessary when selecting an area for sewer line installation.

18. WASTEWATER TREATMENT PLANT LOADING

The amount of organic material, or substrate, entering a wastewater treatment plant is termed the *organic loading rate* (OLR). The organic loading rate to a wastewater treatment plant is expressed as pounds mass of BOD_5 (grams of BOD_5). This value is determined by the characteristics of the water discharged by the population served and should be measured by the utility. Organic loading rates vary based on factors such as the diet of the population served, climate, and the characteristics of inputs to the feed stream. Garbage disposals represent a significant input source for organic loadings. With garbage disposals, the average organic matter contributed per capita per day is 0.24 lbm/day (110 g/d) of suspended solids and approximately 0.17 lbm/day to 0.20 lbm/day (77 g/d to 90 g/d) of BOD_5. Organic loadings for an area are generally expressed in terms of a population equivalent based on an average daily per capita organic contribution.

A wastewater source may be characterized in terms of a *population equivalent*, P_e, a loading rate normalized to a set number of persons. P_e is generally expressed in terms of volume per thousands of persons per day. The average daily per capita contribution of BOD_5 to domestic wastewater is 0.20 lbm/person-day (90 g/person·d). This is the amount of BOD_5 a person contributes on a daily basis to the total wastewater flow entering a wastewater treatment plant. Therefore, the population equivalent can be calculated from Eq. 7.3.

$$P_{e,1000s} = \frac{\text{BOD}_{\text{mg/L}} Q_{\text{L/d}}}{\left(90 \ \dfrac{\text{g}}{\text{person·d}}\right)(1000 \ \text{persons})} \quad [\text{SI}] \quad 7.3(a)$$

$$P_{e,1000s} = \frac{\text{BOD}_{\text{mg/L}} Q_{\text{gal/day}} \times \left(8.345 \ \dfrac{\text{lbm-L}}{\text{MG-mg}}\right)}{\left(10^6 \ \dfrac{\text{gal}}{\text{MG}}\right)(1000 \ \text{persons})} \quad [\text{U.S.}] \quad 7.3(b)$$
$$\times \left(0.20 \ \dfrac{\text{lbm}}{\text{person-day}}\right)$$

Example 7.1

A manufacturing facility discharges its wastewater into a local river. In terms of 1000 persons, determine the population equivalent for the facility, given the following characteristics for the effluent: $Q = 16{,}000$ gal/day and $\text{BOD}_5 = 134$ mg/L.

Solution

Use Eq. 7.3 to calculate the population equivalent for the facility.

$$P_{e,1000s} = \frac{\text{BOD}_{\text{mg/L}} Q_{\text{gal/day}} \left(8.345 \ \dfrac{\text{lbm-L}}{\text{MG-mg}}\right)}{\left(10^6 \ \dfrac{\text{gal}}{\text{MG}}\right)(1000 \ \text{persons})\left(0.20 \ \dfrac{\text{lbm}}{\text{person-day}}\right)}$$

$$= \frac{\left(134 \ \dfrac{\text{mg}}{\text{L}}\right)\left(16{,}000 \ \dfrac{\text{gal}}{\text{day}}\right)\left(8.345 \ \dfrac{\text{lbm-L}}{\text{MG-mg}}\right)}{\left(10^6 \ \dfrac{\text{gal}}{\text{MG}}\right)(1000 \ \text{persons})\left(0.20 \ \dfrac{\text{lbm}}{\text{person-day}}\right)}$$

$$= 0.09 \ \text{gal/person-day}$$

Example 7.2

A local WWTP has an influent BOD_5 of 194 mg/L and a flow rate of 6 MGD. The utility has a per capita BOD_5 loading rate of 0.20 lbm/person-day. What is the population equivalent loading rate for this system?

Solution

Use Eq. 7.3 and the given values to calculate the population equivalent for the facility.

$$P_{e,1000s} = \frac{\text{BOD}_{\text{mg/L}} Q_{\text{MG/day}} \left(8.345 \ \dfrac{\text{lbm-L}}{\text{MG-mg}}\right)}{(1000 \ \text{persons})\left(0.20 \ \dfrac{\text{lbm}}{\text{person-day}}\right)}$$

$$= \frac{\left(194 \ \dfrac{\text{mg}}{\text{L}}\right)\left(6 \ \dfrac{\text{MG}}{\text{day}}\right)\left(8.345 \ \dfrac{\text{lbm-L}}{\text{MG-mg}}\right)}{(1000 \ \text{persons})\left(0.20 \ \dfrac{\text{lbm}}{\text{person-day}}\right)}$$

$$= 48.57 \ \text{gal/person-day}$$

19. ORGANIC LOADING RATE (OLR)

Organic loading rate (OLR) describes the rate at which organic materials are added to a system. Typical loading rates for various wastewater parameters at municipal wastewater treatment plants are reported in Table 7.3. This value is dependent on the hydraulic loading rate (defined in Sec. 5.4) and the amount of organic material in the water source. This term may be used to describe the input of organic material into a wastewater reactor or into a natural system, such as a river or lake.

Table 7.3 Typical Unit Loading Rates for Different Wastewater Parameters Entering Municipal Wastewater Treatment Plants

parameter	loading rate (g/capita·d)
BOD_5	90
COD	180
total suspended solids (TSS)	104
total nitrogen (N)	18
organic nitrogen (N)	9
ammonia nitrogen (N)	9
total phosphorus (P)	4

Reproduced with permission of CRC Press, from *Wastewater Treatment Plants: Planning, Design, and Operation*, Syed R. Qasim, 2nd ed., 1999; permission conveyed through Copyright Clearance Center, Inc.

For reactors (water treatment plants), the OLR is in some instances expressed on a unit area basis as pounds of BOD_5 per unit area per unit time (pounds per square foot per day or kilograms per square meter per day). OLR is calculated from Eq. 7.4.

$$\text{OLR} = \frac{(\text{BOD}_5) Q}{A} \quad [\text{reactors}] \quad 7.4$$

For surface water systems (rivers and lakes), the OLR is generally expressed in terms of mass per unit time (pounds per day or kilograms per day) and is calculated from Eq. 7.5.

$$\text{OLR} = (\text{BOD}_5) Q \quad [\text{surface water systems}] \quad 7.5$$

The OLR for surface water systems is important because it will affect the characteristics of the water. The OLR is likely governed by effluent standards set by a government organization.

Example 7.3

A conventional activated sludge system with an influent of high strength wastewater ($\text{BOD}_5 = 270$ mg/L) has a flow rate of 500,000 gal/day. The rectangular reactor has a width of 50 ft, a length of 100 ft, and a depth of 12 ft. Calculate the organic loading rate for the sludge system. Compare the calculated OLR value to those recommended for a conventional activated sludge reactor.

Wastewater

Solution

Calculate the reactor surface area, A, and volume, V.

$$A = wl = (50 \text{ ft})(100 \text{ ft})$$
$$= 5000 \text{ ft}^2$$
$$V = wld = (50 \text{ ft})(100 \text{ ft})(12 \text{ ft})$$
$$= 60{,}000 \text{ ft}^3$$

Calculate the loading rate in terms of mass BOD_5, per unit area and per unit volume, relative to the activated sludge reactor.

$$\text{OLR}_{\text{per unit area/day}} = \frac{(BOD_5)Q}{A}$$

$$= \frac{\left(270 \ \frac{\text{mg}}{\text{L}}\right)\left(500{,}000 \ \frac{\text{gal}}{\text{day}}\right)}{5000 \text{ ft}^2}$$

$$\times \left(8.345 \times 10^{-6} \ \frac{\text{lbm-L}}{\text{mg-gal}}\right)$$

$$= 0.225 \text{ lbm BOD}_5/\text{ft}^2\text{-day}$$

$$\text{OLR}_{\text{per unit volume/day}} = \frac{(BOD_5)Q}{V}$$

$$= \frac{\left(270 \ \frac{\text{mg}}{\text{L}}\right)\left(500{,}000 \ \frac{\text{gal}}{\text{day}}\right)}{60{,}000 \text{ ft}^3}$$

$$\times \left(8.345 \times 10^{-6} \ \frac{\text{lbm-L}}{\text{mg-gal}}\right)$$

$$= 0.019 \text{ lbm BOD}_5/\text{ft}^3\text{-day}$$

$$\text{OLR}_{\text{per day}} = (BOD_5)Q$$

$$= \left(270 \ \frac{\text{mg}}{\text{L}}\right)\left(500{,}000 \ \frac{\text{gal}}{\text{day}}\right)$$

$$\times \left(8.345 \times 10^{-6} \ \frac{\text{lbm-L}}{\text{mg-gal}}\right)$$

$$= 1126 \text{ lbm BOD}_5/\text{day}$$

The OLR values recommended for a conventional activated sludge reactor are 0.019 lbm BOD_5/ft^3-day to 0.037 lbm BOD_5/ft^3-day (0.3 kg $BOD_5/\text{m}^3\cdot$d to 0.6 kg $BOD_5/\text{m}^3\cdot$d). Since the calculated value falls within the typical range for conventional activated sludge reactors, the design is acceptable.

Example 7.4

The discharge stream from a local WWTP has a BOD_5 concentration of 4.32 mg/L and a flow rate of 3.23 MGD. What is the organic loading rate into the receiving water body?

Solution

Using Eq. 7.5, solve for the OLR into the river.

$$\text{OLR} = (BOD_5)Q$$

$$= \left(4.32 \ \frac{\text{mg}}{\text{L}}\right)\left(3.23 \ \frac{\text{MG}}{\text{day}}\right)\left(8.345 \ \frac{\text{lbm-L}}{\text{MG-mg}}\right)$$

$$= 116.4 \text{ lbm BOD}_5/\text{day}$$

20. DILUTION AND PURIFICATION/ DILUTION RATIOS

Dilution purification, or *self-purification*, is the blending of a contaminated or polluted stream with a more pristine one to produce a stream with acceptable characteristics. The pristine source may be a river, treated effluent, groundwater, or other clean water source. The characteristics of a combined flow, C_f, may be determined by doing a mass balance of the known properties of the mixing streams.

$$C_f = \frac{Q_1 C_1 + Q_2 C_2}{Q_1 + Q_2} \qquad \qquad 7.6$$

Equation 7.6 can be used to determine the BOD, solids content, temperature, pH, dissolved oxygen concentration, or contaminant concentration of mixed streams in municipal and industrial wastewater treatment.

Example 7.5

A 10 MGD WWTP discharges into a river. The effluent DO concentration is 1 mg/L. The river has a flow rate of 100 ft^3/sec and an average DO concentration of 8 mg/L. Assuming complete mix conditions, what is the DO in the river immediately following the discharge point?

Solution

The units for the respective flow rates must be the same, so convert the given WWTP into cubic feet per second.

$$\frac{(10 \text{ MGD})\left(10^6 \ \frac{\text{gal}}{\text{MG}}\right)}{\left(7.48 \ \frac{\text{gal}}{\text{ft}^3}\right)\left(86{,}400 \ \frac{\text{sec}}{\text{day}}\right)} = 15.47 \text{ ft}^3/\text{sec}$$

Use a mass balance approach, Eq. 7.6, to solve for the DO concentration after the two streams mix.

$$C_f = \frac{Q_1 C_1 + Q_2 C_2}{Q_1 + Q_2}$$

$$= \frac{\left(100 \; \frac{\text{ft}^3}{\text{sec}}\right)\left(8 \; \frac{\text{mg}}{\text{L}}\right) + \left(15.47 \; \frac{\text{ft}^3}{\text{sec}}\right)\left(1 \; \frac{\text{mg}}{\text{L}}\right)}{100 \; \frac{\text{ft}^3}{\text{sec}} + 15.47 \; \frac{\text{ft}^3}{\text{sec}}}$$

$$= 7.06 \text{ mg/L}$$

Example 7.6

Runoff from a local slaughterhouse is flowing into a small stream. The runoff has a BOD_5 of 1100 mg/L and a total daily discharge of roughly 275 gal/day. The stream has a background BOD_5 of 4 mg/L and an average daily flow rate of 2700 gal/day. Assuming complete mix conditions at the point where the runoff enters the stream, determine whether the runoff is significantly altering the stream's water quality, as determined by its BOD_5.

Solution

Calculate the stream's BOD_5 after it mixes with the runoff, using the mass balance approach, Eq. 7.6.

$$C_f = \frac{Q_1 C_1 + Q_2 C_2}{Q_1 + Q_2}$$

$$= \frac{\left(275 \; \frac{\text{gal}}{\text{day}}\right)\left(1100 \; \frac{\text{mg}}{\text{L}}\right) + \left(2700 \; \frac{\text{gal}}{\text{day}}\right)\left(4 \; \frac{\text{mg}}{\text{L}}\right)}{275 \; \frac{\text{gal}}{\text{day}} + 2700 \; \frac{\text{gal}}{\text{day}}}$$

$$= 105 \text{ mg } BOD_5/\text{L}$$

To estimate the significance of the increase in the stream's BOD_5, calculate the increase as a percentage of change in the original value.

$$\% \text{ increase} = \frac{C_i - C_o}{C_o} \times 100\%$$

$$= \frac{105 \; \frac{\text{mg}}{\text{L}} - 4 \; \frac{\text{mg}}{\text{L}}}{4 \; \frac{\text{mg}}{\text{L}}} \times 100\%$$

$$= 2525\%$$

An increase of 2525% would indicate that the increase in the stream's BOD_5 is large enough that the runoff will need to be treated prior to its discharge into the stream.

Example 7.7

An industrial center discharges its wastewater directly into a local river at a flow rate of 4 ft^3/sec. The discharge has a DO of 9 mg/L and an ultimate BOD of 44 mg/L. Prior to the discharge point, the river has a BOD_u of 3 mg/L. The flow rate in the river is 15 ft^3/sec, with a velocity of 1 ft/sec. Calculate the total system flow rate and the BOD_u at the industrial discharge point.

Solution

Calculate the flow rate in the river following the industrial discharge.

$$Q_t = Q_1 + Q_2$$

$$= 15 \; \frac{\text{ft}^3}{\text{sec}} + 4 \; \frac{\text{ft}^3}{\text{sec}}$$

$$= 19 \text{ ft}^3/\text{sec}$$

Using Eq. 7.6, calculate the river BOD_u at the discharge point.

$$BOD_{u, \text{post discharge}} = \frac{Q_1 C_1 + Q_2 C_2}{Q_1 + Q_2}$$

$$= \frac{Q_1(BOD_{u1}) + Q_2(BOD_{u2})}{Q_t}$$

$$= \frac{\left(15 \; \frac{\text{ft}^3}{\text{sec}}\right)\left(3 \; \frac{\text{mg}}{\text{L}}\right) + \left(4 \; \frac{\text{ft}^3}{\text{sec}}\right)\left(44 \; \frac{\text{mg}}{\text{L}}\right)}{19 \; \frac{\text{ft}^3}{\text{sec}}}$$

$$= 11.63 \text{ mg/L}$$

Example 7.8

The effluent from a WWTP empties into a pristine river having a flow rate of 16 MGD and a total solids (TS) concentration of 0.85 mg/L. The WWTP discharge rate is 4.9 MGD, and the WWTP TS concentration is 275 mg/L. What is the TS concentration in the river after the WWTP?

Solution

Calculate the TS concentration in the mixed stream using Eq. 7.6.

$$C_f = \frac{Q_1 C_1 + Q_2 C_2}{Q_1 + Q_2}$$

$$= \frac{(16 \text{ MGD})\left(0.85 \ \frac{\text{mg}}{\text{L}}\right) + (4.9 \text{ MGD})\left(275 \ \frac{\text{mg}}{\text{L}}\right)}{16 \text{ MGD} + 4.9 \text{ MGD}}$$

$$= 65.1 \text{ mg/L}$$

21. NUTRIENTS IN WASTEWATER

Wastewater may contain high levels of nutrients (e.g., nitrogen and phosphorus), which can be toxic to aquatic organisms even at very low concentrations (e.g., 0.06 mg/L of NH_3 will damage fish gills). These nutrients may also result in excessive algae growth, which is of particular concern. Algae produce toxins, and their death and consumption by bacteria results in oxygen depletion (i.e., eutrophication, see Sec. 6.14) of the water body. Nutrient removal can be achieved either biologically or by chemical precipitation. Specifically, nitrogen can be removed from wastewater streams through a variety of processes, including membrane separation, ion exchange, volatilization, aerobic biological systems, media filtration, and biological nitrification/denitrification processes (see Sec. 8.10).

At high concentrations, micronutrients are considered to be toxins. Therefore, a careful balance must be maintained in the biological reactors.

PRACTICE PROBLEMS

1. A biological reactor has an influent flow rate of 100,000 gal/day with a BOD_5 of 200 mg/L (medium strength wastewater). The circular reactor has a depth of 8 ft and a diameter of 40 ft. What is the organic loading rate?

(A) 0.06 lbm BOD_5/ft²-day

(B) 0.13 lbm BOD_5/ft²-day

(C) 1.5 lbm BOD_5/ft²-day

(D) 2.7 lbm BOD_5/ft²-day

2. What is the most common technology used in the United States to collect and transport municipal sewage?

(A) pressurized sewers

(B) septic systems

(C) gravity sewers

(D) storm drains

3. What is the minimum required fluid flow rate in sanitary sewers to prevent the accumulation of solid material on the bottom of sewer pipes?

(A) 0.5 ft/sec

(B) 2.0 ft/sec

(C) 10 ft/sec

(D) >20 ft/sec

4. In a full circular sewer pipe, at what depth in the fluid flow field is the fluid velocity the greatest?

(A) at the pipe wall

(B) at the flow centerline

(C) halfway between the pipe wall and the flow centerline

(D) none of the above

5. Effluent from a municipal wastewater treatment plant (WWTP) has a BOD_5 of 79 mg/L and an average daily flow rate of 3100 gal/day. The WWTP discharges into a stream with an average flow rate of 30,000 gal/day and an initial BOD_5 of 4 mg/L. What is the BOD_5 in the stream after mixing with the WWTP treated effluent?

(A) 3.1 mg/L

(B) 5.7 mg/L

(C) 7.6 mg/L

(D) 11 mg/L

SOLUTIONS

1. Calculate the surface area of the reactor.

$$A = \pi r^2 = \pi \left(\frac{D}{2}\right)^2$$
$$= \pi \left(\frac{40 \text{ ft}}{2}\right)^2$$
$$= 1257 \text{ ft}^2$$

Use Eq. 7.4.

$$\text{OLR} = \frac{(\text{BOD}_5)Q}{A}$$

$$= \frac{\left(200 \ \dfrac{\text{mg}}{\text{L}}\right)\left(100{,}000 \ \dfrac{\text{gal}}{\text{day}}\right)}{\times \left(8.345 \times 10^{-6} \ \dfrac{\text{lbm-L}}{\text{mg-gal}}\right)}{1257 \text{ ft}^2}$$

$$= 0.13 \text{ lbm BOD}_5/\text{ft}^2\text{-day}$$

The answer is (B).

2. From Sec. 7.9, there are two general types of municipal wastewater collection systems: gravity and force collection. Conventional gravity sewers are widely used because contents flow naturally, without requiring pumping power. Gravity sewers are used less frequently in areas of flat or hilly terrain where pressurized systems are needed.

The answer is (C).

3. Fluid flow rates less than 2.0 ft/sec allow solids to accumulate on the pipe bottoms and produce hydrogen sulfide, which results in odor problems.

The answer is (B).

4. From Sec. 7.16 and Fig. 7.1, pipe walls provide friction between the pipe and the water flowing within the pipe, which slows the flow. In this problem, the pipe is circular and full of water, so the fluid flow is fastest where it is farthest from the pipe wall, at the flow centerline.

The answer is (B).

5. Use Eq. 7.6 to perform a mass balance on the defined system, which includes the stream and the effluent from the WWTP, to determine the BOD_5 in the mixed water. Assume that the stream and the WWTP effluent are instantaneously and completely mixed.

$$C_{\text{mix}} = \frac{Q_{\text{effluent}}\, C_{\text{effluent}} + Q_{\text{stream}}\, C_{\text{stream}}}{Q_{\text{effluent}} + Q_{\text{stream}}}$$

$$= \frac{\left(3100 \ \dfrac{\text{gal}}{\text{day}}\right)\left(79 \ \dfrac{\text{mg}}{\text{L}}\right) + \left(30{,}000 \ \dfrac{\text{gal}}{\text{day}}\right)\left(4 \ \dfrac{\text{mg}}{\text{L}}\right)}{3100 \ \dfrac{\text{gal}}{\text{day}} + 30{,}000 \ \dfrac{\text{gal}}{\text{day}}}$$

$$= 11.0 \text{ mg BOD}_5/\text{L} \quad (11 \text{ mg/L})$$

The answer is (D).

Wastewater Treatment

Nomenclature

A_s	surface area	ft²	m²
BOD	biochemical oxygen demand	mg/L	mg/L
BOD_5	biochemical oxygen demand after 5 days of incubation	mg/L	mg/L
$CBOD_5$	carbonaceous biochemical oxygen demand after 5 days of incubation	mg/L	mg/L
d	total basin depth	ft	m
d_o	settling zone depth	ft	m
d_p	mean particle diameter of settling material	in	mm
DO	dissolved oxygen	mg/L	mg/L
f	Darcy friction factor	–	–
g	gravitational acceleration	ft/sec²	m/s²
G	flux of solids through a clarifier	lbm/ft²-hr	kg/m²·h
k	constant describing settling material	–	–
L	sedimentation basin length	ft	m
L'	sedimentation zone length	ft	m
MWCO	molecular weight cutoff of a membrane	amu	amu
N	number of persons	–	–
q_o	hydraulic loading rate	gal/day-ft²	m³/d·m²
q_o	overflow rate of sedimentation basin	ft/hr	m/h
q_w	weir overflow rate for sedimentation basin	ft³/hr-ft	m³/h·m
Q	influent flow rate	ft³/hr	m³/h
r	radius	ft	m
SG_p	specific gravity of particle	–	–
t	time	hr	h
t_d	detention time	hr	h
TDS	total dissolved solids	mg/L	mg/L

TSS	total suspended solids	mg/L	mg/L
v	scouring velocity of settling material	ft/sec	m/s
v_h	horizontal velocity of settling particle	ft/sec	m/s
v_s	settling velocity of solid materials	ft/sec	m/s
v_u	bulk downward velocity resulting from underflow	ft/hr	m/h
V	volume of sedimentation basin	ft³	m³
w_r	weir loading rate for clarifier design	gal/day-ft²	m³/d·m²
W	width	ft	m
X	concentration of biomass entering clarifier	mg/L	mg/L
X_u	biomass recycle concentration	mg/L	mg/L

1. NATIONAL POLLUTANT DISCHARGE ELIMINATION SYSTEM

The National Pollutant Discharge Elimination System (NPDES) is a national permit program, first authorized by the Clean Water Act in 1972, that manages water pollution through regulation of point sources. *Point sources* are stationary dispensation locations such as pipes, manmade ditches, and other conveyers of pollution into United States waters. Industrial, municipal, and agricultural facilities must obtain NPDES permits if their waste is discharged into surface waters. Residential homes do not need permits if they are connected to municipal wastewater systems, if they use a septic system, or if they do not otherwise discharge to surface waters. Though the program is national, the permits are administered through the states in most cases. Table 8.1 shows typical effluent standards for industrial and municipal wastewater limitations.

2. SEPTIC TANKS

A *septic system* is a small-scale sewage treatment system, commonly used in areas with no connection to main sewerage pipes (this can include suburbs and small towns, as well as rural areas). In North America, approximately 25% of the population relies on septic tanks. A septic tank generally consists of a tank of 1000–1500 gallons capacity connected to an inlet wastewater pipe at one end and to a septic drain field at the other (see Fig. 8.1). These pipe connections are generally made via a T-pipe, allowing liquid entry and egress without disturbing any scales on the surface. The tank

Table 8.1 NPDES Wastewater Effluent Standards

parameter	maximum permitted
BOD$_5$ (mg/L)	30[*]
oil and grease or TPH (mg/L)	15–55
total suspended solids (mg/L)	30–45
pH	6.0–9.0
temperature (°C)	< 40
color (color units)	2.0
NH$_3$/NO$_3$ (mg/L)	1.0–10
phosphates (mg/L)	0.2
heavy metals (mg/L)	0.1–5.0
surfactants (mg/L)	0.5–1.0 total
sulfides (mg/L)	0.01–0.1
phenol (mg/L)	0.1–1.0
toxic organics (mg/L)	1.0 total
cyanide (mg/L)	0.1

[*]30 day average value; the arithmetic mean of the BOD$_5$ values (by concentration) for effluent samples collected over a period of thirty consecutive days shall not exceed 15% of the arithmetic mean (by concentration) for influent samples collected at approximately the same time during the same period (Clean Water Act).

design typically incorporates two chambers (each of which is equipped with a manhole cover), separated by a dividing wall with openings located about midway between the floor and roof of the tank. Wastewater enters the first chamber, where suspended solids settle out and scum floats to the surface. The settled solids are anaerobically digested, reducing the volume of solids, while the supernatant flows through the dividing wall into the second chamber. There, any remaining suspended solids settle out and the liquid exits to the leach field, or drain field. The tank effluent is distributed through a perforated piping network. The size of the leach field is a function of the wastewater volume and is inversely proportional to the porosity of the drainage field. The entire septic system can operate by gravity alone. Where topographic considerations require, a *lift pump* is included in the system.

The solid waste that remains following anaerobic digestion must be removed after a set amount of time or the septic tank fills up and undecomposed wastewater discharges directly to the drainage field. This time period depends on the volume of the tank relative to the input of solids, the amount of indigestible solids, and the ambient temperature (anaerobic digestion occurs more efficiently at higher temperatures). In general, it is rare for a septic tank system to require emptying more than four times a year. By careful management, many users can reduce emptying to once per year or less. When a tank is emptied, a small residue of about 10% of the sludge should be allowed to remain in the tank to ensure that anaerobic decomposition is rapidly resumed as the tank refills.

3. WASTEWATER TREATMENT

Sewage treatment is the process that removes most of the contaminants from wastewater or sewage and

Figure 8.1 Two-Stage Septic Tank Typically Used for Household Wastewater Treatment

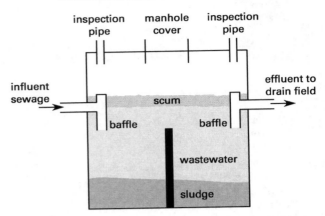

produces both a liquid effluent suitable for disposal to the natural environment and a sludge. Municipal wastewater is most often a combination of liquid flows from sanitary systems, kitchens, businesses, industry, and stormwater runoff, which are collected in a regional sewer system (see Table 8.2). The ultimate composition of a wastewater, and in turn, the processes needed to successfully treat it, will vary according to the characteristics of the locale (climate, population, geography, geology, industry, etc.). For treatment to be effective, sewage must be conveyed to a treatment plant by appropriate pipes and infrastructure in a process that is subject to regulation and controls. At the simplest level, sewage and most wastewaters are treated by separation of solids from liquids, usually by settlement. By progressively converting dissolved materials into solids, usually biological flocs, and settling the solids out, an effluent stream of increasing purity is produced.

Municipal wastewater and sewage may be treated to varying degrees, depending upon the characteristics of the water body into which the treated water is to be discharged. For instance, if the sewage is to be discharged to the ocean, only primary treatment may be required, while if it is to be discharged to a small river, tertiary treatment is necessary. In this respect, treatment plants are generally described as being primary, secondary, or tertiary. In each case, the descriptor term designates the level of treatment the plant employs. Although there are numerous variations of wastewater treatment plants, they all work to achieve the following goals.

- maintain the aesthetic quality (e.g., clarity and odor) of the receiving water body
- remove sludge and scum
- maintain the dissolved oxygen concentration in the receiving water body
- remove/inactivate pathogenic microorganisms (i.e., bacteria and viruses)

4. WASTEWATER TREATMENT PLANT

Wastewater treatment plants are required for treating flows from multiple sources and are typically needed in

Table 8.2 Typical Characteristics of Raw Municipally Generated Wastewater

wastewater characteristic	strong (mg/L)	weak (mg/L)
total solids	1200	350
TDS	850	250
fixed	525	145
volatile	325	105
TSS	350	100
fixed	75	30
volatile	275	70
settleable solids (mL/L)	20	5
BOD_5 at 20°C	300	100
TOC	300	100
COD	1000	250
total nitrogen (as N)	85	20
organic	35	8
free ammonia	50	12
nitrites	0	0
nitrates	0	0
total phosphorus (as P)	20	6
organic	5	2
inorganic	15	4
chlorides	100	30
alkalinity (as $CaCO_3$)	200	50
grease	150	50

urban areas where septic systems are not feasible. The wastewater enters the treatment plant through coarse bar screens that are designed to remove large floating or suspended debris from the influent stream. Sand and other gritty particles are then removed in a grit chamber. These are the processes of *pretreatment*. After grit is removed, water begins *primary treatment* in a large settling basin or tank, called a primary clarifier, where solids settle out and less dense materials (e.g., greases, oils, and light solids) float. Both fractions are removed from the clarifier to produce a waste stream composed of suspended and dissolved organic and inorganic materials. The "clarified" water then flows to the *secondary treatment* process termed *activated sludge*, where the waste is aerated to facilitate the degradation of organics by bacteria (activated sludge is the subject of Chap. 9). Following the activated sludge process, the waste enters a *secondary* (or final) *clarifier* that allows the suspended biological floc to settle out in the form of sludge. A portion of this sludge is then recycled back into the aerated basin (usually termed the *reactor*) while the remainder is removed for disposal. The waste then enters a final disinfection process, called tertiary treatment, prior to being discharged back into the environment. A schematic of the entire process is shown in Fig. 8.2.

Pretreatment

In order to protect pumping systems and to prevent clogging of pipes in wastewater applications, the raw water must be pretreated to remove large objects, sand, grit, oils, and grease. Pretreatment equipment includes bar screens, grit chambers, and comminutors (grinders). Materials removed in bar screens and grit chambers are typically dewatered and deposited in a landfill or other receiving facility.

Primary Treatment

Primary treatment processes remove oils, grease, fats, and coarse (settleable) solids. They do not remove colloidal or dissolved solids. Primary treatment processes consist of clarifiers (sedimentation basins) that typically remove 60% of total suspended solids (TSS) and about 30% of BOD from influent wastewater. Primary clarifiers

Figure 8.2 General Schematic of a Tertiary Wastewater Treatment Plant

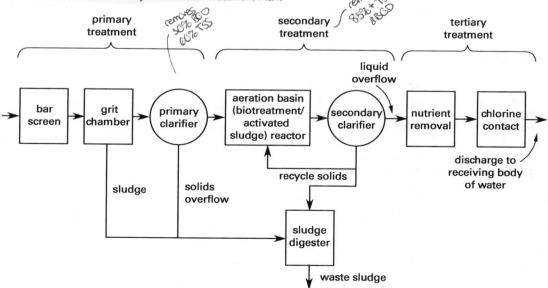

are either circular (see Fig. 8.3) or rectangular tanks. The tanks are large enough that heavier solids can settle and floating material, such as greases, oils, and plastics, can rise to the surface and be skimmed off. The main purpose of the primary clarification stage is to produce both a generally homogeneous liquid capable of being treated biologically and a sludge that can be separately treated or processed. Primary sedimentation tanks are usually equipped with mechanically driven scrapers that continually drive the collected sludge toward a hopper in the base of the tank, where it can be pumped to additional sludge treatment stages.

Figure 8.3 *Typical Circular Clarification Basin for Wastewater Treatment*

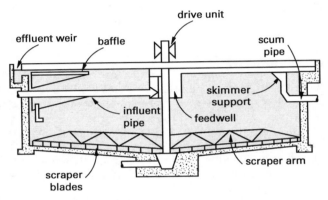

Settled solids are removed in a primary clarifier by a scraper blade rotating around the bottom to guide the sludge into a collection pipe, from which it is removed and pumped into a digester. Scum and/or floating materials on the water surface are pushed (skimmed) by a skimmer arm into an effluent trough for ultimate disposal. In a rectangular clarifier, sludge is collected by a continuously moving collector chain, which acts as a conveyor belt and drags the sludge to a collection point for removal. Water leaves the clarifiers by flowing over a weir, a steel curb with teeth that allows the water to flow slowly.

Secondary Treatment

Secondary treatment consists of two additional steps: biological treatment and secondary clarification. Dissolved and colloidal materials not removed in primary treatment are removed in these two steps. For biological treatment, five types of biological processes are common: aerobic, anoxic, anaerobic, combined (aerobic, anoxic, and anaerobic), and pond (i.e., lagoon) processes (see Table 8.3). The terms suspended and attached are used to distinguish systems in which the biota are suspended in the media, and systems in which the biota are attached to a stationary medium.

The biological treatment is followed by clarification in basins (secondary clarifiers) in which suspended biological solids from the aeration basin are allowed to settle out. Solids removed in the secondary clarifiers are returned to the aeration basins as *returned activated*

Table 8.3 *Types of Biological Treatment Processes Used in Wastewater Treatment*

biological treatment process	suspended design example applications	attached design example application
aerobic	activated sludge, suspended growth, aerated lagoon	trickling filter
anoxic	denitrifying suspended growth reactors	denitrifying packed bed reactors
anaerobic	septic tank	anaerobic biofilter
combined	nitrification/ denitrification	nitrification/ denitrification

sludge to maintain the appropriate population of microorganisms needed to assimilate the organic matter entering the aeration basins, while the rest are wasted (see Fig. 8.4). Approximately 85% or more of the raw influent TSS and BOD_5 is removed once the wastewater has passed through the secondary treatment processes. Performance characteristics of secondary wastewater treatment processes are shown in Table 8.4.

Figure 8.4 *General Layout of Secondary Wastewater Treatment Processes in a Municipal Wastewater Treatment Plant*

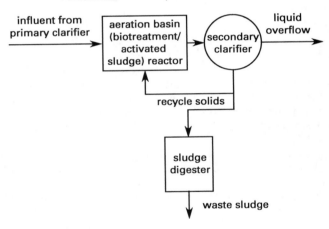

Table 8.4 *Typical Effluent Wastewater Quality Following Secondary Treatment[a]*

wastewater characteristic	maximum average 30 day level	maximum average 7 day level	minimum % removal in 30 days
BOD_5 (mg/L)	30	45	85
suspended solids (mg/L)	30	45	85
pH	6.0–9.0[b] at all times		
$CBOD_5$[c] (mg/L)	25	40	85

[a]according to *Code of Federal Regulations*, Title 40, Part 133 standards
[b]The effluent values for pH must be maintained within the limits of 6.0 to 9.0 unless the publicly owned treatment works demonstrates that inorganic chemicals are not added to the waste stream as part of the treatment process, and contributions from industrial sources do not cause the pH of the effluent to be less than 6.0 or greater than 9.0.
[c]At the option of the NPDES permitting authority, the parameter of carbonaceous BOD_5 ($CBOD_5$) may be substituted for the parameter BOD_5.

Tertiary Treatment

Tertiary or *advanced wastewater treatment* is employed to remove or further reduce the concentrations of various constituents (e.g., BOD, nitrogen, phosphorus, suspended solids, ammonia, toxic compounds, and dissolved inorganic salts). Tertiary treatment includes physical, chemical, and biological processes. Tertiary treatment provides a final stage to raise the effluent quality to the standard required before it is discharged to the receiving water body. More than one tertiary treatment process may be used at any treatment plant.

5. FLOW EQUALIZATION

Flow equalization is aimed at maintaining a nearly constant flow rate through a treatment facility. A secondary objective is to dampen the strength of the wastewater constituents by blending the wastewater streams in an equalization basin. Equalizing the flow rate and dampening the strength of the wastewater maintains the reliability and operational control of wastewater treatment. Flow equalization is recommended for plants with large diurnal variations in influent quality or volume and for plants with capacities less than 0.5 MGD (1.89 MLD). The equalization of flow prevents short-term, high volumes of incoming flow, called *surges*, from forcing solids and organic material out of a treatment process. Flow equalization also controls the flow through each stage of the treatment system, allowing adequate time for the physical, biological, and chemical processes to take place. These processes are most efficient when operated at or near uniform conditions.

There are four commonly used flow equalization techniques in wastewater treatment: alternating flow diversion, intermittent flow diversion, completely mixed combined flow, and completely mixed fixed flow (see Fig. 8.5).

Figure 8.5 *Flow Equalization Systems*

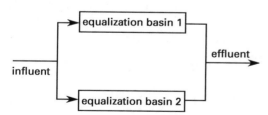

In an *alternating flow* system, one equalization basin collects the total influent flow for a given time, while a second basin is discharging. Mixing is maintained so that effluent concentrations remain constant with relatively constant flow. In an *intermittent flow* system, the waste stream is diverted to an equalization basin for short periods, which is then metered back to the primary flow at a controlled rate. The metering rate is determined by the variance of the influent parameters.

In the *completely mixed combined flow* system, multiple flows are completely mixed to reduce variations in each stream. In the *completely mixed fixed flow* system, the influent flow is completely mixed in a large holding basin directly before the treatment plant.

Equalization basins are typically located downstream of pretreatment facilities, such as bar screens, comminutors, and grit chambers. Where possible, they are also located downstream of the primary clarifiers. These basins may be designed as either in-line or side-line units and are generally divided into two or more separate cells.

Separate basins or unused on-line treatment units, such as primary clarifiers or aeration tanks, may be used for flow equalization during the early period of the plant's design life. If no other unit is available and equalization is performed by a single basin, then a valved bypass pipe is provided around the basin to the downstream portion of the treatment facility. With a diurnal flow pattern, the volume required to achieve the desired degree of equalization can be determined from a cumulative flow volume hydrograph over a representative 24 hr period. The volume required for equalization of flows will generally vary from approximately 20% to 40% of the 24 hr flow for smaller plants and from approximately 10% to 20% of the average daily dry weather flow for larger plants. Figure 8.6 illustrates the method in determining the minimum volume.

Figure 8.6 *Plotting Technique for Determining the Equalization Basin Volume for a Variable Influent Flow Rate*

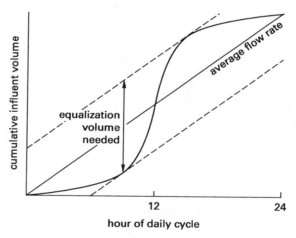

Equalization basins are constructed of earth, concrete, or steel. Earthen basins should be constructed according to the following aerated pond criteria.

- At least 1 ft (0.3 m) of freeboard is provided above the highest possible water level in the basin.

- Corner fillets and hopper bottoms with draw-offs are provided to prevent accumulation of sludge and grit.

- The basins should be capable of being isolated and drained.

- Basins should have provisions for suitable access to facilitate the cleaning and the maintenance of equipment, for cleaning basin walls, and for scum control.

- At least 18 in of freeboard should be provided for aerated basins and at least 24 in of freeboard when mechanical aerators are used.

- Approximately 0.1 hp/1000 gal (0.02 kW/m^3) of mixing power is, as a general rule, required. However, power levels vary with basin geometry. Mixing requirements for normal domestic wastewater range from 0.07 hp/1000 gal (0.01 kW/m^3) to 0.2 hp/1000 gal (0.04 kW/m^3) wetted basin volume.

- Multiple aeration and/or mixing units should be provided to maintain adequate mixing and for continuous operation.

- The air should be supplied at a rate between 1.25 cfm/1000 gal and 2.0 cfm/1000 gal (0.16 L/m^3·s and 0.25 L/m^3·s). Aeration should be sufficient to maintain a minimum of 1.0 mg/L of dissolved oxygen in the basin at all times.

6. PRETREATMENT PROCESSES

Bar Screens

Grates or bar screens remove large objects (e.g., plant debris, litter, etc.) from sewerage. The *bar screen* is made up of metal bars arranged in parallel (see Table 8.5). Materials retained on the screen are removed as needed (i.e., once the screen becomes blocked) and are typically deposited in a landfill. In some cases, the materials are automatically ground up and left in the wastewater flow for later removal.

Grit Chamber

Sand and grit must be removed from the wastewater influent to protect pumps and other sensitive equipment. These materials are removed using *grit chambers*, which are based on a variety of designs and operating schemes. Grit removal devices include horizontal flow grit chambers, aerated grit chambers, vortex-type grit removal systems, detritus tanks, and hydrocyclones. The type of design used is a function of the characteristics and quantity of the grit, head loss requirements, available space, wastewater organic content, and cost. Grit chambers should be located after bar screens, but before pumps and other treatment units. In smaller plants, the material accumulated in the grit chamber may be manually collected. Otherwise, the material is collected mechanically and incinerated or deposited in a landfill. For mechanical removal, a conveyor system with scrapers, buckets, or plows is employed.

Horizontal Flow Grit Chamber

The essential operating principle of a *horizontal flow grit chamber* is to slow the velocity of the incoming wastewater to the point that sand, grit, and stones can settle out while keeping the majority of the organic material

Table 8.5 Typical Design Criteria for Coarse Screening Equipment Commonly Used as Pretreatment Processes in Wastewater Treatment Applications

process	design criteria
trash rack openings[a]	1.5 in to 6.0 in (38 mm to 150 mm)
manual screen openings[b]	1 in to 2 in (25 mm to 50 mm)
mechanically cleaned bar screen openings	0.2 in to 0.6 in (5 mm to 15 mm)
approach velocity (max)	2.0 ft/sec to 3.25 ft/sec (0.6 m/s to 1.0 m/s)
approach velocity (min)[c]	1 ft/sec to 2 ft/sec (0.3 m/s to 0.6 m/s)
continuous screen openings[d]	0.25 in to 1.5 in (6 mm to 38 mm)
approach velocity (max)	2 ft/sec to 4 ft/sec (0.6 m/s to 1.2 m/s)
approach velocity (min)	1 ft/sec to 2 ft/sec (0.3 m/s to 0.6 m/s)
allowable head loss	0.5 ft to 2.0 ft (0.15 m to 0.6 m)

[a]commonly used for combined sewer inflows; opening size dependent on equipment being protected
[b]used at small plants and/or in bypass lines
[c]needed to prevent grit accumulation
[d]effective in the 6–18 mm range

Adapted with permission from ASCE and the Water Environmental Federation from *Design of Municipal Wastewater Treatment Plants*, Vol. 2, 4th ed., 1998.

within the fluid flow. Horizontal flow grit chambers are designed to provide flow velocities within a range from 0.75 ft/sec to 1.25 ft/sec (0.23 m/s to 0.38 m/s). A target velocity of 1.0 ft/sec (0.3 m/s) is employed in most designs. To prevent the resuspension of settled material, the fluid must flow at a velocity less than the scouring velocity, which may be calculated using the *Camp formula*, Eq. 8.1.

$$v = \sqrt{8k\left(\frac{gd_p}{f}\right)(SG_p - 1)} \qquad 8.1$$

k is a dimensionless constant that varies from approximately 0.04 for sand to 0.06 or greater for more cohesive materials. SG_p is the specific gravity of the particles, d_p is the mean particle diameter, and f is the dimensionless Darcy friction factor whose value is generally between 0.02 and 0.03. The detention time of a horizontal flow grit chamber is based on the size and specific weight of the particle to be removed. The chamber must be designed with a sufficient detention period to allow a particle of 0.01 in (0.2 mm) in average diameter to settle from the fluid surface to the chamber bottom before it can leave the chamber through horizontal flow (see Table 8.6).

Table 8.6 *Design Criteria for a Horizontal Flow Grit Chamber*

parameter	design criteria	
	range	commonly used
flow velocity		
(ft/sec)	0.8–1.3	1.0
(m/s)	0.24–0.4	0.3
settling velocity[a,b]		
50-mesh		
(ft/min)	9.2–10.2	9.6
(m/min)	2.8–3.1	2.9
100-mesh		
(ft/min)	2.0–3.0	2.5
(m/min)	0.6–0.9	0.8
head loss (% of channel depth)	30–40	36[c]
inlet-outlet length allowance (%)	25–50	30
detention time (sec)	45–90	60

[a]If particle specific gravity is much less than 2.65, lower velocities should be used.
[b]The mesh size designates the size of particles that may pass through a screen.
[c]for a Parshall flume controlled system

Aerated Grit Chamber ②

In *aerated grit chambers*, grit is removed by directing the flow of the wastewater in a spiral pattern, or vortex, through the tank. This is achieved by introducing air on one side of the grit chamber and directing it to produce a spiral velocity pattern to the fluid flow. The spiral flow pattern accelerates the settling of grit from suspension to the tank bottom. For aerated grit chambers, the minimum detention time is 1 min to 3 min. The exact value of detention time will depend on the density and associated settling velocity of the grit being removed. A denser particle will have a higher settling velocity, requiring a lower detention time. Typical design criteria for an aerated grit chamber are shown in Table 8.7.

Table 8.7 *General Design Characteristics of Aerated Grit Chambers for Wastewater Treatment*

parameter	operating range
transverse velocity at surface (ft/sec)	2–5
depth-width ratio	1.5:1 to 2:1
air supply	
cfm/ft length	3–5
ft³/gal	0.04–0.06
detention time (min)	3–5 (peak flow)
quantity of grit (ft³/MG)	1–10
quantity of scum (ft³/MG)	1–6

Hydrocyclones ③

Hydrocyclone systems may be used to either remove grit from the influent wastewater or to separate grit from grit slurries or sludge. For removing grit from influent wastewater, the fluid is pumped to a head ranging from 12 ft to 30 ft (3.7 m to 9 m). In this process, the grit collects on the sides and bottom of the cyclone system as a result of centrifugal forces.

Example 8.1

An influent sandy grit with an average diameter of 0.02 cm and a specific gravity of 2.65 flows into a horizontal flow grit chamber. Organic material in the influent flow has an average diameter of 0.03 cm and a specific gravity of 1.3, is not sticky, and remains as discrete particles. Determine the minimum and maximum allowable horizontal velocities in the grit chamber.

Solution

To determine the maximum allowable horizontal velocity, solve Eq. 8.1 for the grit scour velocity. Because the grit is principally composed of sand, use a k value of 0.04 and a friction value, f, of 0.03.

$$\mathrm{v} = \sqrt{8k\left(\frac{gd_p}{f}\right)(\mathrm{SG}_p - 1)}$$

$$= \sqrt{\begin{array}{c}(8)(0.04)\left(\dfrac{\left(9.81 \; \frac{\mathrm{m}}{\mathrm{s}^2}\right)\left(100 \; \frac{\mathrm{cm}}{\mathrm{m}}\right)(0.02 \; \mathrm{cm})}{0.03}\right) \\ \times \, (2.65 - 1)\end{array}}$$

$$= 18.6 \; \mathrm{cm/s}$$

To determine the minimum allowable horizontal velocity, solve Eq. 8.1 for the scour velocity of the organics. Since the organic material is not sticky and remains as discrete particles, use a k value of 0.04.

$$\mathrm{v} = \sqrt{8k\left(\frac{gd_p}{f}\right)(\mathrm{SG}_p - 1)}$$

$$= \sqrt{\begin{array}{c}(8)(0.04)\left(\dfrac{\left(9.81 \; \frac{\mathrm{m}}{\mathrm{s}^2}\right)\left(100 \; \frac{\mathrm{cm}}{\mathrm{m}}\right)(0.03 \; \mathrm{cm})}{0.03}\right) \\ \times \, (1.30 - 1)\end{array}}$$

$$= 9.7 \; \mathrm{cm/s}$$

Based on this analysis, the flow through the horizontal grit chamber must have a horizontal velocity less than 18.6 cm/s, but greater than 9.7 cm/s to allow the grit to settle out in the grit chamber while the organics remain suspended and travel to the primary clarifiers. Finally, operating the grit chamber closer to the maximum allowable velocity (18.6 cm/s) will produce a "clean" grit, meaning that it contains little organic material.

7. PRIMARY AND SECONDARY TREATMENT PROCESSES

Clarification Basin

Primary and secondary wastewater treatment systems typically use clarification basins. Clarification and sedimentation processes are used to remove solid materials that have been made to form aggregates through flocculation processes. There are four types of *settling regimes*.

- type I: *Discrete settling*, in which sand or grit-like materials are settled out. The materials readily settle out because of their high density. There is no change in the density or shape of the particles (i.e., they settle out individually).

- type II: *Flocculant* or *uniform settling*, in which the particles change shape and density through aggregation processes. This may be likened to two food particles sticking together. The settled particles tend to be more concentrated than discretely settled particles.

- type III: *Zone settling* or *thickening*, in which a distinct water-solid interface exists. An example of this type of settling is sugar settling to the bottom of a water glass. Secondary clarification in a wastewater treatment train uses zone settling.

- type IV: *Compression settling*, in which the weight of overlying solid material compresses the already settled out materials, forming a relatively dense sludge.

Primary and secondary clarifiers have sloped bottoms to facilitate sludge removal. The depth stated for these basins is most commonly the depth at the basin wall (side). The design of clarification processes is primarily based on the surface loading rate (also called the overflow rate). The *hydraulic (surface) loading rate*, q_o, is calculated in terms of gallons per day-square foot (cubic meters per day·square meter) using Eq. 8.2.

$$q_o = \frac{Q}{A_s} \qquad 8.2$$

Detention time, t_d, also known as *hydraulic resistance time*, is given by Eq. 8.3.

$$t_d = \frac{V}{Q} \qquad 8.3$$

The *weir loading rate*, w_r, is given in gallons per day-foot (cubic meters per day·meter) by Eq. 8.4.

$$w_r = \frac{Q}{L} \qquad 8.4$$

of weir, not tank!

L is the length of the weir, equal to the circumference for circular weirs. Two additional design parameters are the solids flux, G, and the settling velocity, v_s, of the solid materials to be settled out. The *solids flux* is essentially the mass flow of solids through the clarifier per unit area, per unit time.

$$G = q_o X + v_u X \qquad 8.5$$

q_o is the overflow rate for the basin (the critical settling velocity), with units of ft/hr or m/h. X is the biomass, or suspended solids concentration, entering the basin, with units of mg/L. v_u is the bulk downward velocity resulting from the underflow, with units of ft/hr or m/h.

In general, a G value is selected to produce a given or desired biomass recycle concentration, X_u. The settling velocity, v_s, is a function of the solids (density, size, mass) to be removed. The area, A_s, is typically determined experimentally in bench tests. Typical design criteria, including the surface overflow rate, $q_o = Q/A_s$, for rectangular clarification basins are reported in Table 8.8. Values are dependent on the type of settling regime that has been determined for the sedimentation basin.

Table 8.8 *Typical Value Ranges for the Design of Rectangular Wastewater Clarification Basins*

design criteria	settling regime	
	type I	type II
q_o (m³/d·m²)	1.0–2.5	0.6–1.0
d_o (m)	2.5–3.0	3.0–4.0
t_d (h)	2–4	4–6
v_h (m/h)	< 36	< 9
q_w (m³/m·h)	14	6

Chemically Assisted Clarification (Sedimentation Processes)

When gravity-driven clarification is not sufficient to remove suspended solids, chemicals or flocculants may be added to the water flow just before it reaches the sedimentation basin. Flocculants aggregate the suspended particles, increasing their size and mass, which increases their settling velocity. Sedimentation basins for wastewater treatment tend to be rectangular rather than circular. A schematic of a typical rectangular sedimentation basin is shown in Fig. 8.7. Figure 8.8 describes the processes and operational scheme in a sedimentation basin.

d_o is the settling zone depth, given in feet or meters. d is the settling depth, or the total depth of the basin, which includes d_o plus the depth of sludge in the basin, all given in feet or meters. v_s is the settling velocity, and v_h is the horizontal velocity of the settling particle. Typical accepted ratios for the dimensions of sedimentation basins range from 2:1 to 4:1 for length:width ($L{:}W$) and from 10:1 to 20:1 for length:depth ($L{:}d$).

The settling zone is the functional zone in the sedimentation basin, the area in which the particle(s) of interest will settle out of suspension. The inlet zone is the area in

Figure 8.7 *General Design of a Rectangular Sedimentation Basin*

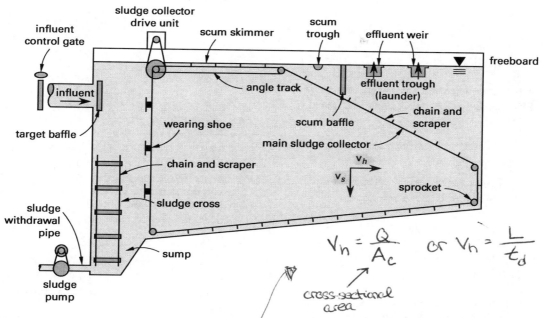

is necessary to minimize the sizes of both the inlet and outlet zones. A large water surface area tends to increase the mixing within the basin due to surface wind and the formation of temperature gradients in the tank.

Example 8.2

A wastewater treatment plant has two primary clarifiers, or settling basins, which receive waste-activated sludge.

Typical design information for two different primary clarifier operational schemes, followed by secondary treatment and receiving waste activated sludge, are given in the table shown.

	primary clarifier followed by secondary treatment	primary clarifier with waste activated sludge return
q_o (m³/d·m²)		
average flow	32.56–48.84	24.42–32.56
peak hourly flow	81.40–122.10	48.84–69.19
t_d (h)	1.5–2.5	1.5–2.5
	(2.0 typical)	(2.0 typical)

The clarifiers are circular, and each has a diameter of 40 ft and a depth of 10 ft. An outlet weir spans the entire periphery. The average influent flow rate is 2 MGD. Calculate the surface loading rate, the hydraulic detention time, and the outlet weir loading rate.

Solution

Because each of the settling basins has the same dimensions, the influent flow rate will be evenly split between the two. The flow going into one basin and all subsequent parameters for that single basin will apply to

Figure 8.8 *Processes and Operational Scheme in a Sedimentation Basin*

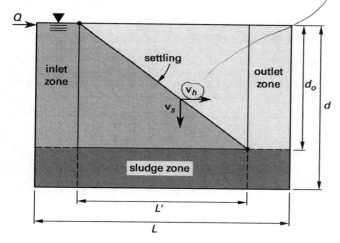

which the water first enters the sedimentation basin, while the outlet zone is the area from which the clarified sewerage exits the sedimentation basin, typically through a weir device. Design equations are similar to those used for clarification basins.

The sludge depth reduces the depth of the functional sedimentation zone of the basin. L' is the length of the functional sedimentation zone. Only the length and depth of the functional zone are used for determining the cross-sectional area of the rectangular basin, A_s, as shown in Eq. 8.6.

$$A_s = L'd_o \qquad 8.6$$

The parameters of q_o, d_o, and t_d for circular sedimentation tanks and clarifiers are comparable to, but not the same as, those values given in Table 8.8. Typically the maximum diameter for a circular tank is 98 ft (30 m). It

both. Calculate the flow rate entering one basin as follows.

$$Q_{\text{basin}} = \frac{Q_{\text{total}}}{\text{number of reactors}} = \frac{2 \text{ MGD}}{2}$$
$$= 1 \text{ MGD}$$

It is important to remember that, while the average flow rate is used in this design, most designs will be based on the peak flow rate.

Using Eq. 8.2, calculate the hydraulic or surface loading rate.

$$A_s = \pi r^2 = \pi (20 \text{ ft})^2$$
$$= 1257 \text{ ft}^2$$

$$q_o = \frac{Q}{A_s} = \frac{1,000,000 \, \dfrac{\text{gal}}{\text{day}}}{1257 \text{ ft}^2}$$
$$= 796 \text{ gal/day-ft}^2$$

The surface loading rate is greater than typical surface loading rates for primary clarifiers that receive waste-activated sludge (785 gal/day-ft^2, at the high end). Therefore, the clarifiers would not likely be able to accept any additional flow (average flow), and the plant would potentially require an additional primary clarifier if more capacity is needed.

Use Eq. 8.3 to calculate the hydraulic detention time.

$$V = A_s d = (1257 \text{ ft}^2)(10 \text{ ft})$$
$$= 12,570 \text{ ft}^3$$

$$t_d = \frac{V}{Q} = \frac{(12,570 \text{ ft}^3)\left(24 \, \dfrac{\text{hr}}{\text{day}}\right)\left(7.48 \, \dfrac{\text{gal}}{\text{ft}^3}\right)}{1,000,000 \, \dfrac{\text{gal}}{\text{day}}}$$

$$= 2.26 \text{ hr}$$

The actual detention time in a clarifier is affected by eddy, wind, thermal currents, and dead spaces. In most cases, fluids pass through the clarifier in less time than the theoretical detention time. Dead spaces and eddy currents have rotational flow. They contribute very little to sedimentation because their inflow and outflow is very small. As a result, the net volume available for settling is reduced and the mean flow-through time for the fluid element is decreased. The magnitudes of the effects of the dead spaces, thermal currents, and the hydraulic characteristics of a basin may be measured using tracer studies. The weir length is the perimeter of the circular basin, because the weir extends around the total basin.

$$L = 2\pi r = (2\pi)(20 \text{ ft})$$
$$= 126 \text{ ft}$$

Using Eq. 8.4, calculate the weir loading rate.

$$w_r = \frac{Q}{L} = \frac{\left(1 \, \dfrac{\text{MG}}{\text{day}}\right)\left(10^6 \, \dfrac{\text{gal}}{\text{MG}}\right)}{126 \text{ ft}}$$
$$= 7937 \text{ gal/ft-day}$$

The weir loading rate is lower than a typical minimum value of 10,000 gal/ft-day. While the calculated values may be acceptable, they do suggest that the design could be improved. For example, reducing the clarifier diameter will increase the weir loading rate. Furthermore, increasing the depth of the clarifier from 10 ft to 12 ft will compensate for the reduced diameter and may increase the detention time.

Example 8.3

A municipality serves a population of 25,000, having a per capita wastewater flow of 120 gal/day. The per capita BOD contribution is 0.20 lbm/day. Determine the volume for a step-aeration process with maximum volumetric BOD loading of 40 lbm BOD/1000 ft^3-day and a minimum aeration period of 6.0 hr. Assume that 35% of the BOD$_5$ is removed in the primary clarifier. Using these values, calculate the required dimensions for two circular final clarifiers.

Solution

Calculate the influent flow rate and the BOD loading rate for the primary clarifier.

$$Q = N Q_{\text{capita}}$$

$$= \frac{(25,000 \text{ persons})\left(120 \, \dfrac{\text{gal}}{\text{person-day}}\right)}{1,000,000 \, \dfrac{\text{gal}}{\text{MG}}}$$

$$= 3 \text{ MGD}$$

$$\text{BOD}_{\text{loading rate}} = N \text{BOD}_{\text{capita}}$$

$$= (25,000 \text{ people})\left(0.20 \, \dfrac{\text{lbm}}{\text{person-day}}\right)$$

$$= 5000 \text{ lbm/day}$$

Calculate the BOD loading rate to the aeration basin following the primary clarifier, assuming that 35% of the BOD$_5$ is removed in the primary clarifier.

$$\text{BOD}_{\text{basin loading rate}} = (1 - R)\text{BOD}_{\text{clarifier loading rate}}$$

$$= (1 - 0.35)\left(5000 \, \dfrac{\text{lbm}}{\text{day}}\right)$$

$$= 3250 \text{ lbm/day}$$

Calculate the necessary aeration basin volume. This depends on the volumetric BOD loading rate or on the aeration time period, whichever requires the larger volume. Therefore, calculate the volume based on the volumetric BOD loading rate or on the aeration time period.

$$V_{\text{BOD loading}} = \frac{\text{BOD}_{\text{basin loading rate}}}{\text{BOD}_{\text{max volumetric loading}}}$$

$$= \frac{3250 \; \dfrac{\text{lbm}}{\text{day}}}{\dfrac{40 \; \text{lbm}}{1000 \; \dfrac{\text{ft}^3}{\text{day}}}}$$

$$= 81{,}250 \; \text{ft}^3$$

$$V_{\text{aeration time period}} = Q t_{\text{aeration}}$$

$$= \frac{\left(3 \; \dfrac{\text{MG}}{\text{day}}\right)\left(10^6 \; \dfrac{\text{gal}}{\text{MG}}\right)(6 \; \text{hr})}{\left(7.48 \; \dfrac{\text{gal}}{\text{ft}^3}\right)\left(24 \; \dfrac{\text{hr}}{\text{day}}\right)}$$

$$= 100{,}267 \; \text{ft}^3$$

The volume required for the aeration time period, 100,267 ft³, is the larger of the two values and controls the design.

Calculate the circular basin dimensions. Use a typical average overflow rate value for an activated sludge plant with an influent flow rate greater than 1 MGD (800 gpd/ft²).

$$q_o = \frac{Q}{A_s}$$

$$A_s = \frac{Q}{q_o}$$

$$= \frac{3{,}000{,}000 \; \dfrac{\text{gal}}{\text{day}}}{800 \; \dfrac{\text{gal}}{\text{day-ft}^2}}$$

$$= 3750 \; \text{ft}^2$$

Because the design stipulates the use of two clarifiers, split the total surface area between two equally sized basins. Therefore, the surface area per clarifier is 1875 ft². Now solve for the basins' diameters.

$$A = \pi r^2$$

$$1875 \; \text{ft}^2 = \pi r^2$$

$$r = 24.43 \; \text{ft} \quad (25 \; \text{ft})$$

$$d = 50 \; \text{ft}$$

Use a side water depth of 11 ft, and calculate the detention time for the basins. Split the feed flow rate between the two basins.

$$Q = \frac{3 \; \dfrac{\text{MG}}{\text{day}}}{2}$$

$$= 1.5 \; \text{MG/day}$$

Equation 8.3 gives the detention time for each basin.

$$t_d = \frac{V}{Q}$$

$$= \frac{(1875 \; \text{ft}^2)(11 \; \text{ft})\left(7.48 \; \dfrac{\text{gal}}{\text{ft}^3}\right)}{\left(1.5 \; \dfrac{\text{MG}}{\text{day}}\right)\left(10^6 \; \dfrac{\text{gal}}{\text{MG}}\right)}$$

$$= 0.10 \; \text{day}$$

Example 8.4

Determine the basin volume for two rectangular primary sedimentation basins, with an average daily influent flow of 15 000 m³/d, a depth of 3.5 m, a maximum surface overflow rate of 60 m³/d·m², and a minimum hydraulic detention time of 1 h.

Solution

Use Eq. 8.6 to solve for the total basin surface area.

$$t_d = \frac{V}{Q}$$

$$= \frac{A_s d}{Q}$$

$$1 \; \text{h} = \frac{A_s(3.5 \; \text{m})\left(24 \; \dfrac{\text{h}}{\text{d}}\right)}{15\,000 \; \dfrac{\text{m}^3}{\text{d}}}$$

$$A_s = 179 \; \text{m}^2$$

Solve for the total basin surface area.

$$q_o = \frac{Q}{A_s}$$

$$60 \; \frac{\text{m}^3}{\text{d·m}^2} = \frac{15\,000 \; \dfrac{\text{m}^3}{\text{d}}}{A_s}$$

$$A_s = 250 \; \text{m}^2$$

The surface area required by the surface overflow rate is larger than the surface area determined from the minimum detention time, and therefore controls the basin design ($A_s = 250 \; \text{m}^2 > 179 \; \text{m}^2$).

Wastewater

Calculate the volume of each basin from the surface area and depth.

$$V_{\text{total}} = A_s d = (250 \text{ m}^2)(3.5 \text{ m})$$
$$= 875 \text{ m}^3$$

Per *Ten States Standards*, each basin is to take half of the average daily flow.

$$V_{\text{basin}} = \frac{V_{\text{total}}}{\text{number of basins}} = \frac{875 \text{ m}^3}{2}$$
$$= 438 \text{ m}^3$$

Biotreatment Reactors

There are four different types of *biotreatment reactors* in wastewater plants. The reactors vary according to the type of flow regime and the conditions within the tank.

1. *Packed bed:* There are two types of packed bed reactors, *fluidized* and *expanded*. In a plug flow reactor, the liquid flows through the reactor as a plug. The stream flows at the same velocity, parallel to the reactor axis with no back mixing and with a uniform residence time. The residence time is equal to the theoretical detention time. The longitudinal position within the reactor is proportional to the time spent within the reactor. Because all of the substrate has the same residence time within the reactor, it all has an equal opportunity for reaction. With respect to reaction time, a plug flow reactor operates like a well-stirred batch reactor. Each volume element behaves as a batch reactor as it passes through the reactor. Any required degree of reaction may be achieved by the use of a plug flow reactor of suitable length and flow rate.

2. *Plug flow reactors* (PFRs): These types of reactors are also known as *continuous flow reactors*. Here there is a continuous influent and effluent from the reactor, and the influent equals the effluent.

3. *Batch reactor:* Batch reactors are filled, a reaction occurs, and then they are emptied. There is no continuous influent or effluent into or out of a batch reactor.

4. *Completely stirred tank reactor* (CSTR) or *complete mix reactor:* Materials entering a completely stirred tank reactor are immediately dispersed homogeneously throughout the reactor volume. The concentration of materials in the effluent is equal to their concentration in the CSTR. The reactor contents exit it in proportion to their statistical population in the reactor. CSTRs are usually circular to facilitate mixing and reduce stagnation zones.

More details about CSTRs and PFRs are discussed in Sec. 9.6 and Sec. 9.11. Activated sludge is the subject of Chap. 9.

8. TERTIARY TREATMENT PROCESSES

Filtration

In wastewater treatment, the purpose of filtration processes is to remove flocculated materials and other solids that were not removed in the secondary clarification process. Filtration processes can take on an array of different designs and operation schemes. *Conventional granular media filtration* processes follow the sedimentation basin in which most large materials have been removed via settling processes. In *direct filtration*, the filtration processes directly follow the flocculation processes. Therefore, in direct filtration, the larger aggregated materials have not been removed prior to the filter.

Filters may be composed of different types of granular media and may be constructed of either uniform media or multimedia. *Uniform media filters* are generally composed of poorly graded sand, while *multimedia filters* may be composed of a media mixture of sand, anthracite coal, and carbon. Common examples of filter media include anthracite coal, beach sand, and garnet sand. The filter media are graded as part of the filter design. The grading allows for stratification within the filter (see Fig. 8.9), according to the densities of the different media (i.e., denser materials settle faster than less dense ones).

Figure 8.9 Schematic of a General Granular Media Filter Showing the Stratification of the Media

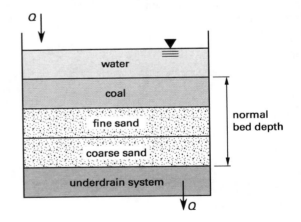

A variety of removal mechanisms are operating in granular media filters, with some mechanisms being more prominent in some filters than others, depending on the type of filter media that is used. The principal removal mechanisms include the following.

- *sieving/straining*—based mostly on size exclusion as the solid materials become trapped in small void spaces or pores in the filter

- *sedimentation*—particles settle out from the flowing water in the small filter pores

- *impaction*—particles collide with and remain on the filter media

- *adsorption*—particles adsorb onto the filter media via favorable or attractive chemical interactions between the two surfaces

Roughing Filters

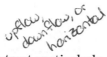
[handwritten: upflow, downflow, or horizontal]

Roughing filters are intended to treat particularly strong or variable organic loads, typically industrial. They are usually tall, circular filters filled with open synthetic filter media to which sewerage is applied at a relatively high rate. The design of the filters allows high hydraulic loading and a rapid flow of air. On larger installations, air is forced through the media using blowers. The resultant mixed liquor is usually within the normal range for conventional treatment processes.

Trickling Filters

In older plants and plants receiving more variable loads, *trickling filter beds* are used. The settled sewerage is spread onto the surface of a deep trickling filter bed made up of coke, limestone chips, or specially fabricated plastic media. Such media must have large amounts of surface area to support the biofilms that form. The mixed liquor is distributed through perforated rotating arms radiating from a central pivot. The distributed mixed liquor trickles through this bed and is collected in drains at the base. These drains also provide a source of air, which percolates up through the bed, keeping it aerobic. Biological films of bacteria, protozoa, and fungi form on the media's surfaces, consuming the organic content. Trickling filter design criteria are reported in Table 8.9.

Table 8.9 *Typical Design and Performance Characteristics of Trickling Filters*

filter type	design BOD$_5$ loading rate	BOD$_5$ removal (%)
low		
(lbm/1000 ft^3-day)	< 25	80–90
(kg/100 m^3·d)	< 40	
intermediate		
(lbm/1000 ft^3-day)	40	50–70
(kg/100 m^3·d)	64	
high		
(lbm/1000 ft^3-day)	40–1000	65–85
(kg/100 m^3·d)	64–160	
roughing		
(lbm/1000 ft^3-day)	100–300	40–65
(kg/100 m^3·d)	160–180	

Filter Cleaning

A typical filter run will last from several hours to several days before the flow through the filter becomes too low or the effluent degrades to a minimum water quality level. Filter effluent quality is generally assessed in terms of turbidity, particle counts, or another parameter. Once the specified threshold is reached, the filter is cleaned through a backwashing process. During backwashing, water is flushed rapidly through the filter to remove the solid material that has become trapped in the filter media. As the water is backwashed through the filter, the total bed depth expands. The extended bed depth is typically 1.25 to 1.75 times deeper than the unexpanded bed depth. The floc and other entrapped materials are less dense than the filter material. During the backwash, the floc and trapped matter reach a higher point in the water column and overflow into a weir in the filter basin. After a specified time, the backwash is turned off and the filter media again resettles into a stratified formation.

Ammonia Removal by Air Stripping

[handwritten: ammonia eutrophication (p. 6-6)]

Ammonia stripping is a desorption process that can be used to remove ammonia from wastewater streams. An example of an ammonia air stripping tower is shown in Fig. 8.10. In ammonia stripping, the ammonia, which is a weak base, reacts with water (an acid relative to ammonia) to form ammonium hydroxide. Lime or another caustic substance is added to the water to raise the pH to 10.8–11.5, converting the ammonium hydroxide ions to ammonia gas (see Eq. 8.7).

$$NH_4^+ + OH^- \rightleftharpoons H_2O + NH_3 \qquad 8.7$$

Ammonia stripping towers are designed to use either cross-flow or countercurrent flow. In a *cross-flow* tower, air enters along the entire depth of the packing media, as the wastewater flows down through it. In a *countercurrent* design, air is drawn through openings at the bottom of the tower as wastewater is pumped through the packing media to the top of the tower. In both designs, the packing media is either wood or plastic. Free ammonia is stripped from the percolated water droplets into the air streams and then discharged to the atmosphere. The effluent ammonia concentration is typically less than 6 mg per cubic meter of air. However, local air regulations must be considered when using ammonia air stripping as a treatment process. Ammonia stripping is most applicable to wastewaters with ammonia contents between 10 mg/L and 100 mg/L. Steam stripping and biological methods (nitrification/denitrification) are more appropriate for treating waters with ammonia contents greater than 100 mg/L. Design criteria for ammonia air stripping towers are shown in Table 8.10.

The amounts of nitrogen and phosphorus removed increase with the levels of treatment (see Table 8.11). Raw wastewater nutrient levels must be reduced to prevent negative impacts (such as eutrophication, see Sec. 6.14) on the receiving water body.

Figure 8.10 *Cross-flow and Countercurrent Designs for Ammonia Air Stripping Towers*

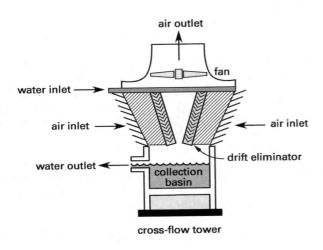

cross-flow tower

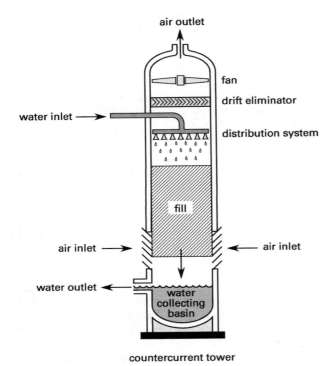

countercurrent tower

Table 8.10 *Design Criteria for Ammonia Air Stripping Towers*

criteria	desirable range
hydraulic loading rate	
(gal/min-ft^2)	1–2
(L/m^3·min)	0.1–0.2
stripping air flow rate	
(ft^3/min-gal)	300–500
(L/m^3·min)	32–54
media packing depth	
(ft)	20–25
(m)	6.1–7.6
pH	10.8–11.5
air pressure drop in water	0.015–0.019
(in per ft of packing media)	

Table 8.11 *Typical Concentrations of Nitrogen and Phosphorus in Various Stages of Treatment*

treatment	total nitrogen (mg/L)	total phosphorus (mg/L)
no treatment (raw wastewater)	15–100 (typically 50)	4–15 (typically 10)
primary treatment	40	7
secondary treatment	25–30	6
biological N removal	5–8	–
biological P removal	–	< 1

Nitrification

Nitrification is a biological process in which ammonia (NH_4^+) is oxidized into nitrite (NO_2^-) and then into nitrate (NO_3^-) (see Fig. 8.11). Ammonia is converted into nitrite by the bacteria *Nitrosomonas*. Nitrite is then converted into nitrate by the bacteria *Nitrobacter*.

$$2NH_4^+ + 3O_2 \rightarrow 2NO_2^- + 4H^+ + 2H_2O \qquad 8.8$$

$$2NO_2^- + O_2 \rightarrow 2NO_3^- \qquad 8.9$$

Nitrification requires an aerobic environment. Nitrification can be achieved in an aerobic biological process at low organic loadings and long hydraulic retention times. Nitrification consumes a large amount of oxygen while also reducing alkalinity (7.2 unit mass alkalinity per unit mass of ammonia oxidized). Key design parameters for nitrification rates are DO concentration, temperature, pH, organic and hydraulic loading rates, and the type of filter media used (if any).

During nitrification, approximately 4.6 unit mass of oxygen is required for every unit mass of ammonia that is oxidized. Nitrification rates are limited at DO concentrations of 0.5 mg/L in suspended growth systems[1] and at DO concentrations of 2.5 mg/L in attached growth systems[2] under steady state conditions. Nitrification rates also depend on the degrees of mass-transport and diffusion resistance, and on the solids retention time. Maximum nitrification rates in attached growth systems occur at DO concentrations between 2 mg/L and 2.5 mg/L. Maximum nitrification rates may not be necessary when there is sufficient contact time available for slower nitrification rates to be sufficient.

---

[1]Suspended growth treatment systems freely suspend microorganisms in water (e.g., activated sludge systems). They use biological treatment processes in which microorganisms are maintained in suspension within the liquid.

[2]In attached growth systems, the microorganisms or biofilm are attached to a growth media. In this configuration, the wastewater will typically flow over the growth media.

Figure 8.11 *Biological Oxidation of Nitrogen (Nitrification) in Wastewater Treatment*

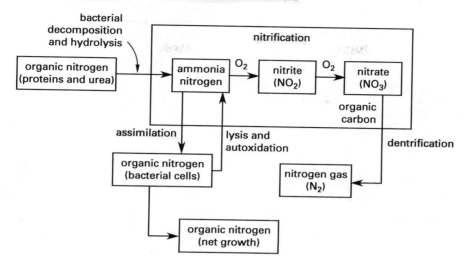

Nitrification rates increase with increasing temperature. Nitrification occurs over a temperature range of 4°C to 45°C. At mixed liquor temperatures below 5°C, the nitrification rate will approach zero. The nitrification rate's dependence on temperature increases with temperature. In nitrification processes, the optimum mixed liquor pH is between 6.5 and 8, though nitrification can occur outside this range. Acidic pH reduces the rate of ammonium oxidation. Therefore, nitrification rates drop as pH drops below 7. pH must be carefully controlled, as the process consumes a significant amount of alkalinity. During nitrification the wastewater is typically introduced at the top of an attached growth reactor. The wastewater then trickles down through the media or trickling filters. The minimum hydraulic time must be set to ensure that all of the media is contacted by the wastewater. Hydraulic and organic loading rates are interdependent, as the raw water influent conditions may be variable. Recirculation in the trickling filters may be used to control the hydraulic flow rate entering the filters and to reduce the influent BOD to the filters. Nitrification efficiency increases with decreased influent BOD. Current nitrification processes used in wastewater treatment include trickling filters, rotating biological contactors, contact reactors, conventional activated sludge processes at low loadings, and two-stage activated sludge systems with separate carbonaceous oxidation and nitrification systems. Loading rates for the different filter media types are reported in Table 8.12.

As for nitrification rates in general, nitrification in trickling filters is dependent on temperature, pH, loading rate, and the wastewater BOD_5. It also depends on the amount of dissolved oxygen, the presence of inhibitors, the filter depth, and the media type. For trickling filters, the BOD_5 organic loading rate should not be greater than 10 lbm/1000 ft^3-day (160 g/m^3·d) Plastic media trickling filters can operate at higher BOD_5 loading rates of up to 23 lbm/1000 ft^3-day (360 g/m^3·d) because

Table 8.12 *Typical Loading Rates for Single Stage Nitrification*

process	nitrification (%)	BOD organic loading rate lbm/1000 ft^3-day	g/m^3·d
trickling filter,	75–85	10–6	160–96
rock medium	85–95	6–3	96–48
trickling filter,	75–85	181–12	288–192
plastic medium	85–95	12–6	192–96

Reprinted with permission from G. Tchobanoglous, F. L. Burton, and H. D. Stensel, and the staff of Metcalf & Eddy, Inc. *Wastewater Engineering: Treatment, Disposal, Reuse.* 3rd ed., copyright © 1991 by the McGraw-Hill Companies.

the plastic media have greater surface area. If two trickling filters are used, then heterotrophic growth occurs in the first filter and nitrification occurs in the second one.[3]

In a *rotating biological contactor* (RBC), the contactor surface supports a biofilm, approximately 1 mm to 4 mm thick, which is responsible for BOD removal. The rotating disks, or media supports, are mounted on a rotating shaft, and provide a large surface area for biomass growth (see Fig. 8.12). The rotating packs of disks, known as the media, are contained in a tank or trough. Polythene, PVC, and expanded polystyrene are commonly used plastics. The shaft is aligned with the flow of wastewater so that the discs rotate at right angles to the flow, with several packs usually combined to make up a treatment train. About 35% to 40% of the disc area is immersed in the wastewater. The first stages of an RBC remove mostly organic materials. Subsequent stages remove ammonia nitrogen (NH_3-N) that is present in the wastewater as a result of nitrification, when the BOD_5 is low enough (< 14 mg/L). RBC performance is negatively affected by low dissolved oxygen in the first stages (BOD removal) and by low pH in the later stages (nitrification). Recirculation in an RBC can improve nitrification by diluting the influent

[3]During heterotrophic growth, carbon is used as the food source.

Figure 8.12 Configuration of a Rotating Biological Contactor (front view)

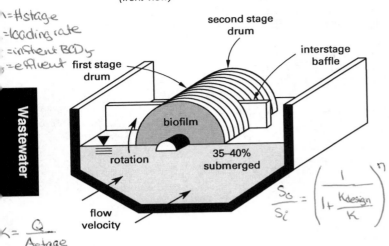

Handwritten notes:
η=#stage
=loading rate
=influent BOD₅
=effluent

$\dfrac{S_o}{S_i} = \left(\dfrac{1}{1 + \dfrac{K_{design}}{K}} \right)^\eta$

$K = \dfrac{Q}{A_{stage}}$

organic carbon. Degradable organic carbon inhibits nitrification at BOD_5 concentrations greater than 15 mg/L to 20 mg/L.

Denitrification

Denitrification is a process in which denitrifying bacteria convert nitrate into nitrogen gas (N_2) by oxidizing an organic substrate using nitrate as an electron acceptor. Denitrification processes follow nitrification in most wastewater treatment schemes (see Fig. 8.13). The organic substrate is supplied by the wastewater. Denitrifying bacteria require an anoxic environment (DO < 0.2 mg/L). The requirements for the denitrification process are the following.

- nitrogen present in the form of nitrates
- an organic carbon source
- an anoxic environment

Because much of the BOD in the wastewater has been removed by the secondary treatment process, additional carbons are typically needed to serve as electron donors for the denitrifying bacteria. Methanol or acetate is typically used for this purpose. The amount of added methanol or acetate is carefully controlled so that it doesn't leave residual BOD in the effluent. The mechanisms used

for biological denitrification are continuous-flow stirred reactors, activated sludge systems, fixed film reactors, and fluidized bed biofilm reactors. For reactions using methanol as the carbon source, the chemical equation is

$$5CH_3OH + 6NO_3^- \rightarrow 3N_2 + 5CO_2 + 7H_2O + 6OH^-$$

8.10

The reaction using acetate is

$$5CH_3COOH + 8NO_3^- \rightarrow 4N_2 + 10CO_2 + 6H_2O + 8OH^-$$

8.11

Phosphorus Removal

Phosphorus may be removed through chemical or biological (activated sludge) processes. Phosphorus can be chemically precipitated out using alum, $Al_2(SO_4)_3$; lime, $Ca(OH)_2$; or ferric chloride, $FeCl_3$. Each of these chemicals precipitates phosphorus as a phosphorus salt (e.g., $AlPO_4$). This precipitate settles out in a clarifier and is collected as sludge. Alum and ferric chloride also form hydroxides in water, $Al(OH)_3$ and $Fe(OH)_3$, and are capable of adsorbing phosphorus to form settleable flocs. Precipitating chemicals are typically added after the activated sludge process, allowing the phosphorus to be removed in the secondary clarifier. However, the resulting chemical sludge is difficult to dispose. The use of chemicals in the treatment process is expensive and makes operation difficult, and often messy.

Enhanced biological phosphorus removal is a biological phosphorus removal process. In this process, specific bacteria called *polyphosphate accumulating organisms* (PAO) are selectively enriched and accumulate large quantities of phosphorus within their cells. When the biomass enriched in these bacteria is separated from the treated water, the bacterial biosolids have a high fertilizer value.

Effluent Polishing

Sand filtration removes much of the residual suspended matter. Filtration over activated carbon removes residual toxins and suspended matter. Wastewater treatment filtration systems are operated in a similar fashion as drinking water treatment filters. The granular media are stratified, and they are back-flushed for cleaning.

Figure 8.13 Schemes for Nitrification/Denitrification in Wastewater Treatment

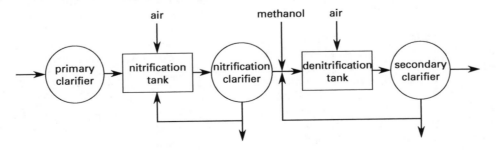

Storage in large artificial ponds or lagoons provides settlement and further biological improvement. These lagoons are highly aerobic, and colonization by native macrophytes, especially reeds, is often encouraged. Small filter-feeding invertebrates such as *Daphnia* and species of *Rotifera* greatly assist in treatment by removing fine particulates.

Constructed wetlands include engineered reed beds and a range of similar methods, all of which provide a high degree of aerobic biological improvement. Such methods can often be used instead of secondary treatment for small communities (see phytoremediation in Sec. 11.18).

Any number of membrane processes may be used in polishing steps in wastewater treatment, depending upon the material to be removed from the effluent stream. Membrane processes are divided according to the *molecular weight cutoff* (MWCO) of the membrane. The following categories of membranes are relevant to wastewater treatment and are listed in order of decreasing MWCO.

- ultrafiltration (UF)
- microfiltration (MF)
- nanofiltration (NF)
- reverse osmosis (RO)

Both UF and MF are generally employed for the removal of suspended materials. NF and RO are used for the removal of dissolved materials (TDS), pathogens, and viruses. Membrane filtration processes are for separation only. Such processes generate a reject stream (concentrate stream) that requires disposal or storage in large artificial ponds or lagoons, or the generated streams may be sent back to the plant head works.

Disinfection of Wastewater Effluent

Prior to discharge of the treated wastewater effluent, it must be disinfected to remove pathogens. The principles and strategies of wastewater disinfection are similar to those outlined in the drinking water section (see Sec. 5.9). Pathogens are destroyed or deactivated, resulting in the termination of their growth and reproduction.

Disinfection can be accomplished using either physical or chemical disinfectants. In water treatment, the disinfectants used must also have a residual effect, which means that they remain active in the water after disinfection (i.e., in the distribution system). The residual disinfectant prevents the recontamination of the water via microorganisms that may be present in the distribution system plumbing and prevents the regrowth of microorganisms that were not wholly killed during the initial disinfection process.

Of the various disinfectant methods discussed in Sec. 5.9, chlorination is the most popular because it is a well-established technology that effectively destroys target organisms through oxidation of the cellular wall. Chlorine can be supplied in many forms, which include chlorine gas, hypochlorite solutions, and other chlorine compounds in solid or liquid form. Chlorination also improves water quality by reducing the concentrations of nitrogen, iron, manganese, sulfide, and certain organic species. In the treatment of wastewater, the effectiveness of the chlorination process is affected by a variety of water quality parameters (see Table 8.13).

Table 8.13 *Wastewater Characteristics Affecting the Performance of Chlorination Disinfection Processes*

wastewater characteristic	effects on chlorine disinfection
ammonia	forms chloramines when combined with chlorine
biochemical oxygen demand	the degree of interference depends on their functional groups and chemical structures
hardness, iron, nitrate	minor effects, if any
nitrite	reduces effectiveness of chlorine and results in the production of THMs*
pH	affects distribution between hypochlorous acid and hypochlorite ions and among various chloramine species
total suspended solids	shields embedded bacteria and increases chlorine demand

*Trihalomethanes (THMs) are a by-product of chlorination processes and are referred to as a disinfection by-product. They are formed when organic material in the wastewater reacts with chlorine. They are the most common type of disinfection by-product.

Adapted from *Wastewater Technology Fact Sheet—Chlorine Disinfection*, EPA, 1999.

Selection of the most appropriate disinfectant for a given utility is dependent on the following factors.

- ability to penetrate and destroy infectious agents under normal operating conditions
- safe and easy handling, storage, and shipping
- absence of toxic residuals and mutagenic or carcinogenic compounds after disinfection
- affordable capital and operation and maintenance (O&M) costs

Parameters Controlling Disinfectant Efficiency

The efficiency of a wastewater disinfection process is determined by a number of different parameters, including the contact time between the disinfectant and the microorganisms, the types and ages of microorganisms present, and the water characteristics. Materials in the water, such as iron, manganese, hydrogen sulfide, and nitrates, often react with disinfectants and reduce their ability to kill the pathogens. The turbidity of the water also affects the disinfection process. Highly turbid waters allow microorganisms to "hide" from chemical disinfectants. *Turbidity* can deplete the disinfectant by causing it to react with nonbiological particles rather than with organisms. For physical disinfection processes, water turbidity may reduce or prevent the passage of light sources and sound waves. Water temperature also influences process efficiency. Increasing the temperature usually increases the speed of reactions between the chemical disinfectant and the microorganisms. Increasing the temperature can also decrease disinfection because the disinfectant may degrade or become volatilized.

Dechlorination

Dechlorination is the process of removing residual chlorine from disinfected wastewater prior to discharge into the environment. This is necessary because chlorine residuals remain in the effluent for some time following disinfection. Sufficient amounts of such residuals in treated wastewater effluent may be harmful to aquatic organisms. Therefore, the residuals must be removed prior to discharge of the effluent into the receiving water body. A variety of methods and processes can be employed for dechlorinating treated wastewater, including chemical addition and activated carbon adsorption. Sulfur dioxide, sodium bisulfite, sodium metabisulfite, and hydrogen peroxide are the most commonly used dechlorinating chemicals. Of these chemicals, sulfur dioxide is most commonly used for dechlorination at public utilities. Sodium metabisulfite and sodium bisulfite are used primarily in small facilities, because they are more difficult to control than sulfur dioxide. Hydrogen peroxide is not frequently used because it is dangerous to handle. Carbon adsorption is an effective dechlorination method. However, it is expensive relative to the other methods. Carbon adsorption is usually used when all of the chlorine residual must be removed from the effluent.

PRACTICE PROBLEMS

1. In pretreatment processes, mechanically cleaned bar screens are used to remove which of the following?

(A) grit

(B) sand

(C) total solids

(D) large debris

2. Approximately how much BOD is typically removed during primary treatment processes in wastewater treatment plants?

(A) 10%

(B) 20%

(C) 30%

(D) > 50%

3. In nitrification processes, what is a typical oxygen requirement (in unit mass O_2 per unit mass NH_4 oxidized) to oxidize a unit mass of ammonia?

(A) 1.5 lbm

(B) 2.6 lbm

(C) 4.6 lbm

(D) 7.4 lbm

4. What is the weir loading rate for a circular primary clarifier followed by an activated sludge process with an influent flow rate of 200,000 gal/day and a total weir length of 8.0 ft?

(A) 15,000 gal/ft-day

(B) 25,000 gal/ft-day

(C) 45,000 gal/ft-day

(D) 55,000 gal/ft-day

5. What is the surface loading rate for a circular primary clarifier that is followed by secondary treatment? The average influent flow rate is 100,000 gal/day, and the diameter of the clarifier is 12 ft.

(A) 500 gal/day-ft^2

(B) 660 gal/day-ft^2

(C) 880 gal/day-ft^2

(D) 900 gal/day-ft^2

SOLUTIONS

1. Section 8.4 details the use and typical design of bar screens in wastewater treatment plants. Bar screens are typically metal bars, with bar spacing varying based on the material to be removed and the type of cleaning used. From Table 8.5, mechanically cleaned bars are typically spaced 0.2 in to 0.6 in apart. As water flows through them, they remove large debris that would otherwise harm downstream pumps and treatment equipment.

The answer is (D).

2. Referring to Sec. 8.4, primary treatment typically consists of physical processes, such as screening and sedimentation. During these processes, approximately 30% of the BOD load in the water is removed as gross (large) solids that settle out and are removed as sludge.

The answer is (C).

3. Section 8.8 details the typical operating parameters of nitrification processes with the primary goal of removing ammonia (NH_4) from the wastewater. During nitrification, approximately 4.6 lbm of oxygen are required for every pound mass of ammonia that is oxidized.

The answer is (C).

4. Using Eq. 8.4, calculate for the weir loading rate, w_r.

$$w_r = \frac{Q}{L} = \frac{200{,}000 \ \frac{\text{gal}}{\text{day}}}{8.0 \ \text{ft}}$$
$$= 25{,}000 \ \text{gal/ft-day}$$

The answer is (B).

5. The reactor surface area is

$$A_s = \pi r^2 = \pi (6 \ \text{ft})^2$$
$$= 113.1 \ \text{ft}^2$$

Using Eq. 8.2, the surface loading rate is

$$q_o = \frac{Q}{A_s} = \frac{100{,}000 \ \frac{\text{gal}}{\text{day}}}{113.1 \ \text{ft}^2}$$
$$= 884 \ \text{gal/day-ft}^2 \quad (880 \ \text{gal/day-ft}^2)$$

The answer is (C).

9 Activated Sludge

Nomenclature

BOD	biochemical oxygen demand	mg/L	mg/L
C_f	solute concentration in the feed water to a membrane	mg/L	mg/L
C_p	solute concentration in the product water from a membrane	mg/L	mg/L
D	dilution rate	day^{-1}	d^{-1}
DO	dissolved oxygen	mg/L	mg/L
f	BOD$_5$/BOD$_u$ ratio	–	–
F:M	food to microorganism ratio	lbm/day-lbm	kg/d
J	water flux through a membrane	gal/ft^2-day	m^3/m^2·d
k_d	decay rate	day^{-1}	d^{-1}
k_m	cell growth constant	mg/mg-day	mg/mg·d
k_{max}	rate of exponential growth of a bacterial culture	hr	h
K_I	inhibition coefficient	mg/L	mg/L
K_s	half-saturation coefficient	mg/L	mg/L
K_t	oxygen transfer coefficient	day^{-1}	d^{-1}
m	mass	lbm/day	kg/d
m	mass of microorganisms in reactor	lbm	kg

MLSS	mixed liquor suspended solids	mg/L	mg/L
MLVSS	mixed liquor volatile suspended solids	mg/L	mg/L
p	transmembrane pressure	lbf/ft^2	Pa
P_x	mass of sludge wasted	lbm/day	kg/d
Q	flow rate	gpd	m^3/d
Q_e	flow rate of effluent out of tank	gpd	m^3/d
Q_o	flow rate into activated sludge tank	gpd	m^3/d
Q_r	flow rate of waste recycled to the head of the aeration tank	gpd	m^3/d
Q_w	flow rate of waste from the recycle line	gpd	m^3/d
r_{O_2}	rate of oxygen utilization	lbm/day	kg/d
r_{Sg}	rate of energy utilization for synthesis	lbm/ft^3-day	kg/m^3·d
r_{Sm}	rate of energy utilization for maintenance	lbm/ft^3-day	kg/m^3·d
r_{Su}	rate of substrate utilization over a set period of time	lbm/ft^3-day	kg/m^3·d
r_X	rate of biomass growth	lbm/ft^3-day	kg/m^3·d
R	recycle ratio	%	%
R	solute rejection from membrane	%	%
R_m	membrane resistance	ft^{-1}	m^{-1}
s	gravimetric solids content	–	–
S	effluent substrate concentration	mg/L	mg/L
S_i	influent substrate concentration	mg/L	mg/L
S_o	initial substrate concentration in reactor	mg/L	mg/L
S^*	maximum substrate concentration	mg/L	mg/L
SDI	sludge density index	g/mL	g/mL
SG$_{sludge}$	specific gravity of sludge	–	–
SRT	solids retention time	day	d
SS	suspended solids	mg/L	mg/L
SVI	sludge volume index	g/ML	mL/g
t	time	day	d
t_h	hydraulic residence time in an ideal plug flow reactor	day	d
T	temperature	°F	°C
TSS	total suspended solids	mg/L	mg/L
U	specific substrate utilization parameter	day^{-1}	d^{-1}
V	reactor volume	ft^3	m^3
V_{air}	volume of air required to meet oxygen demand	ft^3	m^3

V_{settled}	relative volume of settled sludge	n.a.	mL/L or mg/g
W	waste ratio	–	–
X	biomass reactor solids	lbm/gal	kg/m^3
X_e	biomass in effluent stream	lbm/gal	kg/m^3
X_o	biomass in the reactor influent stream	lbm/gal	kg/m^3
X_r	biomass in recycle waste stream	lbm/gal	kg/m^3
X_w	biomass wasted from the recycle line	lbm/gal	kg/m^3
Y	true biomass yield	lbm/lbm	mg/mg
Y_{obs}	observed biomass yield	lbm/lbm	mg/mg

Symbols

β	oxygen saturation coefficient	–	–
η_{BOD}	BOD removal efficiency	–	–
η_{transfer}	gas transfer efficiency	–	–
θ	residency time of bacteria in reactor	day	d
θ_{air}	aeration period during treatment process	hr	h
θ_c	mean cell residence time	day	d
θ_{min}	minimum cell residence time, or washout time	day	d
μ	absolute (dynamic) viscosity of water	lbf-sec/ft^2	Pa·s
μ	specific growth rate coefficient	day^{-1}	d^{-1}
μ^*	maximum specific growth rate coefficient with inhibitory substrate	day^{-1}	d^{-1}
μ_{max}	maximal biomass growth rate coefficient	day^{-1}	d^{-1}
π	osmotic backpressure on the membrane	lbf/ft^2	Pa
ρ	density	lbm/ft^3	kg/m^3

Subscripts

5	5 day
f	feed
p	product
u	ultimate
w	water

1. SLUDGE

In wastewater treatment, the term *sludge* may be used to refer to a number of different entities. Generally, however, sludge describes a mixture of water, organic materials, and inorganic materials that have either been added to, or were already present in, the raw water prior to treatment. Sludge may form as a result of separation processes (filtration) or through the formation and subsequent settling of flocculated materials (chemically or biologically). For biologically mediated processes, sludge may be called *biosolids*, because it is an accumulation of bacteria that have degraded wastes suspended in the water and will ultimately settle out of suspension. Characteristics of sludge produced during different wastewater treatment processes are given in Table 9.1.

The routing of sludge through the stages of wastewater treatment is shown in Fig. 9.1.

2. ACTIVATED SLUDGE

Activated sludge plants use a variety of mechanisms and processes in which dissolved oxygen is consumed to generate a biological floc that removes organic material. *Secondary treatment* is designed to substantially degrade the organic content of the sewage derived from human waste, food waste, soaps, and detergent. To be effective, the biota require both oxygen and a substrate on which to live. The bacteria and protozoa consume biodegradable soluble organic contaminants (e.g., sugars, fats, organic short-chained carbon molecules, etc.), and bind much of the less soluble fractions into floc particles.

Secondary treatment systems are classified as fixed film or suspended growth. In *fixed film systems*, such as roughing filters, the biomass grows on media and the sewage passes over its surface. In *suspended growth systems*, such as activated sludge, the biomass is well mixed with the sewage. Typically, a fixed film system requires a smaller footprint than an equivalent suspended growth system. However, suspended growth systems are better able to cope with rapid changes in the organic loading rate.[1] Suspended growth systems also have higher removal rates for BOD and suspended solids than fixed film systems.

An aeration basin, shown in Fig. 9.2, is followed by a secondary clarifier in wastewater treatment. The secondary clarifier separates the remaining suspended solids to produce two streams: a clarified effluent that is sent to the next treatment process, and a liquid sludge. A portion of the sludge is returned to the aeration basin (*return-activated sludge*), while the remainder is wasted for disposal (*waste-activated sludge*).

Typical design criteria for an activated sludge system are reported in Table 9.2.

3. MIXED LIQUOR SUSPENDED SOLIDS

Mixed liquor is effluent from the primary clarifier and is introduced into the aeration basin and mixed with various microorganisms. *Mixed liquor suspended solids* (MLSS), the waste solids and biomass formed in this process, are given in units of milligrams per liter. The MLSS contains microorganisms, organic matter, water, and degradation products. MLSS concentrations in activated sludge typically range, and should be maintained, between 1000 mg/L and 4000 mg/L. Low concentrations (< 1000 mg/L) result in poor settling. High concentrations (> 4000 mg/L) result both in solids passing

---

[1]Rapid changes in system parameters, including temperature, pH, organic loading rate, and other constituents, is commonly referred to as "shocking" the system. The term *shock* refers to a sudden change in a system parameter that affects the system's biological activity.

Table 9.1 *Typical Physical Characteristics and Quantities of Sludge Produced from Various Wastewater-Treatment Operations and Processes*

treatment operation or process	specific gravity of solids	specific gravity of sludge	dry solids ($lbm/10^3$ gal) range	dry solids ($lbm/10^3$ gal) typical	dry solids ($kg/10^3$ m^3) range	dry solids ($kg/10^3$ m^3) typical
primary sedimentation	1.40	1.02	0.9–1.4	1.25	110–170	150
activated sludge (waste biosolids)	1.25	1.005	0.6–0.8	0.7	70–100[b]	80
trickling filter (waste biosolids)	1.45	1.025	0.5–0.8	0.6	60–100	70
extended aeration (waste biosolids)	1.30	1.015	0.7–1.0	0.8[a]	80–120	100[a]
aerated lagoon (waste biosolids)	1.30	1.01	0.7–1.0	0.8[a]	80–120	100[a]
filtration	1.20	1.005	0.1–0.2	0.15	12–24	20
algae removal	1.20	1.005	0.1–0.2	0.15	12–24	20
chemical addition to primary tanks for phosphorus removal						
low lime (350–500 mg/L)	1.90	1.04	2.0–3.3	2.5[b]	240–400	300[b]
high lime (800–1600 mg/L)	2.20	1.05	5.0–11.0	6.6[b]	600–1300	800[b]
suspended growth nitrification	–	–	–	–	–	–[c]
suspended growth denitrification	1.20	1.005	0.1–0.25	0.15	12–30	18
roughing filters	1.28	1.02	–	–[d]	–	–[d]

[a]assuming no primary treatment
[b]solids in addition to that normally removed by primary sedimentation
[c]negligible
[d]included in biosolids production from secondary treatment processes

Reprinted with permission from G. Tchobanoglous, F. L. Burton, and H. D. Stensel and the staff of Metcalf & Eddy, Inc. *Wastewater Engineering: Treatment, Disposal, Reuse*, 3rd ed., copyright © 1991, by The McGraw-Hill Book Companies.

Figure 9.1 *Generation, Treatment, and Disposal of Municipal Wastewater Sludge*

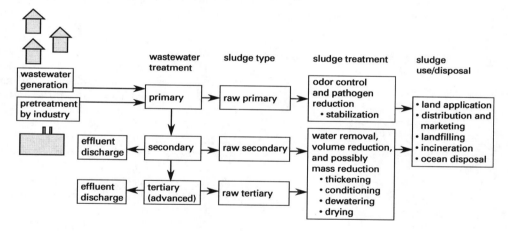

through the clarifier effluent and in excessive DO requirements.

Solids in the MLSS are fractionated according to their volatility. Nonvolatile or fixed solids remain after firing in a furnace at 1022°F (550°C) and are considered to be inert. Volatile solids are those that burn off at or below this temperature. The volatile fraction of the MLSS is designated as the *mixed liquor volatile suspended solids* (MLVSS). The MLVSS concentration represents the concentration, X, of the microorganisms in the reactor. The mass, M, of microorganisms in a reactor of volume V can be determined using Eq. 9.1.

$$M = VX \qquad 9.1$$

4. AERATION SYSTEMS

Extended aeration is most appropriate for small or low-flow wastewater treatment systems. To aerate the mixed liquor, floating or surface aerators are typically housed in a large, shallow pond (see Fig. 9.2). The sludge that forms as a result of aeration is allowed to accumulate at the bottom of the pond until a minimum working volume is reached. After this point, the pond is drained and the sludge is removed for disposal. Floating and surface aeration systems typically have low effluent and overflow rates, between 200 gal/day-ft^2 and 600 gal/day-ft^2 (8 m^3/d·m^2 and 24 m^3/d·m^2).

An *oxidation ditch* is an extended aeration system in the form of a narrow circular or oval channel. Although

Table 9.2 Typical Design Criteria for Activated Sludge Processes

process modification	θ_c (day)	F:M (lbm BOD$_5$/day-lbm MLVSS)	volumetric loading (lbm BOD$_5$/ 10^3 ft^3-day)	MLSS (mg/L)	t_h (hr)	Q_r/Q
conventional	5–15	0.2–0.4	20–40	1500–3000	4–8	0.25–0.75
complete-mix	5–15	0.2–0.6	50–120	2500–4000	3–5	0.25–1.0
step feed	5–15	0.2–0.4	40–60	2000–3500	3–5	0.25–0.75
modified aeration	0.2–0.5	1.5–5.0	75–150	200–1000	1.5–3	0.05–0.25
contact stabilization	5–15	0.2–0.6	60–75	1000–3000[a] 4000–10,000[b]	0.5–1.0[a] 3–6[b]	0.5–1.50
extended aeration	20–30	0.05–0.15	10–25	3000–6000	18–36	0.5–1.50
high rate aeration	5–10	0.4–1.5	100–1000	4000–10,000	2–4	1.0–5.0
Krauss process	5–15	0.3–0.8	40–100	2000–3000	4–8	0.5–1.0
high purity oxygen	3–10	0.25–1.0	100–200	2000–5000	1–3	0.25–0.5
oxidation ditch	10–30	0.05–0.30	5–30	3000–6000	8–36	0.75–1.50
sequencing batch reactor	NA	0.05–0.30	5–15	1500–5000[c]	12–50	NA
deep shaft reactor	NI	0.5–5.0	NI	NI	0.5–5	NI
single-stage nitrification	8–20	0.10–0.25 0.02–0.15[c]	5–20	2000–3500	6–15	0.5–1.50
separate stage nitrification	15–100	0.05–0.20 0.04–0.15[c]	3–9	2000–3500	3–6	0.50–2.00

NA = not applicable; NI = no information.
[a]contact time
[b]solids stabilization unit
[c]total Kjehldahl nitrogen/MLVSS

Adapted with permission from G. Tchobanoglous, F. L. Burton, and H. D. Stensel and the staff of Metcalf & Eddy, Inc. *Wastewater Engineering: Treatment, Disposal, Reuse.* 3rd ed., copyright © 1991, by The McGraw-Hill Book Companies.

Figure 9.2 Typical Layout of an Aeration Basin Used in Wastewater Treatment

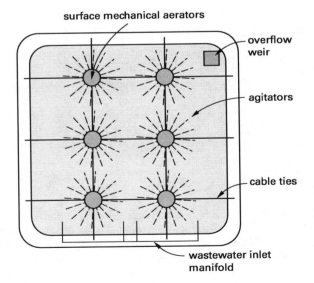

oxidation ditches have relatively large footprints, they have a simple design and are easy to operate. An oxidation ditch combines the processes of oxidation and sedimentation by means of mechanical aeration and final clarification. The ditch effluent is clarified and a portion of the settled sludge is returned to the ditch to maintain a desirable MLSS concentration (ranging from 3000 mg/L to 5000 mg/L). The MLSS concentration is a function of the surface area provided for sedimentation, the rate of sludge return, and the aeration process. The volume of the oxidation ditch should be based on a maximum loading of 15 lbm of BOD$_5$/1000 ft^3 of channel or on the hydraulic retention time, whichever requires the greater volume. Channel depths should be between 3 ft and 5 ft (0.9 m and 1.5 m). A minimum flow-through velocity of 1 ft/sec (0.3 m/s) through the channel cross-sectional area is recommended. The contents in the fluid flow are continuously circulated to maintain good mixing and aeration through brush or diffuser aerators. In order to operate properly, both the fluid velocity and the oxygen concentration must be relatively constant throughout the depth of the fluid channel. This can be achieved through proper channel design (trapezoidal channel cross section is generally recommended) and aerator selection. Oxidation ditches may also be used for nitrogen removal through volatilization. Types of aeration include the following.

- *Step-flow aeration* uses multiple influent points to produce constant aeration throughout the fluid flow.

- In *complete-mix aeration*, the suspension is homogeneous with regard to dissolved oxygen concentration.

- *High rate aeration* systems incorporate both mechanical mixing and aeration to decrease the time required to aerobically degrade a waste (i.e., to increase aeration efficiency). This has the benefit of increasing the BOD load per unit volume of the reactor, so that a higher strength wastewater may be treated in smaller reactors.

- *High purity oxygen aeration* uses pure oxygen, as opposed to compressed air, to aerate a wastewater stream. Such systems are most often used for treating strong sewage and industrial wastewaters.

5. YIELD

Yield is the amount of biomass produced, normalized to the amount of food consumed. It is a measure of the amount of biomass formed per unit of substrate consumed. The units of yield depend on the units used to measure both the biomass and substrate concentrations. Therefore, the units may be in terms of the mass of biomass formed per mass of substrate consumed (gram per gram) or the mass of biomass formed per mass of COD or BOD removed.

There are two types of yield: true growth yield, Y, and observed yield, Y_{obs}. The *true growth yield* is the amount of material that would be formed per unit of substrate consumed if all energy was funneled into biomass growth. The true growth yield represents the proportionality between the rate of biomass growth, r_X, and the rate of energy utilization for synthesis, r_{Sg}. The true growth yield can be approximated in a system where the biomass is growing at the *maximal rate* (μ_{max}), because the substrate utilization for maintenance is small in comparison to that used for growth. A plot of biomass formed versus substrate removed in a batch reactor will give a straight line with a slope equal to Y. The true growth yield is thus normally determined through bench-scale testing using samples of the wastewater to be treated.

Observed yields, Y_{obs}, are generally observed in continuously mixed reactors, such as those used in water treatment plants, and are the amount of biomass actually formed per unit of substrate used. As such, they take into account the energy used for cellular maintenance. The observed yield is always less than the true growth yield. The observed yield is calculated using Eq. 9.2 through Eq. 9.4 and is proportional to the ratio between the rate of biomass growth and the total rate of energy utilization.

$$Y_{obs} = \frac{r_X}{-(r_{Sg} - r_{Sm})} \qquad 9.2$$

$$-r_{Sm} = \mu X \qquad 9.3$$

$$Y_{obs} = \frac{Y}{1 + k_d \theta_c} \qquad 9.4$$

r_{Sm} is the rate of energy utilization for maintenance. θ_c is the mean cell residence time, and k_d is the cell death rate. The general ranges for yields that are observed in common environmental processes follow.

- aerobic processes: $Y_{obs} = 0.4$ mg/mg to 0.8 mg/mg
- anaerobic processes: $Y_{obs} = 0.08$ mg/mg to 0.2 mg/mg

6. KINETICS IN PFRS AND CSTRS

Activated sludge reactors generally take the form of a *completely stirred tank reactor* (CSTR) as shown in Fig. 9.3, or a *plug flow reactor* (PFR) as shown in Fig. 9.4. In a PFR, essentially no back mixing is assumed as the mixture passes through the reactor in the form of a pulse or "plug." As a plug flows through a PFR, the fluid is perfectly mixed in the radial direction but not in the axial direction. Each plug of fluid passing through the reactor is considered to be a separate entity that behaves almost as a batch reactor. An ideal PFR has a fixed residence time, t_h, so that any plug entering the reactor at time t will exit the reactor at time $t + t_h$.

Figure 9.3 Completely Stirred Activated Sludge Wastewater Treatment Process

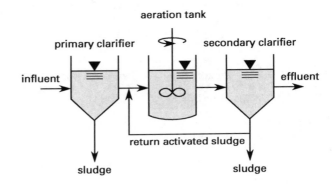

Figure 9.4 Plug Flow Activated Sludge Wastewater Treatment Process

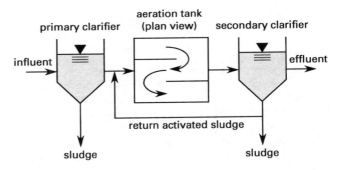

In a CSTR, the influent material is assumed to be completely and homogeneously dispersed throughout the reactor, and the effluent concentration is the same as the concentration in the reactor. Most biological reactors in wastewater treatment are CSTRs. CSTRs are primarily used because they maintain a constant chemical environment at a steady state. This provides for a continuous influent of substrate and a constant effluent quality. Many of the governing design equations are the same for CSTRs and PFRs in activated sludge processes, though there are some differences caused by the inherent differences in operation and flow.

7. BACTERIAL GROWTH KINETICS

Bacteria grow and divide by binary fission, meaning that one cell leads to two, two to four, and so on. This type of multiplication is referred to as exponential growth. If the growth rate is proportional to the biomass concentration, the growth rate is said to be first-order.

The specific growth rate coefficient, μ (per hour), is determined by a number of process variables, principally the concentration of noninhibitory and inhibitory substrates (see Fig. 9.5). The value of μ will be affected by the concentration of a limiting nutrient, provided that all others are provided in excess. At low concentrations of the limiting nutrient, μ increases linearly with limiting substrate concentration, S, in a first-order manner. At high concentrations of S, μ is independent of S in a zero order manner. The *Monod equation*, Eq. 9.5, describes the relationship between the growth rate and the substrate concentration.

$$\mu = \mu_{\text{max}} \left(\frac{S}{K_s + S} \right) \qquad 9.5$$

Figure 9.5 *Specific Growth Rate Curve Showing the Relationship Between the Maximum Growth Rate Constant and the Half Saturation Constant*

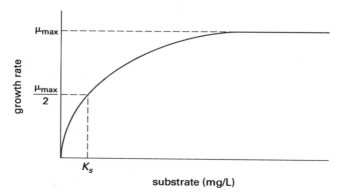

μ_{max} is the maximum *specific growth rate coefficient* (per hour), and K_s is the half-saturation coefficient (milligrams per liter). S is the substrate concentration (milligrams per liter) and is typically the BOD value for the wastewater. If S is much greater than K_s, $\mu \approx \mu_{\text{max}}$. $K_s = S$ when $\mu = \frac{1}{2}\mu_{\text{max}}$ (see Fig. 9.5). μ_{max} is the product of the yield, Y, and the first-order reaction rate constant, k, as shown in Eq. 9.6.

$$\mu_{\text{max}} = kY \qquad 9.6$$

An inhibitory substrate is one that results in a lower value of μ when S is very high. Substrate inhibition generally occurs in industrial treatment systems that use synthetic substrates. The impact of the inhibitory substrate concentration on μ is described by the *Andrews equation*, Eq. 9.7.

$$\mu = \mu_{\text{max}} \left(\frac{S}{K_s + S + \dfrac{S^2}{K_I}} \right) \qquad 9.7$$

K_I is the inhibition coefficient in milligrams per liter. When K_I is very large, the Andrews equation simplifies to the Monod equation. For an inhibitory substrate, μ passes through a maximum, S^*, at S^*. S^* is the maximum substrate concentration and is found from Eq. 9.8.

$$S^* = \sqrt{K_s K_I} \qquad 9.8$$

The theoretical amount of dissolved oxygen that must be added to a biological reactor is called the oxygen utilization rate, r_{O_2}. The oxygen utilization rate is a function of the substrate (food) concentration and the DO removal efficiency. The substrate concentration is the amount of substrate that must be removed by the biological process. The DO removal efficiency is the rate at which the microorganisms in the reactor use dissolved oxygen.

Example 9.1

The amount of BOD removed in a given process may be calculated by measuring the amount of substrate flowing into (S_o) and out of (S) a given reactor. The wastewater feed and effluent for a local WWTP have been carried out, and the pertinent characteristics are summarized in the following table.

quality parameter	value
flow rate, Q (m^3/d)	5600
influent BOD$_5$, S_o (mg/L)	176
effluent BOD$_5$, S (mg/L)	10
recycle solids concentration, X_r (mg/L)	12000
influent TSS, X_o (mg/L)	1800
yield coefficient, Y (g/g)	0.43
endogenous decay rate, k_d (d^{-1})	0.05
half-saturation coefficient, K_s (mg/L)	28.3
oxygen transfer coefficient, K_t (d^{-1})	2.8

Determine the amount of BOD that is removed during aerobic degradation of the waste and how much dissolved oxygen should be added to the reactor to achieve this removal.

Solution

Calculate the amount of substrate (food) that must be removed during the aerobic process.

$$S_o - S = \text{BOD to be removed}$$
$$176 \, \frac{\text{mg}}{\text{L}} - 10 \, \frac{\text{mg}}{\text{L}} = 166 \, \text{mg/L}$$

Calculate the oxygen utilization rate.

$$r_{O_2} = \frac{\left(166 \, \frac{\text{mg DO}}{\text{L}} \right) \left(5600 \, \frac{\text{m}^3}{\text{d}} \right) \left(1000 \, \frac{\text{L}}{\text{m}^3} \right)}{\left(24 \, \frac{\text{h}}{\text{d}} \right) \left(10^6 \, \frac{\text{mg}}{\text{kg}} \right)}$$

$$= 38.7 \, \text{kg DO/h}$$

8. BIOMASS CONCENTRATION

The net amount of biomass produced in a reactor is a summed balance between the new biomass that is formed and the old biomass that dies and is no longer useful. This balance results in the biomass concentration, X, present at any given time in the reactor. For both a PFR and a CSTR, X is calculated according to Eq. 9.9.

$$X = \frac{\left(\frac{\theta_c}{t_h}\right) Y(S_o - S)}{1 + k_d \theta_c} \qquad 9.9$$

k_d is the *death rate* or *endogenous decay rate* (in days), and θ_c is the mean cell residence time (in days).

9. SLUDGE AGE

In biological reactors, a balance exists between the concentration of substrate (waste or food for bacteria) and the concentration of biological organisms (biomass) in the reactor. Consider that the system is substrate limited (e.g., there is a finite amount of food or the system is a batch reactor). The relationship between the concentration of substrate and biomass in a reactor will take on the general behavior shown in Fig. 9.6.

Figure 9.6 is subdivided into four different phases representing the different phases of bacterial population growth in the reactor, described in the following list.

- *Lag phase*—The biomass becomes acclimated to the substrate. Although there is no apparent increase in the biomass concentration, the bacteria may be growing in volume or mass, synthesizing enzymes, proteins, RNA, and so on, and increasing in metabolic activity.

- *Exponential (log) phase*—Biomass growth is not limited by the amount of substrate available because the substrate concentration is substantial. The rate of exponential growth, k_{max}, of a bacterial culture can be expressed as the *generation time*, or the *doubling time*, of the bacterial population.

- *Stationary phase*—Biomass growth is balanced by the availability of the substrate. Exponential biomass growth is limited by the exhaustion of available nutrients, the accumulation of inhibitory metabolites or end products, and the exhaustion of space (i.e., a lack of "biological space").

- *Endogenous phase*—Biomass growth is limited and declines. Death and starvation result from an exhaustion of resources. During the endogenous phase, internal energy reserves are oxidized, and the number of viable bacteria decreases exponentially. This phase is essentially the reverse of growth during the log phase.

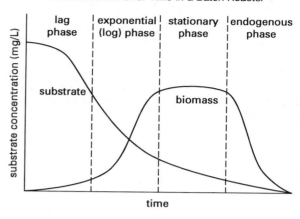

Figure 9.6 *Change in the Concentration of Substrate and Biomass Concentrations Over Time in a Batch Reactor*

The *mean cell residence time*, also called the *sludge age*, θ_c, is the average amount of time that an organism remains in a reactor. θ_c is typically calculated as the total MLSS in the biological reactor, VX, divided by the sum of the solids in the effluent, $Q_e X_e$, and solids wasted, $Q_w X_w$. In conventional aeration processes, the mean residence time, θ_c, ranges from 3 days to 15 days.

Calculation of θ_c is dependent on where the biomass is wasted from the activated sludge system. Equation 9.10 is used when biomass is wasted from the recycle line. This equation may be simplified to Eq. 9.11, because the biomass in the effluent stream, X_e, is approximately equal to zero.

$$\theta_c = \frac{VX}{Q_w X_w + Q_e X_e} \qquad 9.10$$

$$\theta_c = \frac{VX}{Q_w X_w} \qquad 9.11$$

If the cells are wasted from the aeration basin directly (a practice not commonly done), then the biomass in the recycle waste stream, X_r, equals the biomass reactor solids, X, and θ_c is calculated from Eq. 9.12.

$$\theta_c = \frac{V}{Q_w} \qquad 9.12$$

The kinetic model for a continuously stirred tank reactor (CSTR) is more complex than the model for a PFR, because many assumptions must be made about the fluid flow through the reactor. Nevertheless, a CSTR may be modeled according to Eq. 9.13.

$$\frac{1}{\theta_c} = \frac{\mu_{max}(S_o - S)}{S_o - S + (1 + R)K_s \ln\left(\frac{S_i}{S}\right)} - k_d \qquad 9.13 \quad \text{CSTR}$$

S is the effluent substrate concentration (defined in Eq. 9.17). R is the dimensionless recycle ratio, described in Sec. 9.11. S_i is the *influent substrate concentration* found from Eq. 9.14.

$$S_i = \frac{S_o + RS}{1 + R} \qquad 9.14$$

The *rate of substrate utilization*, r_{Su}, is the amount of substrate that is utilized or consumed in a reactor over a set period of time. In a PFR, the r_{Su} is calculated according to Eq. 9.15.

$$r_{Su} = \frac{-\mu_{max}SX}{Y(K_s + S)} \qquad 9.15$$

In CSTRs, the consumption of substrate by the organisms in the reactor is described in terms of the *specific substrate utilization parameter*, U, calculated using Eq. 9.16.

$$U = \frac{-r_{Su}}{X} = \frac{S_o - S}{t_h X} \qquad 9.16$$

The *effluent substrate concentration, S,* for a CSTR may be calculated using Eq. 9.17.

$$S = \frac{K_s(1 + k_d\theta_c)}{\theta_c(\mu_{max} - k_d) - 1} \qquad 9.17$$

10. FOOD TO MICROORGANISM RATIO

Because the biomass reactor solids, X, and the reactor volume, V, are dictated by the design food to microorganism ratio, F:M, Eq. 9.18 may be used in conjunction with recommended values of mean residence times, θ_c, to determine the required flow rate of waste, Q_w, in the recycle line.

The food to microorganism ratio represents the daily mass of substrate, S, supplied to the biomass, X, in the <u>MLSS</u>. F:M (in unit mass of BOD_5 per unit mass of MLSS per day) is

$$\text{F:M} = \frac{S_o Q_o}{XV} \qquad 9.18$$

 Q_o is the flow rate into the activated sludge tank in gallons per day (cubic meters per day). The *hydraulic residence time, t_h,* given in units of days, is the average time that a water molecule will spend in the reactor. t_h is one of the critical parameters for sizing the aeration basin and is found from Eq. 9.19.

$$t_h = \frac{V}{Q_o} \qquad 9.19$$

The inverse of the hydraulic residence time, t_h, is the *dilution rate, D,* defined in Eq. 9.20.

$$D = \frac{1}{t_h} = \frac{Q}{V} \qquad 9.20$$

Substituting for the hydraulic residence time in Eq. 9.18, the F:M may also be expressed as Eq. 9.21.

$$\text{F:M} = \frac{S_o}{t_h X} \qquad 9.21$$

Because S_o, Q_o, and X are typically fixed parameters, a particular reactor volume is selected to achieve the desired F:M ratio. Conversely, the typical design value for the F:M ratio will depend on the type of activated sludge system design used (see Table 9.2 for a range of typical values for various process modifications to activated sludge systems).

At low food to microorganism ratios, the microorganisms are maintained in the endogenous growth phase. This requires rather large reactor volumes or low flow rates, and systems operated in this manner are termed *extended aeration.*

At high food to microorganism ratios, the microorganisms are maintained in the exponential growth phase and are termed *high-rate activated sludge.* Higher MLSS concentrations are maintained, which allows for shorter retention times. At high F:M ratios, the bacteria are highly dispersed and do not settle out well. Therefore, few activated sludge systems operate under these conditions. Exceptions are as follows.

- during start-up
- upon recovery from an upset in which solids washed out significantly
- when a spill of high strength organic material has passed into the aeration basin

Conventional activated sludge processes operate at moderate food to microorganism ratios and the microorganisms are in the declining, or *endogenous*, growth phase. As the F:M ratio decreases, sludge settleability tends to increase, because starving microorganisms flocculate and settle better. Typically, F:M ratios in the range of 0.05 to 0.1 result in BOD removal efficiencies of 95% to 99%. For food to microorganism ratios between 0.1 and 0.2, the removal efficiency may be from 90% to 95%.

Example 9.2

Determine the F:M ratio for a wastewater reactor with an influent flow rate of 40 MLD, an influent BOD_5 concentration of 400 mg/L, a volume of 5×10^6 L, and a mixed liquor suspended solids concentration of 4500 mg SS/L.

Solution

Using Eq. 9.18,

$$F:M = \frac{S_o Q_o}{XV}$$

$$= \frac{\left(40 \ \frac{ML}{d}\right)\left(10^6 \ \frac{L}{ML}\right)\left(400 \ \frac{mg}{L}\right)\left(10^6 \ \frac{mg}{kg}\right)}{\left(4500 \ \frac{mg}{L}\right)(5 \times 10^6 \ L)\left(10^6 \ \frac{mg}{kg}\right)}$$

$$= 0.711 \ \text{kg BOD}_5/\text{kg·d}$$

11. RECYCLE AND WASTE RATIOS

A fraction of the sludge in the secondary clarifier is recycled to the activated sludge reactor to maintain a healthy population of bacteria adapted to the wastewater, a practice called *seeding the reactor*. Recycling reduces the residency time and the reactor volume required to degrade the wastewater. This return rate is described as the dimensionless *recycle ratio*, R, found from Eq. 9.22. Values for R in conventional activated sludge processes range from 0.20 to 0.30. Recycling of sludge leads to a reduction in the hydraulic residence time, t_h, of around 98.5% and a 98.5% reduction in aerator volume. In some conventions, the recycle ratio may also be designated by the symbol α.

The *waste ratio*, W, describes the amount of flow wasted from the secondary clarifier relative to the influent to the activated sludge basin, as shown in Eq. 9.23.

$$R = \frac{Q_r}{Q_o} \qquad 9.22$$

$$W = \frac{Q_w}{Q_o} \qquad 9.23$$

The *recirculation rate*, Q_r, can be determined through mass balances on biomass or mixed liquor suspended solids (MLSS in mg/L), X, in either the clarifier (see Eq. 9.24) or the aeration tank (see Eq. 9.25). In both cases, it is assumed that the influent MLSS concentration, X_o, is equal to 0.

$$Q_r = \frac{XQ_o - X_r Q_w}{X_r - X} \qquad 9.24$$

$$Q_r = Q_o\left(\frac{X}{X_r - X}\right) \qquad 9.25$$

Equation 9.25 should not be applied to systems experiencing high organic loadings (relative to X), as it may no longer be valid to assume that $X_o = 0$ for the aeration basin.

The theoretically required recycle ratio, R, can be determined from the sludge volume index (SVI),

Eq. 9.26 and Eq. 9.27. Because these values are theoretical only, they must be compared with actual experimental measurements to avoid unnecessarily diluting or oversaturating the reactor contents.

$$\frac{Q_r}{Q_o + Q_r} = \frac{V_{\text{settled,mL/L}}}{1000 \ \frac{mL}{L}} \qquad 9.26$$

$$R = \frac{V_{\text{settled,mL/L}}}{1000 \ \frac{mL}{L} - V_{\text{settled,mL/L}}}$$

$$= \frac{1}{\dfrac{10^6}{\text{SVI}_{\text{mL/g}}\text{MLSS}_{\text{mg/L}}} - 1} \qquad 9.27$$

12. WASHOUT TIME

Washout occurs when the microbial residence time, θ, is less than the minimum mean cell residence time, or *washout time*, $\theta_{\min}$ (see Eq. 9.28). Under these conditions, the bacteria are washed out of the reactor at a higher rate than they can reproduce, and the concentration of biomass reactor solids goes to zero, $X = 0$. K_s is the half-saturation coefficient, and k_m is the cell growth constant, also called the BOD degradation rate, in milligrams of BOD per milligrams of cells per day.

$$\theta_{\min} = \frac{W(1 + R)}{W + R}$$

$$= \frac{K_s + S_o}{(Yk_m - k_d)S_o - k_d k_s} \qquad 9.28$$

13. BOD REMOVAL EFFICIENCY

The *removal efficiency* for BOD, or any other constituent in a treatment process, is determined according to Eq. 9.29. In conventional activated sludge processes, the BOD removal efficiency ranges from 85% to 95%.

$$\eta_{\text{BOD}} = \frac{S_o - S}{S_o} \qquad 9.29$$

Example 9.3

A wastewater treatment plant will receive a flow of 10 MGD with a raw wastewater BOD_5 of 250 mg/L. Primary treatment removes 30% of the BOD_5. Calculate the required volume (in gallons) and approximate retention time, t_h, of the activated sludge aeration basin for running the plant as a "high rate" facility (i.e., F:M = 2 lbm/day-lbm) where the aeration basin MLSS concentration will be maintained at 1900 mg MLSS/L.

Solution

Use Eq. 9.29 to calculate the influent BOD_5 to the activated sludge system (i.e., aerobic reactor) following the primary clarifier.

$$\eta_{BOD} = \frac{S_o - S}{S_o}$$

$$
\begin{aligned}
S &= S_o(1 - \eta_{BOD}) \\
&= \left(250\ \frac{mg}{L}\right)(1 - 0.30)\left(8.345 \times 10^{-6}\ \frac{lbm\text{-}L}{mg\text{-}gal}\right) \\
&= 0.0015\ lbm/gal
\end{aligned}
$$

Use Eq. 9.18 to calculate the required volume for the activated sludge reactor.

$$F{:}M = \frac{SQ_o}{XV}$$

$$
\begin{aligned}
X &= \left(1900\ \frac{mg}{L}\right)\left(8.345 \times 10^{-6}\ \frac{lbm\text{-}L}{mg\text{-}gal}\right) \\
&= 0.016\ lbm/gal
\end{aligned}
$$

$$
\begin{aligned}
V &= \frac{SQ_o}{X(F{:}M)} \\
&= \frac{\left(0.0015\ \dfrac{lbm}{gal}\right)\left(10\ \dfrac{MG}{day}\right)\left(10^6\ \dfrac{gal}{MG}\right)}{\left(0.016\ \dfrac{lbm}{gal}\right)\left(2\ \dfrac{lbm}{day\text{-}lbm}\right)} \\
&= 468{,}750\ gal
\end{aligned}
$$

Use the volume calculated from Eq. 9.18 to determine the hydraulic detention time, given by Eq. 9.19.

$$
\begin{aligned}
t_h &= \frac{V}{Q} \\
&= \frac{(468{,}750\ gal)\left(24\ \dfrac{hr}{day}\right)}{\left(10\ \dfrac{MG}{day}\right)\left(10^6\ \dfrac{gal}{MG}\right)} \\
&= 1.13\ hr
\end{aligned}
$$

The detention time is an approximate value because the return activated sludge has not been accounted for.

14. AERATION BASIN PARAMETERS

The *aeration period*, θ_{air}, given in hours, is the theoretical time that the mixed liquor is subjected to aeration and is calculated from Eq. 9.30.

$$\theta_{air} = \frac{V}{Q_o} \qquad 9.30$$

The *rate of oxygen transfer*, m_{O_2}, from the air to the mixed liquor is the amount of oxygen absorbed by the

liquid compared to the amount fed into it through an aeration or oxygenation device. The rate of oxygen transfer is found from Eq. 9.31 and Eq. 9.32.

$$m_{O_2} = K_t D \qquad 9.31$$

$$D = \beta DO_{saturated} - DO_{mixed\ liquor} \qquad 9.32$$

K_t is the oxygen transfer coefficient, and β is the oxygen saturation coefficient (typically 0.8 to 0.9). The oxygen demand is found from Eq. 9.33.

$$m_{O_2} = \frac{Q_o(S_o - S)}{f} - 1.42P_x \qquad 9.33$$

f is a conversion factor (ranging from 0.45 to 0.68) for converting BOD_5 to BOD_u. P_x is the net waste activated sludge produced per day (pounds per day, kilograms per day). The minimum required DO concentration is between 0.2 mg/L and 2.0 mg/L for conventional activated sludge processes. The air delivery rate is greater than the oxygen demand because air contains only approximately 23.2% oxygen by mass, as shown in Eq. 9.34.

$$m_{air} = \frac{m_{O_2}}{0.232\eta_{transfer}} \qquad 9.34$$

$\eta_{transfer}$ is the gas transfer efficiency. The gas transfer efficiency will depend on the type of aeration system used.[2] The required volume of air is calculated from Eq. 9.35.

$$V_{air} = \frac{m_{air}}{\rho_{air}} \qquad 9.35$$

Typical air requirements for conventional aeration are 500 ft^3/lbm BOD_5 to 900 ft^3/lbm BOD_5 (31.2 m^3/kg BOD_5 to 56.2 m^3/kg BOD_5). For F:M ratios greater than 0.3, the air requirement is 1200 ft^3/lbm BOD_5 (75 m^3/kg BOD_5), while for F:M ratios less than 0.3, the air requirement is approximately 1800 ft^3/lbm BOD_5 (112.4 m^3/kg BOD_5).

15. SECONDARY (OR FINAL) CLARIFIERS

Final clarifiers are used in activated sludge processes to separate and thicken sludge, enabling recycling of thickened, activated sludge. Secondary clarifiers are essentially sedimentation basins that follow activated sludge processes. Typically, a fraction of the settled sludge is returned to the aeration basin and the remainder is wasted. The operating principles of the secondary clarifier are the same as those of a sedimentation basin.

---

[2]The following are examples of aeration systems and typical gas transfer efficiencies: fine bubble diffuser (total floor coverage)—20% to 32%; fine bubble diffuser—11% to 15% (side wall installation); jet aerators—22% to 27%; static aerators—12% to 14%; mechanical surface aerators—2% to 4%.

16. SLUDGE MASS AND VOLUME

The volume of sludge produced increases with the level of treatment. Primary treatment produces between 2500 L and 3500 L sludge per 10^6 L of wastewater treated. Biological secondary treatment produces an additional 15 000 L to 20 000 L sludge per 10^6 L of wastewater treated. Use of chemicals for phosphorus removal during tertiary treatment increases sludge volume another 10 000 L sludge per 10^6 L of wastewater treated. The mass of dry sludge produced per unit time, P_x (pounds or kilograms per day), in a biological treatment process can be determined using Eq. 9.36.

$$P_x = \left(Q(S_o - S) \right)\left(\frac{Y}{1 + k_d \text{SRT}} \right) \qquad 9.36$$

The solids retention time is SRT, and k_d is the endogenous decay rate.

The volume and mass of sludge produced during an activated sludge process is primarily dependent on the density and specific gravity of the biomass (see Eq. 9.37). The water content of the sludge is also a significant factor. Sludge mass and volume are determined by its density, ρ_{sludge}, and specific gravity, $\text{SG}_{\text{sludge}}$. For raw sludge, the water content is generally greater than 90%, thus $\text{SG}_{\text{sludge}}$ is approximately 1, the specific gravity of water.

$$m_{\text{wet}} = V\rho_{\text{sludge}} = V\text{SG}_{\text{sludge}}\rho_w \qquad 9.37$$

The sludge volume, V_{sludge}, is best determined from its dried mass. The wet sludge volume can be approximated from Eq. 9.38. Because the sludge is mostly water containing suspended and dissolved solids, the sludge density is approximately equal to that of water at the reported operating temperature.

$$V_{\text{sludge,wet}} = \frac{m_{\text{dry}}}{s\rho_{\text{sludge}}} \qquad 9.38$$

ρ_{sludge} is the density of the sludge, and s is the gravimetric fractional solids content. The mass of solids remaining in a clarifier can be determined from the amount of solids entering and exiting the clarifier. Since most of the solids in sludge are suspended (not dissolved), the amount of dried solids in the sludge may be determined from Eq. 9.39, where the change in suspended solids concentration, ΔSS, and the flow rate are considered.

$$m_{\text{sludge,dry}_{\text{lbm/day}}} = \Delta\text{SS}_{\text{mg/L}} Q_{\text{MGD}} \left(8.345 \ \frac{\text{lbm-L}}{\text{MG-mg}} \right)$$
$$\text{[U.S. only]}$$
$$9.39$$

The volume of the aeration basin, $V_{\text{aeration basin}}$, is a function of the solids retention time (SRT), the mass of sludge produced per day (P_x), and the biomass concentration in the aeration basin (X). The volume calculated using Eq. 9.41 is the working, or wetted, volume and does equal the total reactor volume. The total reactor volume

includes reactor freeboard (i.e., the wall height above the water level).

$$V_{\text{aeration basin}} = \frac{P_x \text{SRT}}{X} \qquad 9.40$$

Example 9.4

The aerobic biological treatment process at a local wastewater treatment plant has an observed yield, Y_{obs}, of 0.60 mg/mg, and an average biomass concentration (XVSS) of 2500 mg/L. Determine the aeration basin volume and the amount of sludge that will be produced on a daily basis in order to treat the raw wastewater, given the following design constraints: a solids retention time, SRT, of 8 days, an endogenous decay coefficient, k_d, of 0.1 day^{-1}, an influent BOD$_5$ of 170 mg/L, an effluent BOD$_5$ of 8 mg/L, and a raw water flow rate, Q, of 1.3 MGD.

Solution

Use Eq. 9.36 to calculate the amount of dry sludge produced from the aerobic biological process.

$$P_x = \left(Q(S_o - S) \right)\left(\frac{Y}{1 + k_d \text{SRT}} \right)$$
$$= \left(\left(1.3 \ \frac{\text{MG}}{\text{day}} \right)\left(170 \ \frac{\text{mg}}{\text{L}} - 8 \ \frac{\text{mg}}{\text{L}} \right) \right)\left(8.345 \ \frac{\text{lbm-L}}{\text{MG-mg}} \right)$$
$$\times \left(\frac{0.60 \ \frac{\text{mg}}{\text{mg}}}{1 + (0.1 \ \text{day}^{-1})(8 \ \text{days})} \right)$$
$$= 586 \ \text{lbm/day}$$

Calculate the aeration basin volume.

$$V_{\text{aeration basin}} = \frac{P_x \text{SRT}}{X}$$
$$= \frac{\left(586 \ \frac{\text{lbm}}{\text{day}} \right)(8 \ \text{days})\left(453{,}592.37 \ \frac{\text{mg}}{\text{lbm}} \right)}{\left(2500 \ \frac{\text{mg}}{\text{L}} \right)\left(3.785 \ \frac{\text{L}}{\text{gal}} \right)}$$
$$= 224{,}723 \ \text{gal}$$

Example 9.5

A wastewater treatment plant treats 3 MGD of municipal wastewater having a BOD$_5$ of 250 mg/L. The plant uses primary treatment followed by an activated sludge process. The BOD$_5$ removal efficiency for the total plant is 96%. The activated sludge system has a design solids retention time (i.e., sludge age) of 8 days, a k_d of 0.1 day^{-1}, and a maximum yield of 0.45 mg biomass/ mg BOD$_5$. Assume a primary clarifier BOD$_5$ removal efficiency of 30% and an operating temperature of 12°C. Assume that the sludge produced in the activated sludge process has a solids content of 5%. Determine the amount of sludge produced per day (dry mass and wet volume) by the activated sludge process at the wastewater treatment plant.

Solution

Determine the BOD_5 concentration entering the activated sludge process. Rearrange Eq. 9.29 and solve for S, the concentration of food consumed, or the BOD_5 concentration that is entering the activated sludge process.

$$\eta_{BOD} = \frac{S_o - S}{S_o}$$

$$\begin{aligned} S &= S_o(1 - \eta_{BOD}) \\ &= \left(250\ \frac{\text{mg BOD}_5}{\text{L}}\right)(1 - 0.30) \\ &= 175\ \text{mg BOD}_5/\text{L} \end{aligned}$$

Determine the BOD_5 concentration in the effluent from the activated sludge system. This value can be calculated using Eq. 9.29 and the BOD_5 removal efficiency that was given for the total (primary treatment + activated sludge) system, 96%.

$$\eta_{BOD} = \frac{S_o - S}{S_o}$$

$$\begin{aligned} S &= S_o(1 - \eta_{BOD}) \\ &= \left(250\ \frac{\text{mg BOD}_5}{\text{L}}\right)(1 - 0.96) \\ &= 10\ \text{mg BOD}_5/\text{L} \end{aligned}$$

The mass of sludge produced in the activated sludge process can be calculated using Eq. 9.36.

$$\begin{aligned} P_x &= \left(Q(S_o - S)\right)\left(\frac{Y}{1 + k_d(\text{SRT})}\right) \\ &= \left(\begin{array}{c} \left(3\ \frac{\text{MG}}{\text{day}}\right)\left(175\ \frac{\text{mg BOD}_5}{\text{L}} - 10\ \frac{\text{mg BOD}_5}{\text{L}}\right) \\ \times \left(8.345\ \frac{\text{lbm-L}}{\text{MG-mg}}\right) \end{array} \right) \\ &\quad \times \left(\frac{0.45\ \frac{\text{mg biomass}}{\text{mg BOD}_5}}{1 + (0.1\ \text{day}^{-1})(8\ \text{days})} \right) \\ &= 1033\ \text{lbm/day} \end{aligned}$$

Using Eq. 9.38, the wet sludge volume is

$$\begin{aligned} V_{\text{sludge,wet}} &= \frac{m_{\text{dry}}}{s\rho_{\text{sludge}}} \\ &= \frac{1033\ \frac{\text{lbm}}{\text{day}}}{(0.05)\left(62.4\ \frac{\text{lbm}}{\text{ft}^3}\right)} \\ &= 331\ \text{ft}^3/\text{day} \end{aligned}$$

(handwritten margin notes:)
① organic cmpds to fatty or amino acids
② amino acids to organic/acetic acids
③ organic acids to methane & CO_2

17. SLUDGE PROCESSING

Following digestion, the sludge is conditioned (thermally and/or chemically) and then dewatered in a variety of ways (e.g., centrifugation, drying beds, pressure filtration, vacuum filtration, filter press, belt press) to produce a cake-like solid for disposal. Polymers or other chemical additives are added to the sludge during the conditioning step to increase the efficiency of the dewatering process. After dewatering, the sludge is removed for disposal. For this purpose, there are a number of options available, including incineration, landfilling, land application, and further treatment (e.g., digestion).

18. SLUDGE STABILIZATION

Biosolids and other sludge produced in the course of wastewater treatment must be treated and ultimately removed for disposal in a safe and effective manner. To effectively meet the requirements for sludge disposal, two principle objectives must be met.

- Regulated and potentially harmful elements in the sludge must be eliminated.

- The volume of sludge to be removed for disposal should be minimized as much as possible.

Regulated and harmful components that may be present in wastewater sludge include toxic organic and inorganic compounds (e.g., heavy metals), and disease-causing microorganisms. Eliminating these harmful elements is known as *sludge stabilization*. The requirement of reducing sludge volume is primarily to reduce the costs associated with sludge disposal. The most common sludge stabilization options include dewatering (see Sec. 9.21), anaerobic digestion, aerobic digestion, and composting. The selection of a wastewater solid treatment method depends on the amount of solids generated and other site-specific conditions. Composting is the preferred option for smaller-scale applications, followed by aerobic digestion, then anaerobic digestion for the larger scale municipal applications.

Anaerobic Digestion

Anaerobic digestion is a bacterial process occurring in the absence of oxygen. The process can either be *thermophilic digestion*, in which sludge is fermented in tanks heated to about 131°F (55°C), or *mesophilic digestion* (86°F to 104°F or 30°C to 40°C), in which sludge is maintained in large tanks for weeks (15 days to 40 days) to allow the sludge to mineralize. Process selection will depend on the amount of sludge, among other considerations. Thermophilic digestion generates *biogas*, which is made up of 60% to 70% methane and carbon dioxide. The methane may be used to heat the digester and/or to help in fueling other plant processes. Methane generation is a key advantage of anaerobic digestion of the sludge. Key disadvantages of anaerobic digestion are the long solids residence times in the digester (up to 30 days) and the high capital costs.

Aerobic Digestion

Aerobic digestion is a bacterial process occurring in the presence of oxygen. Under aerobic conditions, bacteria rapidly consume organic matter and convert it into carbon dioxide. Once there is a lack of organic matter, the bacteria die and are consumed by other bacteria in the digester. This stage of the process, in which solids reduction or sludge digestion occurs, is called *endogenous respiration*. Because aerobic digestion occurs at a faster rate than anaerobic digestion, the capital costs associated with smaller digesters and equipment are lower. However, the operating costs are characteristically higher for aerobic digestion because of the energy costs associated with providing dissolved oxygen to the digester.

Composting

Composting is also an aerobic process. It involves mixing the wastewater solids with sources of carbon, such as sawdust, straw, or wood chips. In the presence of oxygen, bacteria digest both the wastewater solids and the added carbon source. This process is exothermic, resulting in the production of a substantial amount of heat.

19. SLUDGE VOLUME INDEX

The *sludge volume index* (SVI) is an experimentally determined value used to assess the settling characteristics of activated sludge and other biological suspensions (see Table 9.3). The SVI is the volume, in milliliters, occupied by 1 g of a suspension after 30 min of settling. Typically, 1 g (dry weight) of sludge is mixed into 1000 mL of water in a graduated cylinder and allowed to settle for 30 min. The SVI is calculated using Eq. 9.41 and has units of milliliters per gram.

Table 9.3 *Interpretation of Sludge Volume Index (SVI) values*

SVI (mL/g)	settleability of sludge
0–100	good
100–200	acceptable
> 200	poor

The *sludge density index* (SDI) is a second parameter used to characterize biologically produced sludge and is found from Eq. 9.42. The SDI is a measure of the degree of compaction, expressed in grams per milliliter, of the sludge after it has settled in a 1000 mL graduated container for 30 min. The SDI is related to the SVI in that both relate the mass of the sludge to the volume it occupies. SDI and SVI may be used to assess how well the activated sludge will separate from the mixed liquor in the secondary clarifier. The better the sludge settleability, the less volume is required for the settled sludge, and the lower the pumping rate required to keep the solids in circulation.

$$SVI_{mL/g} = \frac{V_{settled,mL/L}}{MLSS_{mg/L}} \qquad 9.41$$

$$SDI_{g/mL} = \frac{MLSS_{g/L}}{SVI_{settled,mL/L}} \qquad 9.42$$

Based on the SVI, the total suspended solids concentration can be determined for the wastewater using Eq. 9.43.

$$TSS_{mg/L} = \frac{\left(1000 \ \frac{mg}{g}\right)\left(1000 \ \frac{mL}{L}\right)}{SVI_{mL/g}} \qquad 9.43$$

Example 9.6

A sample of the influent wastewater has a total suspended solids (TSS) concentration of 4000 mg/L, which settles to a volume of 325 mL in 30 min in a 1 L cylinder. Calculate the SVI and qualitatively assess the settleability of the sludge.

Solution

Substitute the given values into Eq. 9.41.

$$SVI = \frac{V_{settled}}{MLSS}$$

$$= \frac{\left(325 \ \frac{mL}{L}\right)\left(1000 \ \frac{mg}{g}\right)}{4000 \ \frac{mg}{L}}$$

$$= 81.3 \ mL/g$$

Referring to Table 9.3, SVI values below 100 mL/g are considered good for settling sludge and are desired. SVI values greater than 150 mL/g are typically associated with a problem, such as filamentous growth.

20. SLUDGE THICKENING

For the solids generated during chemical and biological processes in secondary and tertiary treatments, concentration (or thickening) is usually the first step toward removal. The primary function of sludge thickening is to reduce the water content and the volume of sludge produced during various treatment processes. Sludge thickening reduces the volume of residuals, improves process operation, and reduces costs for subsequent sludge storage, processing, transfer, end use, or disposal. There are several different methods for sludge thickening, including dissolved air flotation (DAF), centrifugal thickening, gravity thickening, and gravity belt thickening. Performance values for several varieties of sludge thickening processes are given in Table 9.4.

Dissolved air flotation (DAF) separates water from the sludge by attaching air bubbles to the suspended solids (SS). Polymers can be added to the mixture to aid in the flocculation of the SS. Organic polymers used for sludge thickening and dewatering are long-chained, water soluble, synthetic organic chemicals. Polyacrylamide, a non-ionic compound, is the most widely used polymer. The

ea for thickening

$t = \dfrac{Q_d t_s}{h}$

r initial solids $= \dfrac{Xh}{Xu}$

$v = \dfrac{Xh}{Xu}$

thickened solids

Table 9.4 *Typical Solids Concentrations for Sludge Thickening Processes*

$Q_c = \dfrac{Q_d(h-h_u)}{h}$

	solids concentration (%)	
operation	range	typical
gravity thickeners		
primary sludge only	4–10	6
primary and waste-activated	2–6	4
flotation thickeners		
with chemicals	3–6	4
without chemicals	3–6	4
centrifuge thickeners		
with chemicals	4–8	5
without chemicals	3–6	4

Reprinted with permission from G. Tchobanoglous, F. L. Burton, and H. D. Stensel and the staff of Metcalf & Eddy, Inc. *Wastewater Engineering: Treatment, Disposal, Reuse.* 3rd ed., copyright © 1991, by The McGraw-Hill Book Companies.

Wastewater

thickened sludge floats to the surface of the clarifier or basin and is removed by surface skimmers. DAF thickening is commonly used for waste-activated, aerobically digested, contact stabilized, modified activated, or extended aeration sludges without primary settling. DAF is generally not used for primary or trickling filter sludges because they settle more readily by gravity, which is more economical than DAF.

Centrifugation is another sludge thickening technique that is also used for sludge dewatering. In centrifugation, sludge is fed at a constant feed rate into a rotating bowl, typically operating between 1500 rpm and 2000 rpm. The solids are separated from the liquid by centrifugal forces and then are compacted to the bowl wall. The liquid and fine solids exit the unit through the effluent line. The dewatered sludge forms a cake on the sides of the centrifuge and is removed using direct gravity discharge, belt conveyors, screw conveyors, or pumps going out one end of the bowl, while water leaves from the other end. The effluent water may be wasted or returned to the beginning of the plant. The thickened sludge is removed for disposal. In centrifugation, the sludge is thickened to a solids' concentration of around 20% for sludge thickening and 30% to 35% for dewatered sludge. Centrifugation is typically used to thicken waste-activated sludge and other biological sludges from secondary wastewater treatment. It is occasionally used to reduce the volume of stabilized sludges for the purpose of minimizing the volume to transport for ultimate disposal. Centrifugation is not recommended for thickening of primary sludge, which has high settleability and a tendency to contain abrasive material that may damage the centrifuge.

Two solid bowl centrifuge designs may be used: co-current flow and counter-current flow. Differences in the designs pertain to the location of the feed and effluent to and from the bowl, and the internal flow patterns of the liquid and solid phases. In the *co-current* design, the solids travel the full length of the bowl, while the liquids travel in a parallel pattern with the solids phase. Conduits remove the liquid, which then flows over discharge weirs. In the *counter-current* design, feed enters the centrifuge at the junction of cylindrical conical sections. Solids travel to the conical end of the centrifuge while the liquid travels in the opposite direction. The liquid overflows weir plates located at the large diameter end of the centrifuge. Polymer is generally added either into the feed compartment or through an injection port within the centrifuge.

Gravity thickening is a process in which solids are separated from the liquid through gravitational settling (see Table 9.5 for performance of various conventional gravity thickeners). Sludge is pumped into a circular clarifier having a conical bottom, and the sludge thickens through sedimentation. A V-notch weir located at the top of the tank allows the supernatant to return to a clarifier or to the headworks of the treatment plant. In some cases, rakes are used to stir and break up the sludge to enhance the release of water. Gravity thickeners usually thicken sludge to about twice the original solids content, decreasing the volume of fresh sludge by 50%. Solids at the bottom of the tank can reach concentrations of as high as 15% total solids (TS), with typical values between 4% and 6% TS. Suspended solid (SS) concentrations in the supernatant are as low as 200 mg/L. Gravity thickeners are employed principally to thicken primary sludges, lime sludges, combinations of primary and activated sludges, and to a lesser degree, activated sludge only.

Table 9.5 *Typical Performance of Conventional Gravity Thickeners*

sludge origin	feed solids content (%)	thickened solids content (%)
primary treatment (PT)	2–7	5–10
trickling filter (TF)	1–4	3–6
rotating biological contactor (RBC)	1–3.5	2–5
waste-activated sludge (WAS)	0.2–1	2–3
PT + TF	2–5	5–9
PT + RBC	2–5	5–8
PT + WAS	0.5–4	4–7
PT + lime	3–4.5	10–15
PT + (WAS + iron)	1.5	3
PT + (WAS + aluminum)	0.2–0.4	4.5–6.5
anaerobically digested PT + WAS	4	8

Adapted from *Process Design Manual for Sludge Treatment and Disposal*, EPA-625/1-79-011, USEPA, 1979.

In *gravity belt thickening* (GBT), solids are concentrated when free water drains by gravity through a porous horizontal belt. This process is generally carried out along with chemical conditioning of the sludge, typically using a polymer. The polymer is added to aid in dewatering of the sludge. The water drains to a pan below the belt and is pumped back to the beginning of the treatment process. The sludge moves with the belt and is turned continuously by plow-like devices called *chicanes*. The chicanes move the sludge back and forth, allowing the water to fall through the belt to the drain. Near the end of the belt, a ramp contacts the belt at its leading edge, forcing the sludge to go up and over the ramp and causing the sludge to roll back on itself. This rolling motion causes further dewatering of the sludge. GBTs are particularly suitable for the thickening of waste-activated sludge before further processing and for thickening digested sludges as a volume reduction measure before transport. Thickened waste-activated sludge typically has a concentration of solids in the range of 4% to 8%; digested biosolids can be thickened to 10% solids. Equipment and operating variables that influence GBT performance for a particular solid include polymer dose, feed rate, polymer and solids mixing, belt speed, ramp use and angle, belt type, and plow configuration.

21. SLUDGE DEWATERING

If sludge stabilization is not required or has already been carried out, then sludge may be thickened, or *dewatered,* prior to disposal. Dewatering removes a significant amount of water from the solids entering the process, and greatly reduces the volume. Dewatering the sludge from 2% solids to 20% solids reduces the sludge volume by an order of magnitude (60 m^3 to 6 m^3). Operational controls to be considered in the design and operation of dewatering processes include the following.

- dry weight of sludge
- type of sludge
- downstream sludge handling processes
- period of dewatering operation
- pre- and post-dewatering requirements and capacity
- dewatered sludge transport
- sludge conditioning requirements

Typical dewatering processes include lagooning in drying beds to produce a cake that can be applied to land or incinerated; pressing, in which sludge is mechanically filtered, typically through cloth screens, to produce a firm cake; and centrifugation to separate the solid and liquid phases.

Where land is available and labor costs are low, *sand beds* are the simplest and most cost-effective dewatering method. The beds consist of tile drains in gravel, covered by approximately 10 in (25 cm) of sand (see Fig. 9.7). Liquid is lost by seepage into the sand, as well as by evaporation. A typical drying time is 3 months. Sludge is applied on the sand bed and is allowed to dry by evaporation and the drainage of excess water over a period of several weeks, depending on climatic conditions. Bacterial decomposition of the sludge takes

place during the period of the drying process in which the moisture content is sufficiently high. During rainy periods, the process may take a longer time to complete, and this must be taken into account when determining the size of the drying bed.

Figure 9.7 *General Design of a Sludge Drying Bed*

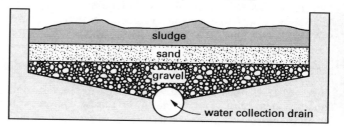

Filter presses apply very high pressures, 70 psi to 220 psi (483 kPa to 1517 kPa), to the sludge cake. Filter presses consist of sets of vertical, juxtaposed recessed plates, pressed against each other by hydraulic jacks at one end of the set. This vertical plate layout forms watertight filtration chambers allowing for removal of the dewatered sludge cake. Sludge is fed into the filtration chamber under pressure. Sludge gradually accumulates in the filtration chamber (on the cloth or membrane) until the final compacted cake is formed. Filtrate is collected at the back of the filtration support and is removed. Following filtration, the chamber is opened and the sludge cake is removed. The filtration capacity for filter presses typically ranges from 3 lbm to 20 lbm (1.5 kg to 10 kg) of solid material per square meter of filtering surface.

22. SLUDGE BULKING AND FOAMING

Sludge bulking and foaming are common problems encountered in activated sludge processes. Bulking occurs when filamentous organisms extend from the flocs into the bulk solution, interfering with settlement and leading to subsequent compaction of the activated sludge. The filaments form weblike structures extending into the bulk liquid and leading to a diffuse floc structure that bridges between the flocs. The sludge becomes extremely voluminous and difficult to settle out in the secondary clarifier. Sludge bulking can result in a host of effluent quality problems, such as solids escaping from the secondary clarifier. Bulking and foaming may also result in sludge accumulations that cause anoxic conditions to develop in the clarifier. The anoxic conditions may result in denitrification in the clarifier, producing nitrogen gas bubbles that cause sludge particles to float.

Poorly compacted sludge has inadequate thickening properties (with respect to gravity settling and dissolved air flotation) and poor dewaterability in centrifuges and belt presses, and results in excessive waste sludge volumes. This may also result in sludge with higher water contents being returned to the aeration tank with a low MLSS, which makes it difficult to maintain a desired operational MLSS in the aeration tank, leading to a fall in effluent quality. Removal of sludge from the clarifier to control the height of the growing sludge blanket results in a declining MLSS concentration in the aeration basin.

Sludge bulking may be controlled or prevented through specific or nonspecific measures, depending on the characteristics of the filamentous organisms. Specific control measures eliminate the operating conditions that favor the growth of filamentous organisms. Specific measures include manipulation of the DO concentration, manipulation of the F:M ratios by controlling the sludge wasting rate, and phosphorus removal prior to the aeration basin. Nonspecific control measures inhibit the growth of filamentous organisms using oxidants such as chlorine, hydrogen peroxide, or ozone.

Activated sludge foaming refers to the formation of dense brown foam in the aeration basin. It is caused by the excessive growth of certain filamentous microorganisms of the genus *Nocardia* and/or possibly *M. parvicella*, which entrain air bubbles from the aeration system and become buoyed to the surface of the aeration tank. Foaming results in a range of maintenance and operational issues. It can reduce available headspace in closed basins, and it can reduce the MLSS in the basin because up to 50% of the total activated sludge may be entrained in the foam. Other problems include odor production in the summer and, if the wasted sludge is processed, foaming in the anaerobic digester.

Foaming is a function of temperature, MLSS concentration, and sludge age. Reducing the sludge age, typically to less than 8 days to 9 days, is the most widely used method for controlling foaming. Antifoaming devices such as water sprays and scum traps may also be used to selectively remove the foam.

23. MEMBRANE BIOREACTORS

A *membrane bioreactor* (MBR) is a variation of the classic activated sludge process that relies on membrane separation of solids instead of using a secondary clarifier to capture and return biomass to the activated sludge process. While the process has been around for many years (as early as 1960), it emerged again in the 1990s with the advent of submerged membranes. The process relies on microfiltration or ultrafiltration membranes to separate the biomass from the treated liquid and produce a crystal clear, filtered effluent. A membrane bioreactor works by submerging a series of filters or membranes in an aeration basin. A vacuum is applied downstream of the membranes to allow for the solid/liquid separation process to occur. The membranes eliminate the need for a secondary clarifier because they act as an absolute barrier. Air is introduced into the system to scour the membranes and to drive the biological treatment.

In general, membrane bioreactors are characterized by relatively short hydraulic retention times (4 hr to 20 hr) and are capable of operating with high biomass concentrations (an MLSS of 10,000 mg/L to 12,000 mg/L). In contrast to the conventional activated sludge process in which the purification stage (aeration tank) and the separation of the biomass from the purified wastewater (settling tank) are carried out separately, membrane bioreactors use a combination of steps from both of these processes. The sedimentation in the final clarifier is generally replaced by the implementation of a membrane filtration. Using membrane technology separates the biomass from the water and considerably improves the quality of the purified wastewater. The use of microfiltration membranes with pore sizes usually between 0.1 μm and 0.4 μm ensures the complete retention of suspended matter and leads to a considerable reduction of the amount of bacteria in the effluent from the treatment process.

Membrane performance is assessed using two parameters: *water flux*, J, in units of gallons per square foot per day (cubic meter per square meter per day) calculated from Eq. 9.44, and *solute rejection*, R (%), found from Eq. 9.45.

$$J = \frac{\Delta p - \Delta \pi}{\mu R_m} \qquad \text{9.44}$$

$$R = 1 - \frac{C_p}{C_f} \qquad \text{9.45}$$

Δp is the pressure applied or transmembrane pressure, $\Delta \pi$ is the osmotic backpressure for the water, R_m is the membrane resistance, μ is the absolute (dynamic) viscosity of the water, C_p is the solute concentration in the product water, and C_f is the solute concentration in the feed water.

24. SLUDGE DISPOSAL

Sludge may be disposed of through land application or by landfilling. There are concerns about sludge incineration because of air pollutants in the emissions. The concerns, along with the high cost of supplemental fuel, make incineration a less attractive, and a less commonly constructed, means of sludge treatment and disposal. There is no process that completely eliminates the requirement for disposal of biosolids. Following either anaerobic or aerobic digestion, it may be possible to dispose of the stabilized sludge through land application. This is an attractive alternative to landfilling, because it is often less expensive and sludge has a number of beneficial land applications (e.g., fertilizer). Land application is possible if the digestion process sufficiently destroys disease-causing microorganisms and parasites. The digested solids can then be safely used as a soil amendment material (with similar benefits to peat) or as a fertilizer for agriculture.

PRACTICE PROBLEMS

1. In biological wastewater treatment systems, "yield" refers to which of the following quantities?

(A) biomass production

(B) BOD_5 removal

(C) COD removal

(D) nutrient production

2. Which of the following parameters most directly determines the specific growth rate of bacteria, μ, in biological wastewater treatment reactors?

(A) pH

(B) total suspended solids

(C) alkalinity

(D) nutrient concentration

3. An activated sludge process produces 1600 lbm of dry sludge per day at an operating temperature of 50°F. If the wet sludge has a solids content of 4.0%, what is the volume of wet sludge produced that must be processed?

(A) 240 ft^3/day

(B) 340 ft^3/day

(C) 540 ft^3/day

(D) 640 ft^3/day

4. What is the effluent BOD_5 from an activated sludge system that treats a wastewater having a BOD_5 of 180 mg/L and a BOD_5 removal efficiency of 93%?

(A) 5 mg BOD_5/L

(B) 10 mg BOD_5/L

(C) 12 mg BOD_5/L

(D) 13 mg BOD_5/L

5. During which of the following biological growth phases is a conventional activated sludge process designed to operate?

(A) lag

(B) exponential

(C) stationary

(D) endogenous

6. What is the F:M ratio for a 770,000 gal conventional activated sludge system with a recycle ratio of 0.25, an influent flow rate of 1 MGD, a MLVSS concentration of 700 mg/L, and an influent BOD_5 of 160 mg/L?

(A) 0.1 lbm/day-lbm

(B) 0.2 lbm/day-lbm

(C) 0.3 lbm/day-lbm

(D) 0.4 lbm/day-lbm

7. What is the percentage BOD_5 removal efficiency for a conventional activated sludge system having a recycle ratio of 0.45, an influent flow rate of 1 MGD, an influent BOD_5 of 235 mg/L, and an effluent BOD_5 of 27 mg/L?

(A) 50%

(B) 66%

(C) 70%

(D) 89%

SOLUTIONS

1. For wastewater applications, "yield" refers to the amount of biomass produced per unit of food consumed. For most applications the food is measured as the BOD_5 concentration.

The answer is (A).

2. A number of process and water quality parameters ultimately determine the specific growth rate coefficient. In most cases, nutrients are present at the lowest concentrations relative to other materials. Therefore, for a given system, the specific growth rate coefficient, μ, for the biological population will most directly be determined by the limiting nutrient concentration (C:N:P).

The answer is (D).

3. At 50°F the density of water is $62.4 \ \text{lbm/ft}^3$, which may be used to approximate the sludge density, as sludge is composed mostly of water. Calculate the wet sludge volume using Eq. 9.38.

$$V_{\text{sludge,wet}} = \frac{m_{\text{dry}}}{s\rho_{\text{sludge}}}$$

$$= \frac{1600 \ \dfrac{\text{lbm}}{\text{day}}}{(0.04)\left(62.4 \ \dfrac{\text{lbm}}{\text{ft}^3}\right)}$$

$$= 641 \ \text{ft}^3/\text{day} \quad (640 \ \text{ft}^3/\text{day})$$

The answer is (D).

4. The effluent BOD_5 concentration from the activated sludge process can be calculated using Eq. 9.29.

$$\eta_{\text{BOD}} = \frac{S_o - S}{S_o}$$

$$S = S_o(1 - \eta_{\text{BOD}})$$

$$= \left(180 \ \frac{\text{mg BOD}_5}{\text{L}}\right)(1 - 0.93)$$

$$= 12.6 \ \text{mg BOD}_5/\text{L} \quad (13 \ \text{mg BOD}_5/\text{L})$$

The answer is (D).

5. From Sec. 9.9 and Fig. 9.6, there are four characteristic phases of growth of a biological system. Sludge settleability improves at lower F:M ratios, which occur when there are more cells than available food, conditions that most readily occur in the beginning of the endogenous phase.

The answer is (D).

6. For a conventional activated sludge system, the F:M ratio should be between 0.2 lbm/day-lbm and 0.4 lbm/day-lbm. Substitute the given values into Eq. 9.18.

$$F:M = \frac{S_o Q_o}{XV}$$

$$= \frac{\left(160 \ \dfrac{\text{mg}}{\text{L}}\right)(1 \ \text{MGD})\left(10^6 \ \dfrac{\dfrac{\text{gal}}{\text{day}}}{\text{MGD}}\right)}{\left(700 \ \dfrac{\text{mg}}{\text{L}}\right)(770{,}000 \ \text{gal})}$$

$$= 0.297 \ \text{lbm/day-lbm} \quad (0.3 \ \text{lbm/day-lbm})$$

The answer is (C).

7. The recycle rate and the flow rate are extraneous information that are not needed for the solution. Substitute the influent and effluent BOD_5 values into Eq. 9.29.

$$\eta_{\text{BOD}} = \frac{S_o - S}{S_o}$$

$$= \frac{235 \ \dfrac{\text{mg}}{\text{L}} - 27 \ \dfrac{\text{mg}}{\text{L}}}{235 \ \dfrac{\text{mg}}{\text{L}}}$$

$$= 0.885 \quad (89\%)$$

The answer is (D).

Topic III: Environmental Impact

Environmental
Impact

10 Hazardous Waste and Pollutants

Nomenclature

A	area of fabric material	ft^2	m^2
AQI	air quality index	–	–
C	concentration	$\mu g/L$	$\mu g/L$
Q	volumetric air flow rate	ft^3/min	m^3/min
TEF	toxic equivalence factor	–	–
TEQ	toxic equivalents	$\mu g/L$	$\mu g/L$
v_f	superficial filtering velocity	ft/min	m/min

Subscripts

BP	breakpoint
Hi	high
Lo	low
p	pollutant

1. THE ENVIRONMENT AND ENVIRONMENTAL ENGINEERING

The natural and constructed environment comprises all living and nonliving things that occur naturally (i.e., non-manufactured) in and on the earth. As manufactured materials, products, and waste streams inevitably make contact and interact with the natural environment, it is necessary to manage and to mitigate the deleterious effects of this interaction on the environment. *Environmental engineering* is the application of science and engineering principles to improve the environment (air, water, and/or land resources), to provide healthy water, air, and land for humans and other organisms, and to remediate polluted sites.

2. ENVIRONMENTAL IMPACT REPORT

Prior to the construction or implementation of large-scale projects, the potential environmental impact of the project must be assessed. Environmental assessments may take a variety of forms. *Environmental impact reports* (EIRs) are studies that look at all the factors in a land development or construction project that may affect the surrounding environment, including the effects on population, traffic, schools, fire protection, endangered species, archeological artifacts, and community beauty. States may require that an EIR be submitted to local governments prior to project approval, unless the governmental body finds there is no possible impact. A finding of no possible impact is called a *negative declaration*. Regardless of the type of EIR prepared, a direct link between impacts and mitigation and remediation measures is generally established in the report.

3. HAZARDOUS AND NONHAZARDOUS WASTE

Waste is specifically defined as any unwanted material: solid, liquid, or gas. There are two classifications of waste set by the United States Environmental Protection Agency (USEPA): hazardous and nonhazardous waste. *Hazardous waste* is waste that is dangerous or potentially harmful to human health or the environment (see Sec. 10.5). It can be a by-product of manufacturing processes or simply a discarded commercial product, such as a cleaning fluid or a pesticide. Hazardous waste can be a solid, liquid, contained gas, or sludge.

Nonhazardous waste includes all solid waste that does not meet the USEPA definition of hazardous waste (see Sec. 10.4) and includes things such as alkaline batteries, cement kiln dust, construction and demolition debris, crude oil and gas waste, fossil fuel combustion waste, municipal solid waste, mining waste, mineral processing waste, organic materials, and scrap tires.

4. PRIMARY LEGISLATION

The disposal of hazardous waste and the remediation of contaminated sites is governed by several key pieces of federal legislation, including the following.

- Resource Conservation and Recovery Act of 1976 (RCRA)

- Comprehensive Environmental Response, Compensation, and Liability Act of 1980 (CERCLA)

- Hazardous and Solid Waste Amendments of 1984 (HSWA)

- Superfund Amendments and Reauthorization Act of 1986 (SARA)

The RCRA is the centerpiece of the United States' efforts to protect groundwater and regulate solid and hazardous waste. It provides a system to track the location of hazardous waste as it is produced, transported, and ultimately disposed. Subtitle C of the RCRA legislation details the requirements for disposing of hazardous waste. Subtitle D provides for the regulation of municipal solid waste disposal and the permitting of municipal landfills.

The HSWA regulates the disposal of hazardous waste and the design and operation of underground storage tanks for petroleum products. The HSWA banned land disposal of hazardous waste, while also substantially reducing the exemption for small hazardous waste generators. Solid waste management units (SWMUs) also became regulated. SWMUs are currently inactive, but were formerly used as hazardous waste sites. Under this legislation, underground storage tanks became regulated by the USEPA, which subsequently mandated design, installation, and maintenance guidelines to prevent leaking.

The CERCLA provides a legal and regulatory basis for cleaning up releases of hazardous substances into the environment. The CERCLA also introduced the concept of hazardous substance reporting. With the CERCLA, the Superfund Tax Act was initiated, which created an excise tax on oil and certain hazardous substances. Money from these taxes was deposited into a fund (known as the Superfund) for the purpose of financing the remediation of non-RCRA hazardous waste sites. The CERCLA is not a true regulatory act because there are no permits or applications. Instead, the CERCLA created a process for identifying releases and cleaning up sites. The National Contingency Plan is a process for identifying and cleaning up designated Superfund sites. It establishes the criteria for placing sites on the Superfund list, studying and evaluating sites, and remediating sites. The National Contingency Plan also led to the National Priority List (NPL), which helps the USEPA prioritize candidate sites for Superfund money.

The USEPA identifies non-RCRA hazardous waste sites, prepares cleanup plans, cleans up, and then sues potentially responsible parties (PRPs) to reimburse the Superfund. The three constituent parts of this process are

1. identification, analysis, and remediation of releases

2. rules of liability associated with remediation of releases

3. general reporting requirements

The SARA enhanced and amplified the liability of PRPs as outlined in CERCLA. This act created the "no reason to know" defense for real estate purchasers, which resulted in new reporting requirements and environmental audits of properties prior to purchase. As a result of this legislation, the list of the top "Toxic 500" companies became public.

5. LEGAL DESIGNATIONS OF HAZARDOUS WASTE

The Resource Conservation and Recovery Act of 1976 (RCRA) and the USEPA define hazardous waste as a waste that appears on one of the following four hazardous wastes lists or exhibits at least one of the following four characteristics.

- *F-list wastes* are wastes from common manufacturing and industrial processes, such as solvents that have been used in cleaning or degreasing operations. Their sources are nonspecific.

- *K-list wastes* are wastes from certain EPA-designated sources, or industries. The EPA names 14 particular industries that can produce K-list waste. These industries include petroleum refinement, wood preservation, and explosives manufacturing, as well as other industries known to use or produce harmful chemicals.

- *P-list* and *U-list wastes* are specific discarded commercial chemical products in an unused form. They may include some pesticides and pharmaceutical products.

Wastes not included in any of the waste lists may still be considered hazardous if they exhibit one of these four characteristics: ignitability, corrosivity, reactivity, and toxicity.

- *ignitability:* Ignitable wastes can create fires under certain conditions, are spontaneously combustible, or have a flash point less than 140°F (60°C).

- *corrosivity:* Corrosive wastes are acids or bases (pH less than or equal to 2, or greater than or equal to 12.5, respectively) that are capable of corroding metal containers, such as storage tanks, drums, and barrels.

- *reactivity:* Reactive wastes are unstable under "normal" conditions. They can cause explosions, toxic fumes, gases, or vapors when heated, compressed, or mixed with water.

- *toxicity:* Toxic wastes are harmful or fatal when ingested or absorbed (e.g., wastes containing mercury or lead).

Hazardous waste is labeled according to its properties, which are grouped into the following four categories by the National Fire Protection Association (NFPA).

- health

- fire

- reactivity
- special hazard (water reactivity, radioactivity, or biohazard)

The NFPA system uses a color coded diamond with four quadrants in which numbers are used to signify the relative hazard in each class.

- health hazard (blue)
- fire hazard (red)
- reactivity hazard (yellow)
- special hazard (white)

The emergency hazards are signaled on a numerical scale of 0 to 4, as follows.

hazard number	level of hazard
0	no unusual hazard
1	minor hazard
2	moderate hazard
3	severe hazard
4	extreme hazard

The U.S. Department of Transportation hazard classes include the following.

- explosive gases (flammable, nonflammable, corrosive, and poisonous)
- flammable liquids
- flammable solids
- oxidizers
- poisons
- radioactive materials
- corrosives

6. POLLUTION

Types of Pollution

Pollution is the release of contaminants into the environment. The principal pollution categories are air, water, soil, noise, light, visual, radioactive, and thermal.

- *Air pollution* is the release of chemicals and particulates, such as carbon monoxide, sulfur dioxide, chlorofluorocarbons (CFCs), and nitrogen oxides, into the atmosphere. Air pollution generally results from the combustion of hydrocarbon-based fuels, but may also result from dust at construction sites and other disturbed areas.

- *Water pollution* is the addition of contaminants to a water body, or the alteration of the properties (temperature, pH, DO concentration) of a water body (lake, river, stream, aquifer, etc). It may result from the direct discharge of contaminants into a water body, or it may occur via urban or agricultural runoff and leaching into groundwater.

- *Soil pollution* results from the release of chemicals through spillage, direct application (accidental or intentional), and leaking storage tanks (above or underground). Among the most significant soil contaminants are hydrocarbons, heavy metals, methyl tertiary-butyl ether (MTBE), herbicides, pesticides, and chlorinated hydrocarbons.

- *Noise pollution* is the addition of high decibel noise to an area and is primarily a concern in urban areas. Common sources of noise pollution are roadways, aircraft, industrial activities, and high-intensity sonar noise.

- *Light pollution* is the artificial illumination of an area and includes light trespass, over-illumination, and astronomical interference.

- *Visual pollution* generally results from the obstruction of a natural view and can result from the presence of overhead power lines, motorway billboards, scarred landforms (e.g., strip mining), open storage of junk, or municipal solid waste.

- *Radioactive pollution* generally results from the loss of radioactive materials during the production or use of radioisotopes, primarily those used in medical imaging. Radioactive pollution may also result from the storage of spent nuclear fuels.

- *Thermal pollution* is the degradation of the ambient temperature of a water body (stream, river, lake, etc.) through the introduction of a manmade wastewater stream. The most common sources of thermal pollution include treated municipal wastewater, industrial wastewater, and power plant cooling tower blow-down (i.e., water used as a coolant). Thermal pollution is a concern due to the relationship between water temperature and the dissolved oxygen concentration, as higher temperatures result in lower dissolved oxygen concentrations.

Air, water, soil, radioactive, and thermal pollution are considered to be potentially harmful, while noise, light, and visual pollution are primarily of aesthetic concern.

Point source pollution is a single identifiable, localized source of air, water, thermal, light, or noise pollution. Examples of point sources of pollution include wastewater discharge from an effluent pipe, light pollution from a street light, air pollution from a smokestack, and thermal pollution from a power plant's outfall. *Nonpoint source pollution* is pollution from a diffuse source that

Environmental Impact

cannot be directly traced back to a single localized point. Nonpoint sources of pollution include runoff from urban or agricultural areas. Pollution from most nonpoint sources is not required to have a National Pollutant Discharge Elimination System (NPDES) permit.[1]

Runoff

The principal environmental issues associated with runoff are the impacts to surface water, groundwater, and soil through transport of water pollutants to these systems. Contaminants of particular concern to surface water systems are petroleum substances, herbicides, pesticides, and fertilizers. In the case of surface waters, the impacts translate to water pollution, because the streams and rivers have received runoff carrying various chemicals or sediments. Surface waters that supply drinking water can be compromised with respect to health risks and drinking water aesthetics (odor, color, and turbidity).

Agricultural runoff is largely associated with soil erosion and sediment loading into surface waters resulting from soil disturbance. Other issues with agricultural runoff are the transport of agricultural chemicals (nitrates, phosphates, pesticides, herbicides, etc.) via surface runoff. Agricultural chemicals in runoff become an issue when chemical use is excessive or poorly timed with respect to high precipitation. The resulting contaminated runoff represents not only a waste of agricultural chemicals, but also an environmental threat to downstream ecosystems. Because agricultural runoff is a nonpoint source of pollution, it is difficult to control and is best reduced through changes in farming practices.

Urban runoff consists of water that has drained from man-made nonporous surfaces in densely populated areas. These surfaces include roads, freeways, sidewalks, roofed structures, parking lots, airports, and industrial sites. Urban surfaces are generally nonporous and cannot filter or biodegrade contaminants like natural soil can. Suspended sediment is the primary pollutant in urban runoff, which also contains oil, grease, pesticides from turf management, road salts, metals, bacteria and viruses from failing septic systems, and toxic chemicals from automobiles, among other substances. The Clean Water Act provides control over urban runoff and storm water discharges through the National Pollution Discharge Elimination System (NPDES) permit program and

---

[1]Most stormwater discharges are considered to be point sources and require coverage by an NPDES permit. Most states are authorized to implement the stormwater NPDES permitting program. The EPA remains the permitting authority in a few states, territories, and in most Native American nations. On June 12, 2006, the EPA published a rule that exempted construction activities at oil and gas sites from the requirement to obtain an NPDES permit for stormwater discharges except in very limited instances. These amendments are consistent with the Energy Policy Act of 2005, which also encourages voluntary application of best management practices for construction activities associated with oil and gas field activities and operations to minimize erosion and control sediment to protect surface water quality.

through management programs developed by the states under the Act. The types of pollutants regulated are

- conventional pollutants, such as those found in household waste (sewage, oil and grease, and detergents)

- toxic pollutants particularly harmful to life, such as organics (pesticides, solvents, and PCBs) and metals (lead, silver, mercury, copper, chromium, and others)

- nonconventional pollutants, such as nitrogen and phosphorus

7. POLLUTANTS

Criteria Pollutants

The USEPA uses the following six criteria pollutants as indicators of air quality. It has established a maximum concentration for each pollutant that, when exceeded, may negatively impact human health.

- ozone

- carbon monoxide

- nitrogen dioxide

- sulfur dioxide

- particulate matter

- lead

The threshold concentrations for criteria pollutants are called *National Ambient Air Quality Standards* (NAAQS). (See Table 10.1.) If these standards are exceeded, the area may be designated as being in nonattainment. A *nonattainment area* is defined in the Clean Air Act and Amendments of 1990 as a locality where air pollution levels persistently exceed the *National Ambient Air Quality Standards*, or a locality that contributes to ambient air quality in a nearby area that fails to meet standards.

Nitrogen Oxides

Nitrogen oxides (NOx) incorporate the sum total of the oxides of nitrogen, the most significant of which are nitric oxide (NO) and nitrogen dioxide (NO_2). NOx are emitted from motor vehicles, power plants, and other combustion operations. NOx gases play a major role in the formation of ozone and nitrogen-bearing particles, which are both associated with adverse health effects. NOx also contribute to the formation of acid rain. Adverse environmental effects of NOx include visibility impairment, acidification of freshwater bodies, increases in levels of toxins harmful to fish and other aquatic life, and changes in the populations of some species of vegetation in wetland and terrestrial systems. Exposure to NO_2 may lead to changes in airway responsiveness and lung function in individuals with pre-

Table 10.1 National Ambient Air Quality Standards for Particle Pollution

pollutant	primary standard level	secondary standard level	averaging time
carbon monoxide, CO	9 ppm (10 mg/m^3)	n.a.	8 h^a
	35 ppm (40 mg/m^3)	n.a	1 h^a
lead, Pb	0.15 μg/m^{3b}	0.15 μg/m^3	rolling 3 month average
nitrogen dioxide, NO$_2$	53 ppb^c	53 ppb	annual (arithmetic mean)
	100 ppb	n.a.	1 h^d
inhalable coarse particles, PM$_{10}$ (2.5–10 μg/m)	150 μg/m^3	150 μg/m^3	24 h^e
fine particles, PM$_{2.5}$ ($\leq$2.5 μg/m)	15.0 μg/m^3	15.0 μg/m^3	annual^f (arithmetic mean)
ozone, O$_3$	35 μg/m^3	35 μg/m^3	24 h^g
	0.075 ppm (2008 std)	0.075 ppm	8 h^h
	0.08 ppm (1997 std)	0.08 ppm	8 h^I
	0.12 ppm	0.12 ppm	1 h^j
sulfur dioxide, SO$_2$	0.03 ppm	0.05 ppm^k	annual (arithmetic mean)
	0.14 ppm	0.05 ppm^k	24 h^a
	75 ppb^l	n.a.	1 h

^{a}Not to be exceeded more than once per year.
^{b}The 1978 lead standard (1.5 μg/m^3 as a quarterly average) remains in effect until one year after an area is designated for the 2008 standard. In areas designated nonattainment for the 1978 standard, however, the 1978 standard remains in effect until implementation plans to attain or maintain the 2008 standard are approved.
^{c}The official level of the annual NO$_2$ standard is 0.053 ppm, equal to 53 ppb, which is shown here for the purpose of clearer comparison to the 1 h standard.
^{d}To attain this standard, the 3 yr average of the 98th percentile of the daily maximum 1 h average at each monitor within an area must not exceed 100 ppb.
^{e}Not to be exceeded more than once per year on average over 3 yr.
^{f}To attain this standard, the 3 yr average of the weighted annual mean PM$_{2.5}$ concentrations from single or multiple community-oriented monitors must not exceed 15.0 μg/m^3.
^{g}To attain this standard, the 3 yr average of the 98th percentile of 24 h concentrations at each population-oriented monitor within an area must not exceed 35 μg/m^3.
^{h}To attain this standard, the 3 yr average of the fourth-highest daily maximum 8 h average ozone concentrations measured at each monitor within an area over each year must not exceed 0.075 ppm.
^{i}(a) To attain this standard, the 3 yr average of the fourth-highest daily maximum 8 h average ozone concentrations measured at each monitor within an area over each year must not exceed 0.08 ppm.
(b) The 1997 standard—and the implementation rules for that standard—will remain in place for implementation purposes as the EPA undertakes rulemaking to address the transition from the 1997 ozone standard to the 2008 ozone standard.
(c) The EPA is in the process of reconsidering these standards.
^{j}(a) The EPA revoked the 1 hr ozone standard in all areas, although some areas have continuing obligations under that standard ("anti-backsliding").
(b) The standard is attained when the expected number of days per calendar year with maximum hourly average concentrations above 0.12 ppm is $\leq$ 1.
^{k}The secondary standard of 0.5 ppm has a 3 hr averaging time.
^{l}To attain this standard, the 3 yr average of the 99th percentile of the daily maximum 1 h average at each monitor within an area must not exceed 75 ppb.

Reprinted from *Code of Federal Regulations, National Ambient Air Quality Standards*, 40 CFR Part 50, USEPA.

existing respiratory illnesses and increases in respiratory illnesses in children (5–12 yr). Long-term exposures to NO$_2$ may lead to increased susceptibility to respiratory infection and may cause alterations in the lung.

Ozone

Ground level ozone (O$_3$) is a primary component in smog and may adversely affect human health. This is in contrast to *stratospheric ozone*, which occurs naturally and provides a protective layer against ultraviolet radiation. O$_3$ is not emitted directly into the air, but is formed by the reaction of volatile organic compounds (VOCs) and NOx in the presence of heat and sunlight. Ground level ozone forms readily in the atmosphere, usually during hot summer weather.

Because O$_3$ is formed in the atmosphere over time, the highest levels of ozone typically occur 30–70 mi (48–113 km) away from the location of the highest ozone precursor (NOx and VOCs) emissions. VOCs are emitted from a variety of sources, including motor vehicles, chemical plants, refineries, factories, consumer and commercial products, and other industrial sources. O$_3$ concentrations are largely affected by climate and weather patterns.

Health effects from O$_3$ exposure include increased susceptibility to respiratory infection, lung inflammation, aggravation of pre-existing respiratory diseases such as asthma, significant decreases in lung function, and increased respiratory symptoms, such as chest pain and cough. Environmental impacts of ground level ozone include reductions in agricultural and commercial forest yields, reduced growth and survivability of tree seedlings, and increased plant susceptibility to disease, pests, and other environmental stresses. In 1997, the USEPA revised the national ambient air quality standards for ozone by replacing the 1 hr ozone standard of 0.12 parts per million (ppm) with a new 8 hr standard of 0.08 ppm.

Acid Rain

Acid rain describes a mixture of wet and dry atmospheric deposition containing higher than normal amounts of nitric and sulfuric acids. The precursors, or chemical forerunners, of acid rain result from both natural sources, such as volcanoes and decaying vegetation, and from manufactured sources, primarily sulfur dioxide (SO_2) and nitrogen oxide (NOx) emissions from fossil fuel combustion. Acid rain is associated with the acidification of soils, lakes, and streams, accelerated corrosion of buildings and monuments, and reduced visibility. In the United States, about two-thirds of all SO_2 and one-quarter of all NOx come principally from coal burning power plants. Acid rain occurs when SO_2 or NOx gases react in the atmosphere with water, oxygen, and other chemicals to form mild solutions of sulfuric or nitric acid.

Wet deposition is acidic rain, fog, mist, and snow. It affects a variety of plants and animals, with the ultimate impact determined by factors including the acidity of the liquid solutions, the chemistry and buffering capacity of the soils involved, and the types of fish, trees, and other living organisms dependent on the water.

Dry deposition is the incorporation of acids into dust or smoke, which then accumulate on the ground, buildings, homes, cars, and trees. Dry deposited acids and particles can be washed from these surfaces by rainstorms, leading to increased acidity in the runoff. About half of the acidity in the atmosphere falls back to earth through dry deposition.

Carbon Monoxide

Carbon monoxide (CO) is a colorless, odorless, and, at high levels, poisonous gas, formed when the carbon in fuel is not burned completely. It is a component of motor vehicle exhaust, which contributes about 60% of all CO emissions nationwide and 95% of all emissions in cities. High concentrations of CO generally occur in areas with heavy traffic congestion. Other sources of CO emissions include refineries and industrial processes and non-transportation fuel combustion, including wood and refuse burning. Peak atmospheric CO concentrations typically occur in winter months when CO automotive emissions are greater and nighttime atmospheric inversion conditions are more frequent. *Atmospheric inversions* occur when pollutants are trapped in a cold layer of air beneath a warmer one. High concentrations of CO enter the bloodstream through the lungs and reduce oxygen delivery to the body, which can cause visual impairment and reduce

work capacity, manual dexterity, learning ability, and performance in complex tasks.

Smog

Smog is essentially a mixture of pollutants and water and is considered air pollution. Classic smog results from large amounts of coal burning in a geographical area and is caused by a mixture of smoke and sulfur dioxide (SO_2). Photochemical smog is a mixture of air pollutants, such as nitrogen oxides (NOx), tropospheric ozone, volatile organic compounds (VOCs), peroxyacyl nitrates, and aldehydes. Photochemical smog components are generally highly reactive and oxidizing. Photochemical smog is caused by a reaction between sunlight and the various chemical components listed. Smog is of principal concern in urban areas.

Lead

Lead (Pb) is a metallic element that occurs naturally in soil, rocks, water, and food. Lead was largely used as an additive to paints and gasoline until the 1970s when its use was significantly reduced. Current sources of Pb are lead-acid car batteries and industrial coatings. Approximately 70% of Pb pollution comes from smelters, power plants fueled by coal, and lead used in the processing of oil shale. Exposure to Pb can occur through inhalation of air containing Pb and ingestion of Pb in food, water, soil, or dust. Excessive Pb exposure can cause seizures, mental retardation, and/or behavioral disorders.

Air Quality Index

The *Air Quality Index* (AQI)[2] was developed by the USEPA to quantify and describe the daily air quality for the general public. The Air Quality Index is divided into six categories of health concern. Each category has a range of AQIs associated with it. A higher AQI indicates lower air quality. The categories are color-coded to simplify identification of hazardous air quality conditions by the public. For example, green corresponds to good air quality, while maroon signalizes hazardous air quality. (See Table 10.2.) The USEPA requires the AQI to be calculated and reported daily for cities with a population exceeding 350,000 people.

The AQI is determined by five of the six criteria pollutants[3] regulated by the Clean Air Act: ground level ozone (O_3), total suspended particulates (TSP), carbon monoxide (CO), sulfur dioxide (SO_2), and nitrogen dioxide (NO_2). Each AQI category is a separate scale for each criteria pollutant that is based on the concentrations deemed permissible by the National Ambient Air Quality Standards (NAAQS). AQI values greater than 100 indicate unhealthy air quality, while values less

---

[2]The USEPA created the Air Quality Index as a replacement for the *Pollutant Standards Index* (PSI) in 1999.

[3]Lead is currently the only criteria pollutant not monitored using the AQI.

Environmental Impact

Table 10.2 Air Quality Index Categories

AQI category	color	AQI range	description
good	green	0 to 50	Air quality is considered satisfactory, and air pollution poses little or no risk
moderate	yellow	51 to 100	Air quality is acceptable. However, for some pollutants, there may be a moderate health concern for a very small number of people who are unusually sensitive to air pollution.
unhealthy for sensitive groups	orange	101 to 150	Members of sensitive groups may experience health effects. The general public is not likely to be affected.
unhealthy	red	151 to 200	Everyone may begin to experience health effects. Members of sensitive groups may experience more serious health effects.
very unhealthy	purple	201 to 300	A health alert is issued that everyone may experience more serious health effects.
hazardous	maroon	301 to 500	A health warning of emergency conditions is issued. The entire population is more likely to be affected.

than 100 indicate satisfactory air quality. The AQI of an air sample can be calculated based on the pollutants present.

Each AQI category has a range of pollutant concentrations associated with it, as shown in Table 10.3. The upper and lower limits of an AQI category's pollutant concentrations are called *breakpoints*. The low and high breakpoint concentrations ($C_{BP,Lo}$ and $C_{BP,Hi}$) are the lowest and highest concentrations acceptable in each category, respectively. For example, an air sample is determined to have a pollutant concentration, $C_p{}^4$, of 8.2 ppm of CO, which falls in Table 10.3 between $C_{BP,Lo}$, 4.5 ppm, and $C_{BP,Hi}$, 9.4 ppm. Therefore, the sample is in the moderate AQI category.

---

[4]It is common practice to round up the pollutant concentration to the nearest whole number.

The AQI of an air sample, AQI_p, is determined using linear interpolation, as shown in Eq. 10.1 and Table 10.3. If an air sample has several pollutants present, the AQI is calculated for each pollutant, and the greatest AQI is selected (i.e., reported) as the daily AQI.

$$AQI_p = \left(\frac{AQI_{Hi} - AQI_{Lo}}{C_{BP,Hi} - C_{BP,Lo}} \right) (C_p - C_{BP,Lo}) + AQI_{Lo}$$

10.1

Radon

Radon is a radioactive noble gas formed by the natural decay of uranium, which is found in nearly all soils. Radon is considered a health hazard because it causes lung cancer. It is colorless, odorless, and tasteless. Radon gas can accumulate in buildings and drinking water. It causes an estimated 20,000 deaths per year in the United States. Radon is a significant contaminant that impacts indoor air quality worldwide, because it enters buildings through cracks and holes in the foundation, or through well water. Radon contamination inside of homes is of particular concern and cannot be remedied once detected. Radon concentrations in the air are typically expressed in units of picocuries per liter (pCi/L) of air. Concentrations can also be expressed in *working levels* (WL) rather than picocuries per liter (1 pCi/L = 0.004 WL).

PCBs

Polychlorinated biphenyls (PCBs) are manufactured mixtures of up to 209 individual chlorinated compounds (congeners). Because of their nonflammability, chemical stability, high boiling point, and electrical insulating properties, PCBs were used in hundreds of industrial and commercial applications, including electrical, heat transfer, and hydraulic equipment. They were also used as plasticizers in paints, plastics, and rubber products, as well as in pigments, dyes, carbonless copy paper, and many other applications. PCBs can be either oily liquids or solids, colorless to light yellow, and have no odor or taste. Some PCBs can exist as air vapor. However, the production of PCBs was stopped in the United States in 1977 because of PCB buildup in the environment, which gave rise to adverse health effects. Products made before 1977 that may still contain PCBs include old fluorescent lighting fixtures, electrical devices containing PCB capacitors, and old microscope and hydraulic oils.

PCB waste is defined by the USEPA as waste containing PCBs as a result of a spill, release, or other unauthorized disposal. The wastes are categorized according to the concentrations of PCBs that they contain and their date of disposal or spillage. PCB contaminated sites are most often cleaned up using incineration processes. Alternative technologies are allowed if the performance of these technologies is equivalent to incineration's *destruction and removal efficiency* (DRE) of 99.9999%. Both stationary

Table 10.3 Air Quality Index Breakpoint Concentrations

AQI	category	O$_3$ (ppm) 8 hr	O$_3$ (ppm) 1 hr[1]	PM$_{10}$ (μg/m^3)	PM$_{2.5}$ (μg/m^3)	CO (ppm)	SO$_2$ (ppm)	NO$_2$ (ppm)
0–50	good	0.000–0.064	–	0–54	0.0–15.4	0.0–4.4	0.000–0.034	–[2]
51–100	moderate	0.065–0.084	–	55–154	15.5–40.4	4.5–9.4	0.035–0.144	–[2]
101–150	unhealthy for sensitive groups	0.085–0.104	0.125–0.164	155–254	40.5–65.4	9.5–12.4	0.145–0.224	–[2]
151–200	unhealthy	0.105–0.124	0.165–0.204	255–354	65.5–150.4	12.5–15.4	0.225–0.304	–[2]
201–300	very unhealthy	0.125–0.374 (0.155–0.404)[3]	0.205–0.404	355–424	150.5–250.4	15.5–30.4	0.305–0.604	0.65–1.24
301–400	hazardous	–[4]	0.405–0.504	425–504	250.5–350.4	30.5–40.4	0.605–0.804	1.25–1.64
401–500	hazardous	–[4]	0.505–0.604	505–604	350.5–500.4	40.5–50.4	0.805–1.004	1.65–2.04

[1]Areas are required to report the AQI based on 8 hr ozone values. However, there are areas where an AQI based on 1 hr ozone values would be more protective. In these cases, the index for both the 8 hr and the 1 hr ozone values may be calculated and the maximum AQI reported.
[2]NO2 has no short-term NAAQS and can generate an AQI only above a value of 200.
[3]The numbers in parentheses are the associated 1 hr values to be used in this overlapping category only.
[4]8 hr O^3 values do not define higher AQI values (> 301). AQI values of 301 or higher are calculated using 1 hr O^3 concentrations.

Adapted from *Guidelines for the Reporting of Daily Air Quality–the Air Quality Index (AQI)*, EPA-454/B-06-001, USEPA.

and mobile incinerators can be used. The choice is often based on cost. Approximately 10 commercial incinerators have been approved for PCB disposal in the United States, and incineration has been used at approximately 65 hazardous waste sites with PCB contaminated soils. Alternative PCB treatment technologies include disposal in chemical waste landfills, high-efficiency boilers, thermal desorption, chemical dehalogenation or dechlorination, solvent extraction, soil washing, vitrification, and bioremediation.

Particulate Matter

Particulate matter (PM), or particle pollution, is a complex mixture of extremely small particles and liquid droplets and is one of the primary forms of air pollution. PM is emitted from numerous industrial, mobile (e.g., cars), residential, and even natural sources. PM is made up of a number of components, including acids (such as nitrates and sulfates), organic chemicals, metals, and soil or dust particles. Particle size is linked to the potential for causing health problems. The USEPA focuses on particles that are 10 μm in diameter or smaller because particles of this size, which generally pass through the throat and nose and enter the lungs, can cause adverse health effects. The USEPA groups particle pollution into two categories: inhalable coarse particles (between 2.5 μm and 10 μm in diameter) and fine particles ($\leq$ 2.5 μm diameter). Inhalable particles are found near roadways and dusty industries, while fine particles are found in smoke and haze. PM is generally designated by putting the maximum size (in numbers of microns) in the subscript. For example, PM$_{2.5}$ designates fine particles.

Sulfur Dioxide

Sulfur dioxide (SO$_2$) is the main sulfur-containing compound produced in the combustion of coal or petroleum. Oxidation of SO$_2$ in the presence of nitrogen oxides (NOx) in the atmosphere produces hydrogen sulfide (H$_2$S), an acid that ultimately produces acid rain. Over 65% of the SO$_2$ released into the air (> 13 million tons/yr) in the United States comes from electrical utilities, especially those that burn coal. Other sources of SO$_2$ are industrial facilities that derive their products from raw materials including metallic ore, coal, and crude oil, or that burn coal or oil to produce process heat (petroleum refineries, cement manufacturing, and metal processing facilities). Health and environmental impacts of SO$_2$ are respiratory illness, visibility impairment, acid rain, plant and water damage, and aesthetic damage to structures.

Dioxins

Dioxins are halogenated organic compounds, most commonly polychlorinated dibenzofurans (PCDFs) and polychlorinated dibenzodioxins (PCDDs). The most toxic dioxin is 2,3,7,8-tetrachlorodibenzo-p-dioxin (TCDD). Dioxins are formed as an unintentional by-product of many industrial processes involving chlorine, such as waste incineration, chemical and pesticide manufacturing, and pulp and paper bleaching. Two dioxins, polybrominated dibenzofurans and dibenzodioxins, have been discovered as impurities in brominated flame retardants. Dioxins are persistent in the environment and readily bioaccumulate up the food chain because they are fat soluble. Dioxins are of particular concern at sites where indirect exposure pathways (e.g., ingestion of fruits/

vegetables, meat, dairy products, and breast milk) are of concern. The toxicity of other dioxins and chemicals, such as PCBs, which act like dioxins, are measured in relation to TCDD using *toxic equivalence factors* (TEFs). *2,3,7,8-TCDD toxic equivalents* (TEQs) are determined from Eq. 10.2 by multiplying the compound concentrations, C, by their respective TEFs and summing them.

$$\text{TEQ} = \sum_{i=1}^{n} C_i(\text{TEF}_i) \qquad 10.2$$

Example 10.1

Determine the 2,3,7,8-TCDD toxic equivalent for a PCB waste mixture that contains the following congeners (given with their respective concentrations and TEFs).

- 3,3',4,4'-TCB (3 μg/L, TEF = 0.0001)
- 2,3',4,4',5-PeCB (2.5 μg/L, TEF = 0.1)
- 2,3,3',4,4',5'-HxCB (0.5 μg/L, TEF = 0.0005)

Solution

Use Eq. 10.2.

$$\begin{aligned}
\text{TEQ} &= \sum_{i=1}^{n} C_i(\text{TEF}_i) \\
&= \left(3 \ \frac{\mu\text{g}}{\text{L}}\right)(0.0001) + \left(2.5 \ \frac{\mu\text{g}}{\text{L}}\right)(0.1) \\
&\quad + \left(0.5 \ \frac{\mu\text{g}}{\text{L}}\right)(0.0005) \\
&= 0.251 \ \mu\text{g/L} \quad (0.3 \ \mu\text{g/L})
\end{aligned}$$

8. EMERGING CONTAMINANTS

In addition to the microorganisms, organic and inorganic chemicals, and other traditional pollutants that are regulated and mitigated by wastewater treatment plants, heavy urbanization and intense agriculture release a wide variety of chemicals into wastewater. These pollutants are referred to as *emerging contaminants* because they are manfactured chemicals in waste streams that are only beginning to be investigated with regard to their sources, pathways, and effects. Substances such as these are of concern because they aren't quarantined and may not be removed from the water by wastewater treatment plants. Studies carried out by the United States Geological Survey have found the following categories of emerging contaminants occurring at low concentrations in U.S. streams and rivers.

- human and veterinary drugs (including steroids, antibiotics, and both prescription and non-prescription drugs)

- natural and synthetic hormones

- detergent metabolites

- plasticizers

- insecticides

- fire retardants

- antioxidants

- fragrances

- polyaromatic hydrocarbons (PAHs)

Endocrine disruptors are emerging contaminants that span several of these categories and have received significant public attention.[5] These chemicals interfere with normal endocrine function in animals and/or humans. They include natural estrogens of human and/or animal origin, synthetic estrogens found in birth control pills and other pharmaceuticals, and chemicals such as dioxins or other chlorinated hydrocarbons that chemically resemble hormones. Chemicals suspected of acting as endocrine disruptors are found in insecticides, herbicides, fumigants, and fungicides that are used in agriculture as well as in the home. Industrial workers can be exposed to chemicals such as detergents, resins, and plasticizers with endocrine disrupting properties. Endocrine disruptors enter the air or water as a by-product of many chemical and manufacturing processes, and when plastics and other materials are burned. Further, studies have found that endocrine disruptors can leach out of plastics, including the type of plastic used to make hospital intravenous bags. Many endocrine disruptors are persistent in the environment and accumulate in fat, so greatest exposures can come from eating fatty foods and fish from contaminated water. Studies are being done to understand the links between these chemicals, human fertility, and cancers.

Though the presence of emerging contaminants is known, much research is still needed to understand their effects, particularly the effects of chronic exposure to these chemicals on human and ecological health. Besides direct toxicity, interactive effects from complex mixtures of these compounds in the environment may prove to have a significant impact. Most studies have been done on the water itself. However, for those chemicals that are known to be released but are not detected in water samples, there may be an affinity for sorption to sediment. Understanding the fate and degradation pathways of these chemicals is important to predicting their effects on the environment.

9. AIR POLLUTION PREVENTION

Pollution prevention, also commonly referred to as *source reduction*, was promulgated in 1990 by the Federal Government in the Pollution Prevention Act. *Pollution prevention* is the process of reducing or eliminating waste at the source; the modification of

---

[5]Congress passed the Food Quality Protection Act in 1996, requiring that the EPA initiate an Endocrine Disruptor Screening Program to screen pesticide chemicals and environmental contaminants for their potential to affect the endocrine systems of humans and wildlife.

waste-producing processes that enter a waste stream; promoting the uses of nontoxic or less toxic substances; implementing conservation techniques; and re-using materials rather than wasting them. This act also detailed pollution prevention practices (such as recycling, green engineering, and sustainable agriculture) that improve efficiency in the use of energy, water, and other natural resources. Soil and water pollution, and their associated remediation techniques, are addressed in Chap. 11. Air pollution is addressed in the following sections of this chapter.

Air pollution generally results from the combustion of hydrocarbon-based fuels, but may also result from dust at construction sites and other disturbed areas. Particle removal is dependent on particle size, but the calculation of collection efficiency is based only on the mass percentage collected. A variety of processes have been developed to remove particulate matter from gas streams, including cyclones, wet scrubbers, electrostatic precipitators, and baghouses. Electrostatic precipitators and baghouses are among the most efficient technologies for removing particulate matter and are thus generally used for air pollution prevention. Particulate matter removal efficiencies of various pollution control devices are given in Table 10.4.

Table 10.4 Removal Efficiencies of Various Particulate Matter Control Devices

control device	removal efficiency (%)			
	$<1 \ \mu m$	$1–3 \ \mu m$	$3–10 \ \mu m$	$>10 \ \mu m$
electrostatic precipitator	96.5	98.3	99.1	99.5
fabric filter	100	99.8	>99.9	>99.9
venturi scrubber	>70.0	99.5	>99.8	>99.8
cyclone	11	54	85	95

Cyclones

A cyclone separator is a type of inertial or impingement separator (see Fig. 10.1). Inertial devices, which are primarily used for the collection of medium-sized and coarse particles, use the inertial properties of the particles to separate them from the carrier gas stream. Cyclones are low-cost, low-maintenance centrifugal collectors, which are typically used to remove particulates in the size range of 10 μm to 100 μm. The fine dust-removal efficiency of cyclones is typically below 70%. Therefore, cyclones are often used as a preliminary to other particulate matter removal mechanisms.

Wet Scrubbers

Wet scrubbers use a liquid spray (scrubbing liquid), usually water, primarily to remove gaseous emissions from gas streams. Particulate control (removal of particulate matter and fine dust) is usually a secondary function. A number of different types of wet scrubbers exist, including venturi scrubbers, jet (fume) scrubbers, and spray towers or chambers. Venturi scrubbers use

Figure 10.1 Typical Gas Cyclone

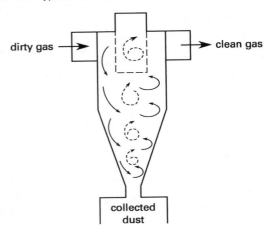

large amounts of scrubbing liquid and electric power, and operate with high-pressure drops. Jet or fume scrubbers rely on the kinetic energy of the liquid spray. The typical removal efficiency of jet or fume scrubbers (for particles smaller than 10 μm) is lower than that of venturi scrubbers. Spray towers can handle larger gas flows with minimal pressure drops. Therefore, they are often used as pre-coolers. Wet scrubbers may contribute to corrosion, so removal of water from the effluent scrubber gas may be required. There are two effluent streams from the wet scrubber, the cleaned gas stream and the dirty liquid scrubbing spray. Wet scrubbers are employed under the following circumstances.

- The contaminant cannot be removed easily in a dry form.

- Soluble gases are present.

- Wettable particles are present.

- The contaminant will undergo some subsequent wet process (e.g., recovery, wet separation or settling, or neutralization).

Typical gas flow rates range from 700 ft^3/min to 100,000 ft^3/min (20 m^3/min to 3000 m^3/min).

Spray Towers (Gas Absorption)

A spray tower is a type of wet scrubber that consists of empty cylindrical or rectangular chambers in which the gas stream flows upward through a bank or series of banks of spray nozzles and contacts liquid droplets generated by the spray nozzles. These devices have low gas pressure drops and nearly all of the contacting power is derived from the liquid stream. The necessary contacting power is determined by the liquid pressure and flow rate. Physical absorption depends on properties of the gas stream and liquid solvent (e.g., density and viscosity), and the specific characteristics of the materials in the gas and the liquid stream (e.g., diffusivity, equilibrium solubility). These properties are temperature dependent, and lower temperatures generally favor absorption of gases by the solvent. Absorption is also enhanced by greater contacting surface area, higher liquid-gas ratios, and higher concentrations in the gas stream. Chemical absorption may be limited by the reaction rate, although the rate limiting step is typically the physical absorption rate, not the chemical reaction rate.

Venturi Scrubber

A *venturi scrubber* accelerates the waste gas stream to atomize the scrubbing liquid and to improve gas-liquid contact inside of a duct (see Fig. 10.2). A narrowing section built into the duct forces the gas stream to accelerate as the duct narrows and then expands, increasing its velocity and turbulence. The scrubbing liquid is sprayed either into the gas stream before the gas encounters the venturi throat, in the throat, or upward against the gas flow in the throat. The scrubbing liquid is then atomized into small droplets by the turbulence in the throat and droplet-particle interaction is increased. After the throat section, the mixture decelerates and further impacts between droplets occur, causing the droplets to agglomerate. Once the particles are captured by the liquid, they and the excess liquid droplets are separated from the gas stream by an entrainment section. Venturi scrubbers generally utilize a vertical down-flow of gas through the venturi throat and incorporate a wet approach entry section to avoid a dust buildup. An adjustable throat allows for adjustment of the gas velocity and pressure drop and a wet or flooded elbow located below the venturi throat and ahead of the entrainment separator reduces wear by abrasive particles.

Figure 10.2 *Venturi Scrubber*

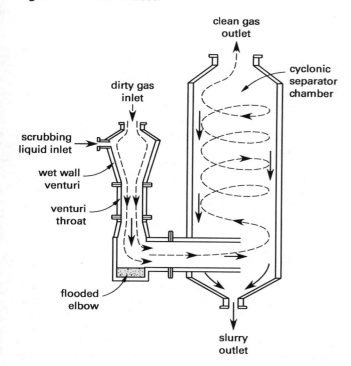

Fiber Bed Scrubbers

In *fiber bed scrubbers*, a wet waste gas passes through fiber beds made of spun glass, fiberglass, or steel. Fine fiber bed scrubbers are best suited for the removal of mists, as they are subject to rapid clogging by particulate matter. To minimize clogging in the process of particulate matter removal, the fiber beds must be composed of coarse fibers and have a high void fraction. The fiber beds are often sprayed with the scrubbing liquid so particles can be collected by deposition on droplets and fibers. For removing particulate matter, the scrubber design may include several beds and an impingement device. The final fiber bed in this design is generally dry to facilitate removal of any liquid droplets still entrained in the gas stream.

Mechanical Scrubbers

Mechanical scrubbers are devices in which a power-driven rotor produces a fine spray for increasing contact between the gas and liquid streams. The liquid droplets are the principal collecting bodies for the dust particles, and the rotor acts as a turbulence producer. An entrainment separator must be used to prevent carryover of spray. Mechanical scrubbers are usually preceded by a prefilter to remove coarse dust and larger materials. This type of scrubber relies almost exclusively on inertial interception for particulate matter collection.

Impingement Plate Scrubbers

Impingement plate scrubbers are composed of a vertical chamber with horizontally mounted plates inside a hollow shell (see Fig. 10.3) and operate as countercurrent particulate collection devices. The scrubbing liquid flows down the tower while the gas stream flows upward. Contact between the liquid and the particle-laden gas occurs on the plates. The plates are equipped with openings (e.g., pores, slots, or valve-like openings) that allow the gas to pass through. The simplest impingement plate scrubber is the sieve plate, which has round perforations. The scrubbing liquid flows over the plates and the gas flows up through the holes. The gas velocity prevents the liquid from flowing down through the perforations. Gas-liquid-particle contact is achieved within the froth generated by the gas passing through the liquid layer. In all types of impingement-plate scrubbers, the scrubbing liquid flows across each plate and down the inside of the tower onto the plate below. After the bottom plate, the liquid and collected particles flow out of the tower. Impingement plate scrubbers are most efficient at removing particles larger than approximately 1 μm.

Condensation Scrubbers

The *condensation scrubber* is a recently developed form of wet scrubber, where particulates act as nuclei for the formation of liquid droplets. To achieve this, the gas stream must be saturated with water before being scrubbed (see Fig. 10.4). Upon saturation, steam is

Figure 10.3 Impingement Plate Scrubber

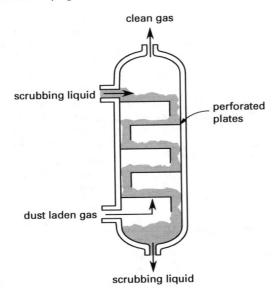

Figure 10.5 Dry Scrubber

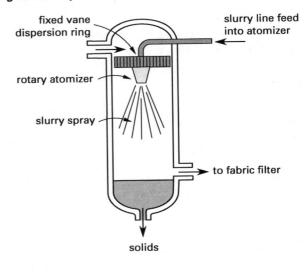

Figure 10.4 Process Flow for a Condensation Scrubber

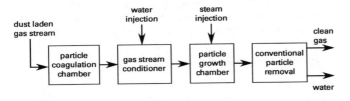

hydrochloric acid, polyaromatic hydrocarbons (PAHs), hydrofluoric acid, heavy metals, and sulfur dioxide.

10. ELECTROSTATIC PRECIPITATORS

Electrostatic precipitators remove particulate matter using an electrostatic field to impart a charge onto the particles, then collect them on oppositely charged plates (see Fig. 10.6). The collection plates are cleaned periodically to remove the collected dust into a *hopper* (a particle or dust collection device that usually consists of cloth-like bags or tubes) for disposal. Removal efficiencies for well-designed systems can exceed 99.9% of the total inlet dust loading. Precipitators are especially efficient in collecting fine particulate matter and can also remove approximately 99% of trace emissions of some toxic metals. An advantage of electrostatic precipitators is their tolerance for higher temperatures. Unlike fabric filters, they can operate with lower pressure drops. A potential drawback to precipitators is that once they have been designed and constructed for a particular combination of gas conditions and dust type, they cannot easily be adapted to other conditions. The key design parameters for precipitators are the gas flow, the drift velocity, and the area of the collection plates. Process performance is dictated by several parameters including the fly ash loading rate, the resistance of the fly ash, and the sulfur content of the fuel.

11. BAGHOUSES

A *baghouse* is a collection of fabric bags in a housing, operated by forcing dirty air through the bags. The bags vary in the types of fabrics used, size, gas flow orientation, and the method used to clean the bags. The bags are usually made of cotton, wool, synthetic, or glass fibers, and there may be hundreds of bags within one structure. This kind of fabric filtration is a well-known

injected into the gas stream, leading to a condition of supersaturation and to the condensation of water on the particulate matter. The large condensed droplets are then removed by one of several conventional devices, such as a high efficiency mist eliminator.

Dry Scrubbers

Dry scrubbers are a widely used flue-gas desulfurization technology. In dry scrubbers, an alkaline slurry is sprayed into a hot gas stream placed upstream from the particulate control device (see Fig. 10.5). As the slurry droplets evaporate, sulfur dioxide absorbs into the droplets and reacts with the dissolved and suspended alkaline material. Dry scrubbers generate a waste stream that is dry, and therefore, easier to handle than the sludge generated in a wet scrubber. This technology is limited to a flue gas volume generated from 200 MWe (megawatts electric) and smaller plants. Therefore, it is used in small to medium-sized coal fired plants. A removal efficiency of over 90% for sulfur dioxide has been achieved using dry scrubbers. Some advantages of dry scrubbers are low waste disposal costs, low water consumption, and low pressure drops. Common contaminants that are removed using dry scrubbers are

Figure 10.6 *Typical Layout for an Electrostatic Precipitator*

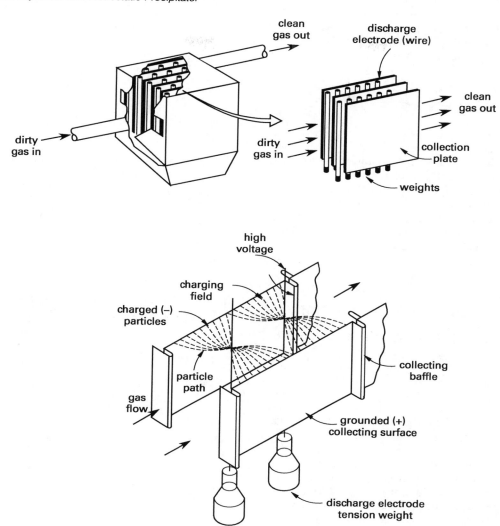

and widely used method for separating dry particles from a stream of gases (usually air or combustion gases). The dusty gas flows into and through the fabric, leaving the dust on the inside of the bag. The cleaned gas exits through the bag to the other side, then out of the baghouse. The fabric does some filtering of the dust, but it is more important in its role as a support medium for the layer of dust that quickly accumulates on it. This dust layer actually does the highly efficient filtering of small particles for which baghouses are known. The advantages of baghouses include their high collection efficiencies for small particles, versatility, and reasonable pressure drops. The disadvantages of baghouses are their large footprint, humidity limitations, potential for combustion, and susceptibility to degradation at high temperatures by corrosive chemicals.

Given the inherently high efficiency (approximately 100%) of baghouses for particle removal, their design is based on the correct selection of the fabric and method of cleaning the bags. These criteria are based on the characteristics of the process stream and the *superficial filtering velocity*, v_f, or the *air-to-cloth ratio*, which may be calculated using Eq. 10.3.

$$v_f = \frac{Q}{A} \qquad \text{10.3}$$

v_f is the superficial filtering velocity given in unit length per minute, Q is the volumetric air flow rate given in unit volume per minute, and A is the area of fabric material in unit area. The key design parameters are the gas volumetric flow rate, the superficial filtering velocity, the type of fabric, the net cloth area, and the number of compartments. Generally, the superficial filtering velocity must be low (≤ 2 ft/sec (≤ 0.61 m/s)) to allow for good filtration and to prevent large pressure drops.

12. FLUIDIZED BED COMBUSTORS

A *fluidized bed combustion system* is composed of a heated bed of sand-like material suspended in a rising column of air. These devices are capable of combusting many types and classes of fuels. They vastly improve the combustion efficiency of high moisture content fuels. The sand material acts as an abrasive to the fuel particles and enhances the combustion process by removing the carbon dioxide and char layers that normally form around the fuel particle. This allows oxygen to reach the combustible material much more readily and increases the rate and efficiency of the combustion process.

13. SELECTIVE CATALYTIC REDUCTION

Selective catalytic reduction (SCR) systems work by chemically reducing nitrogen oxides (e.g., NO and NO_2) to nitrogen gas (N_2). In selective catalytic reduction processes, a gaseous or liquid reductant is added to the flue gas stream and is absorbed onto a catalyst. The reductant reacts with the nitrogen oxide in the flue gas to form water (H_2O) and nitrogen gas (N_2). In a lean gas stream, it is necessary to add a reductant to the system to enable this reaction. The reduction system is categorized according to the source of the reductant used. In general, either ammonia (or more commonly, urea) or a hydrocarbon (lean NOx reduction) serves as the reductant. In lean systems, no reductant may be added because the hydrocarbon is already present. SCR units typically achieve approximately 80% NOx reduction.

1. How many pollutants are listed as criteria pollutants by the EPA and are used to assess air quality?

(A) 2
(B) 6
(C) 10
(D) 13

2. Solid and hazardous waste is primarily regulated by the Federal Government under what piece of legislation?

(A) RCRA
(B) CWA
(C) CAA
(D) CERCLA

3. The EPA organizes hazardous waste into how many different lists used to describe its origin?

(A) 1
(B) 2
(C) 3
(D) 4

4. A waste is defined as ignitable when it has a flash point below what temperature?

(A) 35°C
(B) 40°C
(C) 60°C
(D) 100°C

5. Los Angeles has measured an 8 hour ground level ozone (O_3) concentration of 0.086 ppm. Given the following table of breakpoint information for ground level ozone, what was the Air Quality Index (AQI) value for the city during the 8 hour period?

AQI	category	O_3 (ppm) 8 hr	O_3 (ppm) 1 hr
0–50	good	0.000–0.064	–
51–100	moderate	0.065–0.084	–
101–150	unhealthy for sensitive groups	0.085–0.104	0.125–0.164
151–200	unhealthy	0.105–0.124	0.165–0.204

(A) 85
(B) 104
(C) 117
(D) 135

6. Cheyenne, WY has measured the following concentrations for two air pollutants: $PM_{2.5}$ at 44.4 $\mu g/m^3$ and CO at 9.3 ppm. Given the following table of breakpoint information for ground level ozone, what was the Air Quality Index (AQI) value for the city following the monitoring period?

AQI	category	$PM_{2.5}$ ($\mu g/m^3$)	CO (ppm)
0–50	good	0.0–15.4	0.0–4.4
51–100	moderate	15.5–40.4	4.5–9.4
101–150	unhealthy for sensitive groups	40.5–65.4	9.5–12.4
151–200	unhealthy	65.5–150.4	12.5–15.4

(A) 65

(B) 74

(C) 99

(D) 106

SOLUTIONS

1. From Sec. 10.7, the six criteria pollutants are ozone, carbon monoxide, nitrogen dioxide, sulfur dioxide, particulate matter, and lead.

The answer is (B).

2. From Sec. 10.4, the EPA tracks hazardous waste as it is produced, transported, and disposed of and regulates Subtitle D and C landfills through the Resource Conservation and Recovery Act (RCRA), which was enacted in 1976.

The answer is (A).

3. From Sec. 10.5, the EPA organizes hazardous waste into the following lists: F-list (nonspecific source), K-list (source-specific), P-list (specific commercial chemical products), and U-list (specific commercial chemical products).

The answer is (D).

4. From Sec. 10.5, a waste may be considered hazardous when it displays one of four different characteristics: ignitability, corrosivity, reactivity, and toxicity. To be considered ignitable, the waste must ignite at or below a temperature of 60°C.

The answer is (C).

5. Refer to the data table given in the problem statement to determine the AQI category corresponding to 0.086 ppm. Use the ground level ozone column for an 8 hr monitoring period. 0.086 ppm falls into the unhealthy for sensitive groups category, which has an AQI range of 101–150. The breakpoint concentration range for this category is 0.085–0.104 ppm. Therefore, the $C_{BP,Lo}$ value is 0.085 ppm with a corresponding AQI_{Lo} of 101. Similarly, the $C_{BP,Hi}$ value is 0.104 ppm with an AQI_{Hi} of 150. From Eq. 10.1, the AQI is

$$AQI_p = \left(\frac{AQI_{Hi} - AQI_{Lo}}{C_{BP,Hi} - C_{BP,Lo}}\right)(C_p - C_{BP,Lo}) + AQI_{Lo}$$

$$= \left(\frac{150 - 101}{0.104 \text{ ppm} - 0.085 \text{ ppm}}\right)$$
$$\times (0.086 \text{ ppm} - 0.085 \text{ ppm}) + 101$$
$$= 103.6 \quad (104)$$

The answer is (B).

6. To calculate the AQI value for the city during the monitoring period, calculate the AQI of both pollutants. The highest AQI will determine the city's daily AQI.

Refer to the data table given in the problem statement. A PM$_{2.5}$ concentration of 44.4 μg/m^3 falls into the 40.5–65.4 μg/m^3 range, unhealthy for sensitive groups (AQI = 101–150). Therefore, $C_{BP,Lo}$ is 40.5 μg/m^3 with an AQI$_{Lo}$ of 101, and $C_{BP,Hi}$ is 65.4 μg/m^3 with an AQI$_{Hi}$ of 150. From Eq. 10.1, the AQI of PM$_{2.5}$ is

$$
\begin{aligned}
AQI_{p,PM2.5} &= \left(\frac{AQI_{Hi} - AQI_{Lo}}{C_{BP,Hi} - C_{BP,Lo}}\right)(C_p - C_{BP,Lo}) + AQI_{Lo} \\
&= \left(\frac{150 - 101}{65.4 \ \frac{\mu g}{m^3} - 40.5 \ \frac{\mu g}{m^3}}\right) \\
&\quad \times \left(44.4 \ \frac{\mu g}{m^3} - 40.5 \ \frac{\mu g}{m^3}\right) + 101 \\
&= 106
\end{aligned}
$$

Calculate the AQI of CO. The reported concentration of 9.3 ppm falls into the moderate range of 4.5–9.4 ppm (AQI = 51–100). Therefore, $C_{BP,Lo}$ is 4.5 ppm with an AQI$_{Lo}$ of 51, and $C_{BP,Hi}$ is 9.4 ppm with an AQI$_{Hi}$ of 100. From Eq. 10.1, the AQI of CO is

$$
\begin{aligned}
AQI_{p,CO} &= \left(\frac{AQI_{Hi} - AQI_{Lo}}{C_{BP,Hi} - C_{BP,Lo}}\right)(C_p - C_{BP,Lo}) + AQI_{Lo} \\
&= \left(\frac{100 - 51}{9.4 \text{ ppm} - 4.5 \text{ ppm}}\right)(9.3 \text{ ppm} - 4.5 \text{ ppm}) \\
&\quad + 51 \\
&= 99
\end{aligned}
$$

Since the PM$_{2.5}$ AQI is larger, the AQI for Cheyenne during the monitoring period is 106.

The answer is (D).

11 Environmental Remediation

Nomenclature

a	width of rectangular study area	ft	m
A	area	ft^2	m^2
b	length of rectangular study area	ft	m
C_{in}	inlet volatile organic compound concentration	mg/L	mg/L
C_{out}	outlet volatile organic compound concentration	mg/L	mg/L
h	total head of dewatered aquifer	ft	m
H	Henry's law constant	atm	kPa
H_D	dimensionless Henry's law constant	–	–
H_m	Henry's law constant	atm·L/ mol	atm·L/ mol
HTU	height of a transfer unit of a stripping tower	ft	m
i	hydraulic gradient	ft/ft	m/m
K	soil permeability or hydraulic conductivity	ft/sec	m/s
n	porosity	–	–
N	number of wellpoints	–	–
NTU_{OG} or N_{NOG}	number of transfer units controlled by gas resistance	–	–
NTU_{OL} or N_{OL}	number of transfer units controlled by liquid resistance	–	–
p_{vap}	vapor pressure	atm	atm
Q	groundwater flow	ft^3/sec	m^3/s
Q_d	pump flow per unit length of ditch	ft^3/sec	m^3/s
Q_G	volumetric gas flow	ft^3/min	m^3/min
Q_L	volumetric water flow	–	–
r_e	equivalent radius of influence for rectangular areas	ft	m
r_o	radius of influence	ft	m
R	stripping factor	–	–
S_w	solubility of gas in water	mol/L	mol/L
t	saturated thickness	ft	m
T	transmissivity	ft^2/day	m^2/d
V	volume	ft^3	m^3
x	mole fraction	–	–
z	total packing height, effective height of stripping tower	ft	m

Subscripts

s	solid
tot	total
vap	vapor

1. INTRODUCTION

The most appropriate technique for the remediation of contaminated sites depends on factors such as cost, site conditions, contaminant properties, and other mitigating factors. However, the process is generally determined by selecting the *best available technology* (BAT), as required by government legislation.

Remediation is the restoration of a contaminated environmental site to a condition that is not a threat to the health of humans or other life forms. In some cases, a great deal of effort is spent to clean up contaminated sites. Remediation may be carried out either in situ (in the ground) or ex situ (after removing the contaminated soil, fluid, or vapor from the ground). In situ treatments take place without great disturbance of the site. With ex situ treatments, the contaminated soil, fluid, or vapor is withdrawn and treated above ground or off-site. Therefore, ex situ treatments generally result in greater site disturbance and are usually more costly. Numerous products and technologies exist for remediating groundwater, surface water, soils, air, and sediments, including the following.

- incineration
- activated carbon absorption

Environmental Impact

- ion exchange

- soil washing

- chemical precipitation

- oxidation

- encapsulation of wastes into cement and other binders

- electrolytic treatment

- vitrification

- biological treatment (bioremediation)

2. ENVIRONMENTAL CONTAMINANTS

Remediation strategies are designed based on the state or phase of the target contaminant(s). *Nonaqueous phase liquids* (NAPLs) are immiscible (i.e., undissolved) hydrocarbons, which are common groundwater contaminants. NAPLs have characteristically low solubilities in water. NAPLs comprise either a single chemical or a complex mixture of several chemicals. For example, a spill of trichloroethylene (TCE) will produce a single component NAPL comprised of TCE, while a gasoline spill will yield an NAPL comprising a number of different aromatic and aliphatic hydrocarbons. Predicting the behavior of a mixed NAPL is much more complex than predicting the behavior of a single component NAPL. At very high concentrations, NAPLs may saturate the water and remain mostly as a pure product.

NAPLs exhibit different behaviors in the subsurface than in dissolved contaminant plumes. For instance, dissolved contaminant plumes are invisible to the naked eye, and their transport is governed by advection and dispersion. Conversely, NAPL plumes form a visible and separate oily phase, and their transport is governed by gravity, buoyancy, and capillary forces, in addition to advection and dispersion. When an NAPL is released in the subsurface, it is forced into pore structures in the soil or aquifer matrix by the hydrostatic pressure on the continuous body of the NAPL. Under pressure, the NAPL can enter very small pores and fractures in the subsurface as long as the original NAPL source (e.g., the waste pond or leaking underground storage tank) provides continuous pressure. As the supply is exhausted, pressure on NAPL is removed, and small ganglia of NAPL break off. *Ganglia* are relatively small groups of material that have separated from a larger mass. The location of the ganglia is dependent on the density of the NAPL.

Light nonaqueous phase liquids (LNAPLs) are a fraction of NAPLs that are less dense than water. Examples of LNAPLs are fuel hydrocarbons (i.e., petroleum products such as crude oil, gasoline, and benzene) deposited by spills or by accidental releases of gasoline, kerosene, and diesel fuel. LNAPLs can serve as a source of contamination for soil gas (vapor phase) and groundwater (aqueous phase), depending on their volatility and solubility in water. Of primary concern are the LNAPL dissolution products, which include the BTEX compounds of benzene, toluene, ethylbenzene, and xylene. LNAPLs are generally thought to be more manageable than the heavier fractions of NAPL because they do not penetrate very deeply into the water table and tend to be relatively biodegradable. The ganglia of LNAPLs generally reside in the unsaturated zone.

Dense nonaqueous phase liquids (DNAPLs) are the fraction of NAPL that are denser than water, including chlorinated hydrocarbons (e.g., trichloroethylene, TCE). DNAPLs are of concern because they are used in a wide variety of industrial applications, such as degreasing, dry cleaning, metal stripping, chemical manufacturing, wood-treating operations, and transformer oils. DNAPLs can serve as a source of contamination for both soil gas (vapor phase) and groundwater (aqueous phase). DNAPL dissolution products (e.g., the chlorinated aliphatic hydrocarbons PCE, DCE, and TCE, and vinyl chloride) are mostly encountered in groundwater systems.

When a DNAPL is released at the surface, it migrates downward through the unsaturated zone, continuing until either a low permeability region or stratigraphic barrier is reached or the original supply of NAPL is exhausted. The greater density of DNAPLs relative to LNAPLs results in a much broader and deeper zone of contamination. When the supply of NAPL is depleted, the DNAPL breaks up into residual ganglia that reside in both unsaturated and saturated zones. Chlorinated solvents, which are DNAPLs, are not easily biodegraded and persist for a long time in the subsurface. The physical properties of chlorinated solvents allow them to move through very small fractures in the soil. All of these characteristics make DNAPLs a formidable remediation challenge and make aquifers contaminated with large quantities of DNAPL nearly impossible to restore to original condition using current proven groundwater cleanup technologies (see Fig. 11.1).

Figure 11.1 *Evolution and Migration of a DNAPL Contaminant Plume in a Groundwater Aquifer*

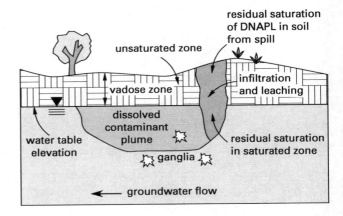

3. REMEDIATION STRATEGIES FOR NAPLS

It is nearly impossible to remove all residual NAPL once it enters and becomes trapped in the soil matrix.

Locating and accessing the NAPL is difficult because it spreads widely in isolated ganglia. There are few successful cases where sufficient NAPL has been removed from an aquifer's subsurface to restore contaminated water to drinking water quality. NAPL that remains trapped in the soil matrix acts as a continuing source of dissolved contaminants to the groundwater. The most commonly applied remediation strategies for NAPLs are natural attenuation (passive remediation), containment (a physical barrier), excavation, pump and treat, soil vapor extraction (SVE), and in situ bioremediation.

4. GROUNDWATER DEWATERING

Many civil and environmental projects require work below the groundwater table (i.e., groundwater elevation). Specifically, for environmental purposes such as groundwater remediation, it may be necessary to either lower the groundwater elevation or to extract groundwater to the surface so that it may be treated. This process is known as pump and treat. These actions are broadly categorized as *groundwater dewatering* processes. Dewatering processes remove water using wells and pumping systems. Following removal, the water may be treated to remove any number of contaminants, then discharged to a surface water body. In some instances where a discharge permit cannot be attained as a result of environmental impacts, it may be necessary to truck the water off-site for disposal. Dewatering of cofferdams and trenches is a common practice during the construction of bridges, culverts, and public utilities, such as water mains, sewers, and storm drains.

Dewatering involves creating a drawdown curve (or cone of depression) below the base of the excavation. Soil permeability, depth of the water table (groundwater elevation), and the depth and geometry of the excavation are important factors. A variety of techniques are available for dewatering. In construction, trench dewatering (single stage or multistage dewatering) is often used (see Fig. 11.2). Trenches are dug around the perimeter of the excavation site and pumps are used to draw down the water table in the area. Single stage systems have wells at a single elevation, while multistage systems have wells at multiple elevations. In general, high pumping rates are required for excavations that are deep, relative to the groundwater elevation, and for highly permeable soils (e.g., gravel or sand). Conversely, lower pumping rates are needed for shallow excavations (just below the water table) and less permeable soils (e.g., clay or silt).

The dewatering rate (extraction capacity) for any number, N, of wells with individual capacity Q_d can be estimated using Eq. 11.1 and Eq. 11.2.

$$t^2 - h^2 = \frac{NQ_d}{\pi K}(\ln r_o - \ln r_e) \qquad 11.1$$

Figure 11.2 *Single Stage Trench Dewatering Process*

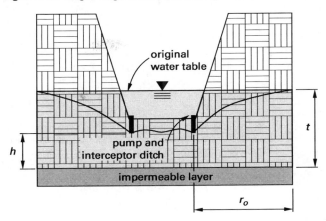

Sichardt's equation, Eq. 11.2, is used to calculate the *radius of influence* for an unconfined aquifer.

$$r_o = 3000(t - h)\sqrt{K_{\mathrm{m/s}}} \qquad \text{[SI]} \quad 11.2$$

r_o is the radius of influence for a single well, t is the depth to an impermeable layer from the original water table (commonly called the saturated thickness), h is the total head of the dewatered aquifer, and K is the soil's permeability or hydraulic conductivity in m/s. Most dewatering activities are set out over a rectangular area. The equivalent radius of influence, r_e for rectangular areas is calculated using Eq. 11.3.

$$r_e = \sqrt{\frac{ab}{\pi}} \qquad 11.3$$

a is the width and b is the length of the rectangular study area. If the study area is circular, then r_e is equal to zero. If wells are spaced evenly along an interceptor ditch (trench), the well capacity, Q_d, may be expressed in terms of volume per length of ditch.

Example 11.1

Dewatering is required at a site containing an unconfined aquifer with a reported saturated zone thickness of 60 m. The closest contaminant plume is 300 m away. In order to excavate an area 30 m long and 15 m wide, the groundwater table is lowered 5 m to the base of the excavation using a series of wells located along the perimeter of the rectangular project area. A total of 25 wells, each connected to a pumping system, is used. The hydraulic conductivity of the aquifer is 30 m/d. Determine whether the dewatering plan will be impacted by the contaminant plume, and what the total discharge will be from the sum of the well pumps.

Solution

The total head of the dewatered aquifer, h, is

$$h = 60 \text{ m} - 5 \text{ m} = 55 \text{ m}$$

From Eq. 11.2, the radius of influence is

$$r_o = 3000(t - h)\sqrt{K}$$

$$= (3000)(60 \text{ m} - 55 \text{ m})\sqrt{\frac{30 \ \frac{\text{m}}{\text{d}}}{86\,400 \ \frac{\text{s}}{\text{d}}}}$$

$$= 280 \text{ m}$$

Use Eq. 11.3 to calculate the equivalent radius of dewatering for the rectangular project area.

$$r_e = \sqrt{\frac{ab}{\pi}} = \sqrt{\frac{(15 \text{ m})(30 \text{ m})}{\pi}}$$

$$= 12.0 \text{ m}$$

Add the radius of influence and the equivalent dewatering radius.

$$r_o + r_e = 280 \text{ m} + 12.0 \text{ m}$$

$$= 292 \text{ m}$$

The radius of influence from the center of the dewatered area is 292 m. Therefore, the dewatering plan will not be influenced by the contaminant plume because it is located 300 m from the project center.

Using Eq. 11.1, the total flow rate from all of the well pumps is

$$t^2 - h^2 = \frac{NQ_d}{\pi K}(\ln r_o - \ln r_e)$$

$$(60 \text{ m})^2 - (55 \text{ m})^2 = \frac{NQ_d}{\pi \left(30 \ \frac{\text{m}}{\text{d}}\right)}$$

$$\times (\ln 280 \text{ m} - \ln 12.0 \text{ m})$$

$$NQ_d = \left(17\,226 \ \frac{\text{m}^3}{\text{d}}\right)\left(264 \ \frac{\text{gal}}{\text{m}^3}\right)$$

$$= 4{,}547{,}664 \text{ gal/day}$$

The flow rate from a single well is

$$Q_d = \frac{NQ_d}{N} = \frac{4{,}547{,}664 \ \dfrac{\text{gal}}{\text{day}}}{25 \text{ wells}} = 181{,}907 \text{ gal/day}$$

5. ADSORPTION PROCESSES

Adsorption is the process where layers of atoms or molecules of one substance are attracted onto the surface structure of another substance. *Adsorption processes* are designed to remove a given contaminant or set of contaminants from a system through adsorption from one phase to another. In an adsorption process, the substance being adsorbed loads into the surface structure of the host substance.

Adsorption is of considerable industrial importance and is a major part of many different processes throughout the chemical and processing industries, including chemical and biochemical reaction, purification, filtration, gas and liquid processing, and catalysis. Adsorption processes may be applied to the treatment of both air and water pollution sources. Because adsorption is a complex process, the correct design and implementation of its operation is critical.

6. ACTIVATED CARBON ADSORPTION

Activated carbon is a commonly used adsorbent, due to its large surface area. It can be made from a variety of carbonaceous materials, including coal, wood, and coconut shells, that are activated in a high temperature and controlled oxidation process. Once the adsorption capacity of the activated carbon is depleted, it must be removed from service and replaced with fresh carbon to maintain a constant level of removal efficiency.

Activated carbon adsorption is used in the remediation field mainly for removing volatile organic compounds (VOCs; see Sec. 11.16) such as hydrocarbons, solvents, toxic gases, and organic-based odors. During the adsorption process, molecules of a contaminated gas are attracted to, and accumulate on, the surface of the activated carbon. Chemically impregnated activated carbons may also be used to remove certain inorganic pollutants, such as hydrogen sulfide, mercury, and radon. Activated carbon is a particularly attractive treatment option because it can achieve nearly 100% removal efficiencies and can be applied to either single component emissions or complex mixtures of pollutants. Activated carbon is generally used in concert with soil vapor extraction processes, where the vapor phase is removed from the ground and passed through the activated carbon column before being reinjected or vented to the atmosphere.

Activated carbon may be regenerated or reactivated by heating it to approximately 1500°C (2732°F) in order to burn off adsorbed materials. Reactivation restores most of the original carbon adsorption capacity.

7. CHEMICAL OXIDATION

Chemical oxidation is a remediation process that can be conducted either in situ or ex situ. However, it is more efficient and more commonly done in situ. Chemical oxidation processes have been used to treat a range of volatile organic compounds (VOCs), including dichloroethylene (DCE), trichloroethylene (TCE), perchloroethylene (PCE), and benzene, toluene, ethylbenzene, and xylene (BTEX), as well as semi-volatile organic compounds (SVOCs), including pesticides, polycyclic aromatic hydrocarbons (PAHs), and polychlorinated biphenyls (PCBs). In situ chemical oxidation delivers chemical oxidants to contaminated media in order to destroy the contaminants by converting them to innocuous compounds commonly found in nature. These processes have been successfully used to remediate an extensive variety of hazardous wastes in groundwater, sediment, and soil. The oxidants are readily available, and treatment time is usually measured in months rather than years, making the process economically feasible. Oxidants that are generally used in remediation include oxygen, potassium permanganate, and hydrogen peroxide. Advanced oxidation processes, including ultraviolet (UV) radiation and ozone, have more recently been developed and used for groundwater remediation ex situ.

In situ chemical oxidation can be applied in conjunction with other treatments such as pump-and-treat and air sparging to degrade more resilient compounds. It is less costly and disruptive than other traditional soil treatments, such as excavation and incineration. In situ chemical oxidation may be used in applications in which the effectiveness of bioremediation is limited by the range of contaminants and/or climate conditions. The volume and chemical composition of individual treatments are based on the contaminant levels and volume, subsurface characteristics, and pre-application laboratory test results. Delivery vehicles for in situ chemical oxidation processes are variable. The oxidant can be injected through a well or injector head directly into the subsurface, mixed with a catalyst and injected, or combined with an extract from the site and then injected and recirculated. When hydrogen peroxide is used, stabilizers may be needed due to the compound's volatility.

8. ELECTROKINETICS

Electrokinetics is an in situ process in which a low-intensity direct electrical current is applied across electrode pairs that are implanted in the ground on each side of a contaminated area. The current causes the contaminants to migrate to the electrodes via a phenomenon known as *electro-osmosis* and *ion migration*. The direction and other migration properties of contaminants are determined by the electrochemical charge of the contaminant (positive or negative). Electrokinetic remediation processes have been used to separate and extract heavy metals, radionuclides, and organic contaminants from saturated or unsaturated soils, sludges, and sediments. Electrokinetics is particularly appropriate for low-permeability soils. However, it is also useful for high-permeability soils and for a wide range of contaminants, as long as they have an electrochemical charge.

9. EX SITU SOIL WASHING

Ex situ soil washing is a water-based process for scrubbing soils off-site to remove contaminants. In this process, contaminants are removed either by dissolving and

suspending them in the wash solution, or by concentrating them into a smaller volume of soil through particle size separation, gravity separation,[1] and attrition scrubbing.[2] Contaminant dissolution or suspension is promoted by chemical treatment of the soil or through modification of the wash solution's pH. Particle size separation is viable because most organic and inorganic contaminants tend to bind, either chemically or physically, to clay, silt, and organic soil particles. The silt and clay, in turn, are attached to sand and gravel particles by physical processes, primarily compaction and adhesion. Separating the fine particles from the coarser sand and gravel particles effectively separates and concentrates the contaminants into a smaller volume of soil that can be further treated or disposed. The clean, larger fraction can be returned to the site for continued use. Soil washing is generally a short- to medium-term process (compared to the long-term process of natural attenuation).

10. IN SITU SOIL WASHING

In situ soil washing flushes a contaminated soil and/or groundwater zone in situ with a wash solution down a gradient. The wash solution is then extracted down-gradient of the injection point and treated off-site to remove the accumulated contaminants. Wash solutions may consist of surfactants, cosolvents, acids, bases, solvents, or plain water. This process requires that the soil be moderately to highly permeable to ensure sufficient contact between the wash solution and the contaminated areas. In situ washing is applicable for the removal of a variety of organic contaminants (e.g., nonaqueous phase liquids) and may be used to remove some inorganic contaminants.

11. IN SITU VITRIFICATION

In situ vitrification is a remediation process in which the contaminants are permanently encased inside a large block of glass. In situ vitrification uses four large graphite electrodes that are inserted into the ground in a square pattern. The vitrification depth is limited by the

---

[1]*Gravity separation* separates soil particles based on differences in density. It typically involves feeding soil into a vessel in which a fluid (commonly water) is slowly flowing upward. The particles fall through the water at speeds that vary with their size, density, and, to a lesser extent, shape. If the flow rate of the water is slowly increased, the slower sinking particles flow upward with the fluid flow and are carried from the container. Intermediate particles will remain stationary, and the largest or densest particles will continue to migrate downward. The next smallest sizes of particles are removed by increasing the flow.

[2]*Attrition scrubbing* is an ex situ remediation process that consists of mechanical agitation of soil and water mixtures in a tank to remove contaminants bound to the external surfaces of particles. This is not a size separation process. Attrition scrubbing is often used to pretreat raw soil before size or gravity separation processes. Attrition scrubbers break up soil agglomerates into individual particles and "scrub" oxide or other coatings from the particles, facilitating subsequent classification by particle size. Soil scrubbing is accomplished mostly by particle-to-particle abrasion or attrition, but also by the interaction between equipment parts (e.g., paddles or propellers) and the particles.

length of the graphite electrodes and the availability of power. As the electrodes are driven into the ground, electricity is applied to them. The electricity arcs from one electrode to another and, as it passes through the soil, heat is produced. This heat reduces the soil into a molten form. As the ground liquefies, the electrodes move deeper and increase the amount of molten soil. When the electrodes have reached a maximum depth, the power is removed and the electrodes are left to be encased in the vitrified soil. As the molten soil solidifies into glass, the graphite electrodes become entombed.

Vitrification occurs at the glass transition temperature.[3] Temperatures in this process range from 3000°F to 3300°F (1600°C to 1800°C). Individual blocks of glass that weigh as much as 1400 tons (1300 tonnes) have been formed. Depths exceeding 20 ft (6 m) have been achieved. Adjacent processes can fuse the vitrified blocks together to form a single contiguous monolith. Organic pollutants are pyrolyzed into gases. The gases rise to the surface, where they are collected by a gas hood over the site and treated. Inorganics or heavy metals are encased in the vitrified soil.

12. IN SITU STABILIZATION

In situ stabilization makes the targeted contaminants innocuous and/or immobile by injecting stabilizers into the contaminated soil or groundwater. The contaminants are physically bound or enclosed within a stabilized mass (solidification), or their mobility is greatly reduced through chemical reaction (stabilization). This technique requires that the soil be moderately to highly permeable. Stabilization and solidification may be used for treating a variety of organic and inorganic contaminants.

13. PERMEABLE REACTIVE BARRIERS

Permeable reactive barriers are created from trenches that are backfilled with reactive media. The reactive media provide passive treatment of any contaminated groundwater passing through them. Permeable reactive barriers are typically used in concert with passive barriers, passive treatment walls, reactive treatment walls, or trenches. A permeable reactive trench is located to intercept a contaminant plume. It is backfilled with media such as zero-valent iron, microorganisms, zeolite, activated carbon, peat, bentonite, limestone, and sawdust. The reactive material removes or degrades the contaminant(s) through sorption, precipitation, or oxidation processes. Reactive barriers are applicable to a wide range of organic and inorganic contaminants. The choice of the reactive material depends on the properties

---

[3]The *glass transition temperature* is the temperature below which the physical properties of amorphous materials vary in a manner similar to those of a crystalline phase (glassy state), and above which amorphous materials behave like liquids (rubbery state). A material's glass transition temperature is the temperature below which molecules have little relative mobility.

of the contaminants to be removed. The geologic materials surrounding the trench must be relatively permeable, and a relatively shallow aquitard must be present to provide a "basement" to the system. Groundwater flow is preferred for moving the contamination, and the groundwater quality must support the desired reaction without imposing an additional load on the reactive media or creating undesirable by-products.

14. AQUIFERS

An *aquifer* is an underground layer of water-bearing permeable rock or unconsolidated materials (i.e., gravel, sand, silt, or clay) that can transmit enough water to supply a well under normal hydraulic gradients. In characterizing aquifers, the geologic materials are classified as consolidated rock or unconsolidated sediment. Consolidated rock generally consists of materials like sandstone, shale, granite, and basalt. Unconsolidated sediment contains granular material like sand, gravel, silt, and clay. There are four major aquifer classifications.

- alluvium (sand, gravel, and silt)
- sedimentary bedrock (consolidated sediments)
- glacial sediments (unconsolidated materials)
- igneous or metamorphic bedrock

Important aquifer characteristics include the porosity of the aquifer's geologic material and the aquifer's hydraulic conductivity, transmissivity, and water storage capacity. Each of these characteristics is linked to the type of material(s) that compose the aquifer.

Porosity, n, is a dimensionless index of the relative pore volume in a soil sample, calculated from Eq. 11.4.

$$n = \frac{V_{\text{tot}} - V_s}{V_{\text{tot}}} = \frac{V_{\text{voids}}}{V_{\text{tot}}} \qquad 11.4$$

V_{tot} is the total sample volume, V_s is the volume of solid materials in the sample, and V_{voids} is the volume of voids or pores in the sample. Porosity determines the amount of water that can be stored in a given aquifer. Groundwater movement through an aquifer is a function of the size of the spaces within the soil or rock, how well these spaces are connected, and the pressure gradient of the water table. Groundwater flow is described using Darcy's Law, Eq. 11.5.

$$Q = KAi \qquad 11.5$$

K is the hydraulic conductivity of the aquifer in feet per year or meters per year, A is the cross-sectional area of the aquifer, and i is the hydraulic gradient. K has units of length per unit time, which is the same as velocity. K is a function of the properties of both the materials that make up the aquifer and the fluid passing through it. For example, viscous fluids like crude oil will move through an aquifer at a slower rate than less viscous fluids like water. Aquifer transmissivity is a measure of the amount of water that can be transmitted horizontally through a unit width by the full saturated thickness of the aquifer under a hydraulic gradient equal to one. The transmissivity, T, in unit area per day, is the product of the hydraulic conductivity and the saturated thickness of the aquifer and is found from Eq. 11.6.

$$T = tK \qquad 11.6$$

t is the saturated thickness of the aquifer in unit length. For multilayer aquifers, the total transmissivity is the sum of the transmissivity of each of the layers.

Aquifer characteristics, such the static water level, location of water-bearing zones, and types of geologic materials present, may be determined through drilling. These characteristics, along with pump test data, are used to characterize the aquifer depth, thickness, and the nature of the aquifer (confined or unconfined). The type(s) and locations of geologic formations and water-bearing zones within the aquifer can indicate whether it is a shallow aquifer (the water table is located near the land surface) or a deep aquifer (the water table is located far below the land surface), and whether the aquifer is composed of consolidated or unconsolidated materials.

A *confined aquifer* has a layer of porous material above it that has a low permeability or a low K value, resulting in the groundwater being under pressure. The low permeability material is referred to as the *confining layer*. Confined aquifers are generally located beneath unconfined aquifers. An *unconfined aquifer* does not have a confining layer above it, so its movement is less restricted and it is under little to no pressure.

Pump tests are used to determine the hydraulic conductivity and transmissivity of an aquifer. Pump tests make it possible to estimate the effect that different pumping rates or the addition of one or more pumping wells have on water levels. These tests consist of pumping a well at a certain rate and recording the drawdown of the water level in the pumping well and in observation wells located around the pumping well over a certain time period.

15. THERMAL DESORPTION

Thermal desorption processes use heat to volatilize and desorb (physically separate) contaminants from a feedstock such as soil, sludge, or sediments. The feedstock may be heated directly or indirectly, at which time some destruction of the contaminant(s) may occur. However, contaminants are typically first separated from the feedstock, then treated further. The vaporized hydrocarbons are generally treated in a secondary unit, such as an afterburner, catalytic oxidation chamber, condenser, or carbon adsorption unit, prior to discharge to the atmosphere. Afterburners and oxidizers destroy the organic constituents. Condensers and carbon adsorption units trap organic compounds for subsequent treatment or disposal. After treatment, feedstock is transferred to a stockpile area and tested for residual contamination.

Depending on the site requirements, the feedstock will be treated further, disposed off-site, or used to backfill site excavations.

Thermal desorption units are commonly divided into high temperature and low temperature units. Low temperature units operate between 200°F and 600°F (93°C and 316°C) and are used to treat halogenated and nonhalogenated volatile organic compounds (VOCs) and petroleum hydrocarbons. High temperature units, which operate between 600°F and 1000°F (316°C and 537°C), are used to treat VOCs, semi-volatile organic compounds (SVOCs), polyaromatic hydrocarbons (PAHs), polychlorinated biphenyls (PCBs), pesticides, coal tar wastes, creosote, paint wastes, and mixed wastes.

16. AIR SPARGING

Air sparging is the process of injecting a gas (air or oxygen) under pressure into one or more wells installed within the saturated zone to volatilize contaminants dissolved in groundwater. The contaminants may be present as a nonaqueous phase liquid or adsorbed to the soil matrix. The volatilized contaminants migrate upward and are removed upon reaching the vadose zone, typically through soil vapor extraction. The *vadose zone*, also known as the unsaturated zone, is the portion of the subsurface between the ground level and the top elevation of the groundwater table, or saturated zone. Air sparging is most appropriate for removing volatile organic contaminants from moderately to highly permeable geologic materials.

17. VOLATILIZATION/VOC REMOVAL

Volatility

Volatilization is the conversion of a solid or liquid to a gas or vapor by application of heat, reduction in pressure, chemical reaction, or a combination of these processes. The volatility of a substance is its ability to escape into the vapor phase and is accounted for in terms of its *Henry's law constant*,[4] denoted as either H or K_H (having units of atm or kPa). *Henry's law* states that, at constant temperature, the concentration of a gas in a liquid of given volume is directly proportional to the partial pressure of the gas in equilbrium with the liquid. Accordingly, the solubility of a gas in a liquid at a given temperature is proportional to the partial pressure of the gas above the liquid. Henry's law constant is the mathematical description of Henry's law and may be estimated using Eq. 11.7. p_{vap} is the vapor pressure of the gas at a known temperature (atm), and x is the mole fraction of the gas (i.e., $x =$ moles of gas/moles of solution or water).

$$H \approx \frac{p_{vap}}{x} \qquad 11.7$$

---

[4]Henry's constant is available for most gases and liquids and may be found in reference tables as a function of temperature.

Henry's law constant, H_m, in atm·m^3/mol or atm·L/mol, can be calculated using the relationship shown in Eq. 11.8. S_w is the solubility of the gas in water in moles per liter, and p_{vap} is the vapor pressure in atmospheres. Table 11.1 gives values of H_m for several common chemical compounds.

$$H_m = \frac{p_{vap}}{S_w} \qquad 11.8$$

Table 11.1 *Henry's Constants for Selected Volatile Organic Compounds (VOCs) at 20°C*

compound	Henry's law constant, H_m (atm·L/mol)
trichloroethylene	500
perchloroethylene	800
1,1,1-trichloroethane	700
chloroform	180
methylene chloride	125
o-dichlorobenzene	71
carbon tetrachloride	1183
methyl-ethyl ketone	1.7
methyl-isobutyl ketone	7.1
ethylbenzene	389
1,1,2,2-tetrachloroethane	20

Vapor pressure is defined as the pressure of a vapor in equilibrium with its other phases and is generally used to describe a liquid's tendency to evaporate. A substance with a high vapor pressure at normal temperatures is often referred to as being volatile. Vapor pressure is inversely proportional to the boiling point of a substance. In other words, the higher the substance's vapor pressure at a given temperature, the lower its boiling point.

Volatile organic compounds (VOCs) are organic compounds that have high vapor pressures under normal conditions, so they tend to vaporize into the atmosphere. A wide range of carbon-based molecules, such as aldehydes, ketones, and hydrocarbons, are VOCs. The term may refer both to well-characterized organic compounds and to mixtures of variable composition. Common examples of chemical mixtures that behave as VOCs include paint thinners, dry-cleaning solvents, and some constituents of petroleum fuels (e.g., gasoline and natural gas). VOCs contribute to air pollution, particularly in urban and industrial areas, and are therefore an environmental concern.

In air pollution studies, VOCs are generally divided into two categories, methane-based (MVOCs) and nonmethane-based VOCs (NMVOCs). Methane (CH$_4$) is a greenhouse gas that contributes to global warming. Other hydrocarbon VOCs are also significant greenhouse gases, because they play a role in creating ozone and in prolonging the life of methane in the atmosphere (although the effect varies depending on local air quality). Within the NMVOCs, the aromatic compounds benzene and xylene are suspected carcinogens, which

may lead to leukemia after prolonged exposure. Some VOCs also react with nitrogen oxides in the air in the presence of sunlight to form ozone. Ozone is beneficial in the upper atmosphere because it absorbs UV light, protecting humans, plants, and animals from exposure to dangerous solar radiation. However, ozone is harmful in the lower atmosphere because it causes respiratory problems and damages structures.

VOCs are released from water sources because they are driven to reach equilibrium between their concentrations in the water and in the surrounding atmosphere. This property may be exploited for removing VOCs from waste streams through processes such as air stripping and gas stripping.

Air Stripping

In *air stripping*, the VOCs are removed from an aqueous solution by greatly increasing the surface area of the contaminated water that is exposed to air. Types of aeration methods include packed towers, diffused aeration, tray aeration, and spray aeration. In *gas stripping*, a second gas is introduced into the VOC liquid solution. The VOC transfers to the stripping gas and is removed from the water because of the inherent vapor pressure difference between the water and the VOC in the presence of the stripping gas. It is necessary that the stripping gas being introduced is free of contaminants, especially the contaminant to be removed, in order to maximize the removal efficiency.

In air stripping, the *stripping factor*, R, is a function of the gas-to-water ratio required to effectively remove a volatile organic compound (VOC) having a given Henry's constant. The ratio of the volumetric air flow, Q_G, to the volumetric water flow, Q_L, is the dimensionless air-to-water ratio, Q_G/Q_L. The required Q_G/Q_L ratio is dependent on the *dimensionless Henry's constant*, H_D,[5] of the VOCs to be removed and the hydraulic stability of the column. The air flow rate will be dependent on the pressure drop across the stripping tower. The stripping factor is determined according to Eq. 11.9.

$$R = \frac{H_D Q_G}{Q_L} \qquad 11.9$$

For stripping to occur, R must be greater than one. When no off-gas treatment is required, an R value between 10 and 15 is sufficient in most cases. Lower R values (3 to 7) and Q_G/Q_L ratios are necessary when off-gas treatment is required.

Stripping tower design is based on the transfer unit method. A *transfer unit* is a measure of the difficulty of the mass transfer operation. The transfer unit depends on the solubility and concentrations of VOCs

---

[5]The dimensionless Henry's law constant is calculated by dividing the Henry's law constant, H_m, having units of atm·m^3/mol or atm·L/mol by the universal gas constant (8.205×10^{-5} atm·m^3/mol·K) and temperature (K).

in the feed stream. The *number of transfer units* is expressed as NTU_{OG} or NTU_{OL} (alternatively, N_{OG} or N_{OL}), depending on whether the gas or liquid resistance controls the tower design. The NTU value depends on the influent VOC concentration, C_{in}, the desired effluent VOC concentration, C_{out}, and the flow rates that go into the R value (see Eq. 11.9). Equation 11.10 is used to determine the value of NTU_{OL} for an air stripper.

$$NTU_{OL} = \left(\frac{R}{R-1}\right)\ln\left(\left(1-\frac{1}{R}\right)\left(\frac{C_{in}}{C_{out}}\right)+\frac{1}{R}\right) \qquad 11.10$$

For R values greater than 12, Eq. 11.10 is simplified to Eq. 11.11.

$$NTU_{OL} = \ln\frac{C_{in}}{C_{out}} \qquad [R>12] \qquad 11.11$$

The total packing height, z, is the effective height of the tower. It is calculated from the NTUs and the *height of the transfer unit*, HTU, as shown in Eq. 11.12. The HTU value is a function of the type of packing media used in the tower. The HTU relates the stripping factor, liquid load, and the packing efficiency. In common practice, to determine the HTU, one can use a correlation or experimental data adapted to the conditions of the design that are also applicable to the packing being considered.

$$z = (HTU_{OG})(NTU_{OG})$$
$$= (HTU_{OL})(NTU_{OL}) \qquad 11.12$$

A 10% safety factor is commonly used in calculating the effective tower height. Tower heights greater than 40 ft (12 m) have several drawbacks, including increased capital, problems with media compression, and higher operation and maintenance costs. Because the removal efficiency improves little with increasing height, tower heights should generally not exceed 30 ft to 40 ft (9 m to 12 m).

Hydraulic and Pneumatic Fracturing

In order to increase soil permeability to liquids and vapors, hydraulic and pneumatic fracturing are used. These processes enhance the fracture networks within the soil substrata, accelerating contaminant removal. Fracturing methods are particularly applicable to vapor extraction, biodegradation, and thermal treatments. In *hydraulic fracturing,* water is injected under high pressure into the bottom of a borehole to cut a notch. From this notch, a slurry of water, sand, and thick biodegradable gel is pumped under high pressure into the borehole to expand the fracture. As the gel degrades, it leaves a large and highly permeable sand-filled lens. Similarly, *pneumatic fracturing* involves the injection of highly pressurized air into consolidated sediments to expand existing fractures and create a secondary fracture network. Pneumatic fracturing is most appropriate for unconsolidated sediments or bedrock.

Blast Enhanced Fracturing

At sites with fractured bedrock formations, *blast enhanced fracturing* is used to improve the rate and predictability of the recovery of contaminated groundwater. Highly fractured areas are created through detonation of explosives in boreholes. This technique is distinguished from hydraulic or pneumatic fracturing in that it involves explosives and is not generally conducted in the overburden. *Overburden* is the soil and/or rock layers that lie above the soil or rock formation of interest.

Directional Wells

Directional wells are drilled directly at any non-vertical inclination for monitoring or remediating groundwater. This type of well is especially useful when contaminant plumes cover a large area, have a linear geometry, or when surface obstructions are present. For remediation purposes, directional wells can be used in groundwater and nonaqueous phase liquid extraction, air sparging, soil vapor extraction, in situ bioremediation, in situ flushing, permeable reactive barriers, and hydraulic and pneumatic fracturing.

18. BIOLOGICAL PROCESSES

Biofiltration

Biofiltration is a pollution control technique that uses microorganisms to filter or chemically oxidize contaminant compounds. In biofiltration, vapor-phase organic contaminants are passed through a bed of porous media onto which they adsorb. Microorganisms in the media then degrade the contaminants. Specific strains of bacteria may be introduced into the filter media and optimal conditions provided to preferentially degrade specific compounds. Biofiltration is commonly used to treat wastewaters, surface water runoff and, most commonly, polluted air. Biofilters are highly efficient at removing a number of common air pollutants, including aliphatic and aromatic hydrocarbons, alcohols, aldehydes, organic acids, acrylate, carbolic acids, amines, and ammonia. Biofilter media may consist of compost, activated carbon, bulking agents, buffering agents, and inorganic additives.

Biofilter media have a dual role. One role is to filter and adsorb contaminants from the feed stream. The other role is to provide an environment for biological growth. Biofilter media support a range of organisms, including various types of bacteria, fungi, and macroscopic life, such as worms and nematodes. A well-designed biofilter is self-sustaining and includes a balance of organisms.

Biofiltration has several advantages over conventional activated carbon adsorption processes. Bioregeneration maintains a constant maximum adsorption capacity as well as a stationary and relatively short mass transfer zone. The filter does not require regeneration, and the required bed length is greatly reduced relative to carbon beds. These features reduce capital and operating expenses. Additionally, contaminants are destroyed and not just separated, as with carbon adsorption processes.

Bioventing

Bioventing is a remediation process that stimulates the natural in situ biodegradation of degradable contaminants in the vadose zone. Soils in the capillary fringe and saturated zone are not affected. Air is supplied in situ at low flow rates to oxygenate deprived microorganisms in the soil. This stimulates biodegradation and minimizes the volatilization of any VOCs that may be present. Air is removed from the soil via tangential boreholes to off-gas treatment facilities before it is released into the atmosphere.

Bioventing has been successfully used to remediate soils contaminated by petroleum hydrocarbons, nonchlorinated solvents, some pesticides, and other organic chemicals. Bioventing is most often used at sites with mid-weight petroleum products (i.e., diesel fuel and jet fuel), because lighter products (i.e., gasoline) tend to volatilize readily and can be removed more rapidly using soil vapor extraction. The success of the process is a function of the soil characteristics (stratification and structure), biodegrability of the target contaminants, and the indigenous bacterial population in the soil. Soil permeability determines the rate at which oxygen can be supplied to the subsurface microorganisms. The biodegradability of the petroleum constituents determines both the rate at which and the degree to which the constituents will be metabolized by the microorganisms.

The advantages of bioventing are that it

- uses readily available equipment
- is easy to install
- has minimal disturbance to site operations
- works in inaccessible areas (e.g., under buildings)
- has a short treatment time (6 mo to 2 yr under optimal conditions)
- is easily combined with other technologies (e.g., air sparging and groundwater extraction)
- may not require costly off-gas treatment

The disadvantages of bioventing are that

- high contaminant concentrations may initially be toxic to microorganisms
- it is not applicable for certain site conditions (e.g., low soil permeabilities, high clay content, or insufficient delineation of subsurface conditions)

- it cannot always achieve very low cleanup standards
- it only treats unsaturated-zone soils—other methods may be needed to treat saturated-zone soils and groundwater

Bioventing is not appropriate for sites with groundwater tables located within 3 ft (1 m) of the land surface. Special considerations must be taken for sites with a groundwater table located within 10 ft (3 m) of the land surface because groundwater upwelling can occur within wells that are under vacuum. This problem can be avoided by using injection wells instead of extraction wells to induce air flow.

Natural Attenuation

Natural attenuation is the process of allowing a contaminant or host of contaminants to degrade naturally over time. This process relies on a variety of physical, chemical, and biological processes that, under favorable conditions, act without human intervention to reduce the mass, toxicity, mobility, volume, or concentration of contaminants in soil or groundwater. Contaminant degradation and movement are often closely monitored.

Phytoremediation

Phytoremediation is the use of plants to remediate or remove contaminants from affected sites in situ. Phytoremediation includes the following processes.

- *rhizofiltration*—absorption, concentration, and precipitation of heavy metals by plant roots
- *phytoextraction*—extraction and accumulation of contaminants in harvestable plant tissues, such as roots and shoots
- *phytotransformation*—degradation of complex organic molecules to simple molecules, which are incorporated into plant tissues
- *phytostimulation* or *plant-assisted bioremediation*—stimulation of microbial and fungal degradation by release of exudates/enzymes into the root zone
- *phytostabilization*—absorption and precipitation of contaminants, principally metals, by plants

Periodic harvesting of the plants may or may not be required, depending on the types of contaminants to be removed. Phytoremediation is applicable to a wide range of organic and inorganic contaminants. It is most appropriate for sites where large volumes of groundwater with relatively low concentrations of contaminants must be remediated to strict standards. Phytoremediation is most effective where groundwater is within 10 ft (3 m) of the ground surface, and soil contamination is within 3 ft (1 m) of the ground surface.

PRACTICE PROBLEMS

1. Volatility describes what characteristic of a given chemical compound?

(A) treatability

(B) evaporation

(C) density

(D) reactivity

2. What is the equivalent radius of influence for a rectangular groundwater-dewatering site that is 400 ft long, 200 ft wide, and has a groundwater-pumping rate of 1000 gal/min?

(A) 100 ft

(B) 120 ft

(C) 160 ft

(D) 180 ft

3. If the mean air temperature is 20°C, what is the stripping factor for an air stripping tower that is designed to remove chloroform (Henry's law constant $= 2.79 \times 10^{-3}$ atm·m^3/mol) and uses an air-to-water ratio of 100:1?

(A) 11.6

(B) 12.0

(C) 12.5

(D) 13.0

4. Bioventing is most commonly used in remediation processes to enhance what?

(A) chemical oxidation processes

(B) in situ biodegradation processes

(C) soil vitrification

(D) volatilization processes

5. Assuming that the system in question has a stripping factor of 15, what is the number of transfer units required if a water has an influent ethylbenzene concentration of 50 mg/L and the effluent concentration of ethylbenzene is required to be less than or equal to 0.02 mg/L?

(A) 8

(B) 10

(C) 12

(D) 14

SOLUTIONS

1. Volatility describes the tendency of a given compound to evaporate.

The answer is (B).

2. From Sec. 11.4 and using Eq. 11.3, the equivalent radius of influence for a rectangular groundwater dewatering site is calculated using the width and length of the site. The groundwater-pumping rate given in the problem statement is extraneous data and is not used to solve the problem.

$$r_e = \sqrt{\frac{ab}{\pi}}$$

$$= \sqrt{\frac{(200 \text{ ft})(400 \text{ ft})}{\pi}}$$

$$= 159.6 \text{ ft} \quad (160 \text{ ft})$$

The answer is (C).

3. From Sec. 11.17, the stripping factor for a volatile compound like chloroform is calculated using the dimensionless Henry's constant and the air-to-water ratio. The Henry's law constant for chloroform is made dimensionless by dividing the given value $(2.79 \times 10^{-3}$ atm·m^3/mol) by the product of the universal gas constant $(8.205 \times 10^{-5}$ atm·m^3/mol·K) and the temperature in kelvins.

$$H_D = \frac{2.79 \times 10^{-3} \ \dfrac{\text{atm·m}^3}{\text{mol}}}{\left(8.205 \times 10^{-5} \ \dfrac{\text{atm·m}^3}{\text{mol·K}}\right)(20°C + 273°)}$$

$$= 0.116$$

From Eq. 11.9, the stripping factor is

$$R = \frac{H_D Q_G}{Q_L} = \frac{(0.116)(100)}{1}$$

$$= 11.6$$

The answer is (A).

4. From Sec. 11.23, bioventing is a remediation process in which air is supplied in situ to areas above the water table to oxygenate oxygen-poor areas in order to stimulate the growth of microorganisms and the biodegradation of contaminants.

The answer is (B).

5. From Sec. 11.17, for stripping factors greater than 12, the number of transfer units required to reduce the contaminant concentration to the desired level can be calculated using Eq. 11.11.

$$NTU_{OL} = \ln \frac{C_{in}}{C_{out}}$$

$$= \ln \left(\frac{50 \ \frac{mg}{L}}{0.02 \ \frac{mg}{L}} \right)$$

$$= 7.8 \quad (8)$$

The answer is (A).

APPENDIX 1.A
Acronyms and Abbreviations

abbreviation	acronym
ABS	acrylonitrile-butadiene-styrene plastic
AQI	Air Quality Index
ASP	activated sludge process
BAT	best available technology
BTEX	compounds composed of benzene, toluene, ethylbenzene, and xylene
CERCLA	Comprehensive Environmental Response, Compensation, and Liability Act
CFC	chlorofluorocarbons
CSTR	completely stirred tank reactor
DAF	dissolved air floatation
DCE	dichloroethylene
DDT	insecticide dichlorodiphenyltrichloroethane
DNAPL	dense nonaqueous phase liquid
DRE	destruction and removal efficiency
EIR	environmental impact report
EPA	United States Environmental Protection Agency
FC	fecal coliform
FEMA	Federal Emergency Management Agency
FS	fecal streptococcus
GBT	gravity belt thickener
gpcd	gallons per capita per day
HDPE	high density polyethylene
HSWA	Hazardous and Solid Waste Amendments
LD_{50}	lethal dose; concentration from which 50% of exposed population will die
LNAPL	light nonaqueous phase liquid
MF	microfiltration
msl	mean sea level
MSW	municipal solid waste
MTBE	methyl tertiary-butyl ether
MVOC	methane-based volatile organic compound
NAPL	nonaqueous phase liquid
NF	nanofiltration
NFPA	National Fire Protection Association
NMVOC	nonmethane-based volatile organic compound
NO_X	nitrogen oxide gases
NPDES	National Pollutant Discharge Elimination System
NRCS	Natural Resources Conservation Service
NSDWR	National Secondary Drinking Water Regulations
NTU	nephelometric turbidity units
NTU	number of transfer units
PAH	polycyclic aromatic hydrocarbons or polyaromatic hydrocarbons
Pb	lead
PCB	polychlorinated biphenyls
PCDD	polychlorinated dibenzodioxin
PCDF	polychlorinated dibenzofuran
PCE	perchloroethylene
PFR	plug flow reactor
PM	particulate matter
PM_{10}	particulate matter with a size between 2.5 μm and 10 μm
$PM_{2.5}$	particulate matter with a size less than 2.5 μm
PRP	potentially responsible parties
PSI	Pollutants Standards Index
PT	primary treatment
PVC	polyvinyl chloride
RBC	rotating biological contactor
RCRA	Resource Conservation and Recovery Act

(continued)

APPENDIX 1.A *(continued)*
Acronyms and Abbreviations

abbreviation	acronym
RDF	refuse derived fuel
RO	reverse osmosis
SARA	Superfund Amendments and Reauthorization Act
SCR	selective catalytic reduction
SRF	solid recovered fuel
SVE	soil vapor extraction
SVOC	semi-volatile organic chemicals
TC	total coliform
TCDD	2,3,7,8-tetrachlorodibenzo-p-dioxin
TCE	trichloroethylene
TEF	toxic equivalence factors
TEQ	2,3,7,8-TCDD toxic equivalent
TF	trickling filter
THMs	trihalomethanes
TKN	total Kjehldahl nitrogen
TOC	total organic carbon
TSP	total suspended particulates
UF	ultrafiltration
USACE	United States Army Corps of Engineers
USDA	United States Department of Agriculture
USDC	United States Department of Commerce
USEPA	United States Environmental Protection Agency
USGS	United States Geological Survey
USNWS	United States National Weather Service
UV	ultraviolet
VOC	volatile organic compound
WAS	waste-activated sludge
WHPA	wellhead protection areas
WL	working level
WWTP	wastewater treatment plant

Appendices

APPENDIX 1.B
Flow Rate and Velocity Unit Conversions

multiply	by	to obtain
gallon/minute	0.06309	liter/second
	0.004419	acre-foot/day
	0.002228	cubic foot/second
	0.001440	million gallon/day
	63.09×10^{-6}	cubic meter/second
liter/second	15.85	gallon/minute
	0.07005	acre-foot/day
	0.03531	cubic foot/second
	0.02282	million gallon/day
	0.0001	cubic meter/second
acre-foot/day	226.3	gallon/minute
	14.28	liter/second
	0.5042	cubic foot/second
	0.3259	million gallon/day
	0.01428	cubic meter/second
cubic foot/second	448.8	gallon/minute
	28.32	liter/second
	1.983	acre-foot/day
	0.6463	million gallon/day
	0.02832	cubic meter/second
million gallon/day	694.4	gallon/minute
	43.81	liter/second
	3.068	acre-foot/day
	1.547	cubic foot/second
	0.04382	cubic meter/second
cubic meter/second	15,850	gallon/minute
	1000	liter/second
	70.04	acre-foot/day
	35.31	cubic foot/second
	22.83	million gallon/day

Appendices

APPENDIX 1.C
Volumetric Unit Conversions

multiply	by	to obtain
cubic inch	0.01639	liter
	0.004329	U.S. gallon
	5.787×10^{-4}	cubic foot
	2.143×10^{-4}	cubic yard
	0.1639×10^{-4}	cubic meter
	0.0013×10^{-5}	acre-foot
liter	61.02	cubic inch
	0.2642	U.S. gallon
	0.03531	cubic foot
	0.001308	cubic yard
	0.001	cubic meter
	810.6×10^{-9}	acre-foot
U.S. gallon	231.0	cubic inch
	3.785	liter
	0.1337	cubic foot
	0.004951	cubic yard
	0.003785	cubic meter
	3.068×10^{-6}	acre-foot
cubic foot	1728	cubic inch
	28.32	liter
	7.481	U.S. gallon
	0.03704	cubic yard
	0.02832	cubic meter
	22.96×10^{-6}	acre-foot
cubic yard	46,660	cubic inch
	764.6	liter
	202.0	U.S. gallon
	27	cubic foot
	0.7466	cubic meter
	619.8×10^{-6}	acre-foot
cubic meter	61,020	cubic inch
	1000	liter
	264.2	U.S. gallon
	35.31	cubic foot
	1.308	cubic yard
	810.6×10^{-6}	acre-foot
acre-foot	75.27×10^{6}	cubic inch
	1,233,000	liter
	325,900	U.S. gallon
	43,560	cubic foot
	1.613	cubic yard
	1233	cubic meter

APPENDIX 2.A
Physical Properties of Water at Atmospheric Pressure
(U.S. customary units)

temperature	density	specific weight	absolute (dynamic) viscosity	kinematic viscosity	vapor pressure
(°F)	(slug/ft^3)	(lbf/ft^3)	(lbm-sec/ft^2)	(ft^2/sec)	(lbf/in^2)
32	1.940	62.416	0.374×10^{-4}	1.93×10^{-5}	0.09
40	1.940	62.423	0.323×10^{-4}	1.67×10^{-5}	0.12
50	1.940	62.408	0.273×10^{-4}	1.41×10^{-5}	0.18
60	1.939	62.366	0.235×10^{-4}	1.21×10^{-5}	0.26
70	1.936	62.300	0.205×10^{-4}	1.06×10^{-5}	0.36
80	1.934	62.217	0.180×10^{-4}	0.929×10^{-5}	0.51
90	1.931	62.118	0.160×10^{-4}	0.828×10^{-5}	0.70
100	1.927	61.998	0.143×10^{-4}	0.741×10^{-5}	0.95
120	1.918	61.719	0.117×10^{-4}	0.610×10^{-5}	1.69
140	1.908	61.386	0.0979×10^{-4}	0.513×10^{-5}	2.89
160	1.896	61.006	0.0835×10^{-4}	0.440×10^{-5}	4.74
180	1.883	60.586	0.0726×10^{-4}	0.385×10^{-5}	7.51
200	1.869	60.135	0.0637×10^{-4}	0.341×10^{-5}	11.52
212	1.847	59.843	0.0593×10^{-4}	0.319×10^{-5}	14.70

Adapted from *Design of Roadside Channels with Flexible Linings*, Hydraulic Engineering Circular No. 15, 3rd ed., Table A.7, 2005, U.S. Federal Highway Administration.

Appendices

APPENDIX 2.B
Physical Properties of Water at Atmospheric Pressure
(SI units)

temperature	density	specific weight	absolute (dynamic) μ viscosity	kinematic ν viscosity	absolute vapor pressure
(°C)	(kg/m³)	(N/m³)	(Pa·s)	(m²/s)	(Pa)
0	1000	9810	1.79×10^{-3}	1.79×10^{-6}	611
5	1000	9810	1.51×10^{-3}	1.51×10^{-6}	872
10	1000	9810	1.31×10^{-3}	1.31×10^{-6}	1230
15	999	9800	1.14×10^{-3}	1.14×10^{-6}	1700
20	998	9790	1.00×10^{-3}	1.00×10^{-6}	2340
25	997	9781	8.91×10^{-4}	8.94×10^{-7}	3170
30	996	9771	7.97×10^{-4}	8.00×10^{-7}	4250
35	994	9751	7.20×10^{-4}	7.24×10^{-7}	5630
40	992	9732	6.53×10^{-4}	6.58×10^{-7}	7380
50	988	9693	5.47×10^{-4}	5.53×10^{-7}	12 300
60	983	9643	4.66×10^{-4}	4.74×10^{-7}	20 000
70	978	9594	4.04×10^{-4}	4.13×10^{-7}	31 200
80	972	9535	3.54×10^{-4}	3.64×10^{-7}	47 400
90	965	9467	3.15×10^{-4}	3.26×10^{-7}	70 100
100	958	9398	2.82×10^{-4}	2.94×10^{-7}	101 300

Adapted from *Design of Roadside Channels with Flexible Linings*, Hydraulic Engineering Circular No. 15, 3rd ed., Table A.6, 2005, U.S. Federal Highway Administration.

APPENDIX 5.A
Selected Ten States Standards

3.1. Surface Water

A surface water source includes all tributary streams and drainage basins, natural lakes, and artificial reservoirs or impoundments above the point of water supply intake. A source water protection plan enacted for continued protection of the watershed from potential sources of contamination shall be provided as determined by the reviewing authority.

3.1.1. Quantity

The quantity of water at the source shall be adequate to meet the maximum projected water demand of the service area as shown by calculations based on a one in 50 year drought or the extreme drought of record, and should include consideration of multiple year droughts.

4.1. Clarification

Clarification is generally considered to consist of any process, or combination of processes, which reduces the concentration of suspended matter in drinking water prior to filtration.

4.1.1. Presedimentation

Detention time: Three hours detention is the minimum period recommended; greater detention may be required.

4.1.2. Coagulation

Coagulation shall mean a process using coagulant chemicals and mixing by which colloidal and suspended material are destabilized and agglomerated into settleable or filterable flocs, or both. The engineer shall submit the design basis for the velocity gradient (G value) selected, considering the chemicals to be added and water temperature, color, and other related water quality parameters. For surface water plants using direct or conventional filtration, the use of a primary coagulant is required at all times.

4.1.2.a. Mixing: The detention period should not be more than 30 seconds with mixing equipment capable of imparting a minimum velocity gradient (G) of at least 750 ft/sec-ft. The design engineer should determine the appropriate G value and detention time through jar testing.

4.1.3. Flocculation

Flocculation shall mean a process to enhance agglomeration or collection of smaller floc particles into larger, more easily settleable or filterable particles through gentle stirring by hydraulic or mechanical means.

4.1.3.b. Detention: The detention time for floc formation should be at least 30 minutes with consideration to using tapered (i.e., diminishing velocity gradient) flocculation. The flow-through velocity should be not less than 0.5 ft/min nor greater than 1.5 ft/min.

4.1.3.c. Equipment: Agitators shall be driven by variable speed drives with the peripheral speed of paddles ranging from 0.5 ft/sec to 3.0 ft/sec. External, non-submerged motors are preferred.

4.1.3.d. Piping: Flocculation and sedimentation basins shall be as close together as possible. The velocity of flocculated water through pipes or conduits to settling basins shall be neither less than 0.5 ft/sec nor greater than 1.5 ft/sec. Allowances must be made to minimize turbulence at bends and changes in direction.

4.1.4. Sedimentation

Sedimentation shall follow flocculation. The detention time for effective clarification is dependent upon a number of factors related to basin design and the nature of the raw water. The following criteria apply to conventional sedimentation units.

4.1.4.a. Detention time: Detention time shall provide a minimum of four hours of settling time. This may be reduced to two hours for lime-soda softening facilities treating only groundwater. Reduced sedimentation time may also be approved when equivalent effective settling is demonstrated or when overflow rate is not more than 0.5 gal/min-ft^2 (1.2 m/h).

4.1.4.c. Outlet weirs and submerged orifices shall be designed as follows.

4.1.4.c.1. The rate of flow over the outlet weirs or through the submerged orifices shall not exceed 20,000 gal/day-ft (250 m^3/d·m) of the outlet launder.

4.1.4.c.2. Submerged orifices should not be located lower than 3 ft below the flow line.

4.1.4.c.3. The entrance velocity through the submerged orifices shall not exceed 0.5 ft/sec.

4.1.4.d. Velocity: The velocity through settling basins should not exceed 0.5 ft/min. The basins must be designed to minimize short-circuiting. Fixed or adjustable baffles must be provided as necessary to achieve the maximum potential for clarification.

4.1.5 Solids Contact Unit

Units are generally acceptable for combined softening and clarification where water characteristics, especially temperature, do not fluctuate rapidly, flow rates are uniform, and operation is continuous. Before such units are considered as clarifiers without softening, specific approval of the reviewing authority shall be obtained. Clarifiers should be designed for the maximum uniform rate and should be adjustable to changes in flow which are less than the design rate and for changes in water characteristics. A minimum of two units are required for surface water treatment.

4.1.5.9. Detention Period

The detention time shall be established on the basis of the raw water characteristics and other local conditions that affect the operation of the unit. Based on design flow rates, the detention time should be

4.1.5.9.a. two to four hours for suspended solids contact clarifiers and softeners treating surface water, and

(continued)

APPENDIX 5.A *(continued)*
Selected Ten States Standards

4.1.5.9.b. one to two hours for the suspended solids contact softeners treating only groundwater.

4.1.5.12. Weirs or Orifices

The units should be equipped with either overflow weirs or orifices constructed so that water at the surface of the unit does not travel over 10 ft horizontally to the collection trough.

4.1.5.12.a. Weirs shall be adjustable, and at least equivalent in length to the perimeter of the tank.

4.1.5.12.b. Weir loading shall not exceed

4.1.5.12.b.1. 10 gal/min-ft of weir length (120 L/min·m) for units used for clarifiers, and

4.1.5.12.b.2. 20 gal/min-ft of weir length (240 L/min·m) for units used for softeners.

4.2 Filtration

Acceptable filters shall include, upon the discretion of the reviewing authority, the following types.

4.2.a. rapid rate gravity filters (4.2.1),

4.2.b. rapid rate pressure filters (4.2.2),

4.2.c. diatomaceous earth filtration (4.2.3),

4.2.d. slow sand filtration (4.2.4),

4.2.e. direct filtration (4.2.5),

4.2.f. deep bed rapid rate gravity filters (4.2.6),

4.2.g. biologically active filters (4.2.7),

4.2.h. membrane filtration (see Interim Standard on Membrane Technologies), and

4.2.i. bag and cartridge filters (see policy statement on Bag and Cartridge Filters for Public Water Systems).

4.2.1. Rapid Rate Gravity Filters

4.2.1.6. Filter material: The media shall be clean silica sand or other natural or synthetic media free from detrimental chemical or bacterial contaminants, approved by the reviewing authority, and having the following characteristics.

4.2.1.6.a. a total depth of not less than 24 in and generally not more than 30 in,

4.2.1.6.b. a uniformity coefficient of the smallest material not greater than 1.65, and

4.2.1.6.c. a minimum of 12 in of media with an effective size range no greater than 0.45 mm to 0.55 mm.

4.2.1.6.d. Types of filter media

4.2.1.6.d.1. Anthracite: Filter anthracite shall consist of hard, durable anthracite coal particles of various sizes. Blending of non-anthracite material is not acceptable. Anthracite shall have an

4.2.1.6.d.1.a. effective size of 0.45 mm to 0.55 mm with uniformity coefficient not greater than 1.65 when used alone,

4.2.1.6.d.1.b. effective size of 0.8 mm to 1.2 mm with a uniformity coefficient not greater than 1.7 when used as a cap,

4.2.1.6.d.1.c. effective maximum size of 0.8 mm when used as a single media on potable groundwater for iron and manganese removal only (effective sizes greater than 0.8 mm may be approved based upon onsite pilot plant studies or other demonstration acceptable to the reviewing authority).

4.2.1.6.d.1.d. a specific gravity greater than 1.4,

4.2.1.6.d.1.e. an acid solubility less than 5%, and

4.2.1.6.d.1.f. a Mho's scale of hardness greater than 2.7.

4.2.1.6.d.2. Sand: Sand shall have

4.2.1.6.d.2.a. an effective size of 0.45 mm to 0.55 mm,

4.2.1.6.d.2.b. a uniformity coefficient not greater than 1.65,

4.2.1.6.d.2.c. a specific gravity greater than 1.4, and

4.2.1.6.d.2.d. an acid solubility less than 5%.

4.2.1.6.d.4. Granular activated carbon (GAC): Granular activated carbon as a single media may be considered for filtration only after pilot or full scale testing and with prior approval of the reviewing authority.

4.2.1.6.e. Support media

4.2.1.6.e.2. Gravel: Gravel, when used as the supporting medium, shall consist of cleaned and washed, hard, durable, rounded silica particles and shall not include flat or elongated particles. The coarsest gravel shall be $2\frac{1}{2}$ in in size when the gravel rests directly on a lateral system, and must extend above the top of the perforated laterals. Not less than four layers of gravel shall be provided in accordance with the following size and depth distribution.

size	depth
$2\frac{1}{2}$ in to $1\frac{1}{2}$ in	5 in to 8 in
$1\frac{1}{2}$ in to $\frac{3}{4}$ in	3 in to 5 in
$\frac{3}{4}$ in to $\frac{1}{2}$ in	3 in to 5 in
$\frac{1}{2}$ in to $\frac{3}{16}$ in	2 in to 3 in
$\frac{3}{16}$ in to $\frac{3}{32}$ in	2 in to 3 in

(continued)

APPENDIX 5.A *(continued)*
Selected Ten States Standards

4.2.3.8. Filtration

4.2.3.8.a. Rate of filtration: The recommended nominal rate is 1.0 gal/min-ft^2 of filter area (2.4 m/h) with a recommended maximum of 1.5 gal/min-ft^2 (3.7 m/h). The filtration rate shall be controlled by a positive means.

4.3. Disinfection

Chlorine is historically the preferred disinfecting agent. Disinfection may be accomplished with gas and liquid chlorine, calcium or sodium hypochlorites, chlorine dioxide, ozone, or ultraviolet light. Disinfection is required for all surface water supplies, groundwater under the direct influence of surface water, and for any groundwater supply of questionable sanitary quality or where other treatment is provided. Disinfection with chloramines is not recommended for primary disinfection. The required amount of primary disinfection needed shall be specified by the reviewing authority. Continuous disinfection is recommended for all water supplies. Consideration must be given to the formation of disinfection by-products (DBP) when selecting the disinfectant.

4.3.1. Chlorination Equipment

4.3.1.2. Capacity: The chlorinator capacity shall be such that a free chlorine residual of at least 2 mg/L can be maintained in the water once all demands are met after contact time of at least 30 minutes when maximum flow rate coincides with anticipated maximum chlorine demand. The equipment shall be of such design that it will operate accurately over the desired feeding range.

4.3.3. Residual Chlorine

4.3.3.a. Minimum free chlorine residual in a water distribution system should be 0.2 mg/L. Minimum chloramine residuals, where chloramination is practiced, should be 1.0 mg/L at distant points in the distribution system.

7.3 Distribution System Storage

7.3.1. Pressures

The maximum variation between high and low levels in storage structures providing pressure to a distribution system should not exceed 30 ft. The minimum working pressure in the distribution system should be 35 psi (240 kPa) and the normal working pressure should be approximately 60 psi to 80 psi (410 kPa to 550 kPa). When static pressures exceed 100 psi (690 kPa), pressure reducing devices should be provided on mains or as part of the meter survey on individual service lines in the distribution system.

8.2. System Design

8.2.1 Pressure

All water mains, including those not designed to provide fire protection, shall be sized after a hydraulic analysis based on flow demands and pressure requirements. The system shall be designed to maintain a minimum pressure of 20 psi (140 kPa) at ground level at all points in the distribution system under all conditions of flow. The normal working pressure in the distribution system should be approximately 60 psi to 80 psi (410 kPa to 550 kPa) and not less than 35 psi (240 kPa).

Selections from *Recommended Standards for Water Works, Policies for the Review and Approval of Plans and Specifications for Public Water Supplies*, 2007 ed.

APPENDIX 5.B

National Primary Drinking Water Regulations Code of Federal Regulations (CFR), Title 40, Ch. I, Part 141, October 2003

microorganisms	MCLG[a] (mg/L)[b]	MCL or TT[a] (mg/L)[b]	potential health effects from ingestion of water	sources of contaminant in drinking water
Cryptosporidium	0	TT[c]	gastrointestinal illness (e.g., diarrhea, vomiting, cramps)	human and animal fecal waste
Giardia lamblia	0	TT[c]	gastrointestinal illness (e.g., diarrhea, vomiting, cramps)	human and animal fecal waste
heterotrophic plate count	n/a	TT[c]	HPC has no health effects; it is an analytic method used to measure the variety of bacteria that are common in water. The lower the concentration of bacteria in drinking water, the better maintained the water is.	HPC measures a range of bacteria that are naturally present in the environment.
Legionella	0	TT[c]	Legionnaire's disease, a type of pneumonia	found naturally in water; multiplies in heating systems
total coliforms (including fecal coliform and *E. coli*)	0	5.0%[d]	Not a health threat in itself; it is used to indicate whether other potentially harmful bacteria may be present.[e]	Coliforms are naturally present in the environment as well as in feces; fecal coliforms and *E. coli* only come from human and animal fecal waste.
turbidity	n/a	TT[c]	Turbidity is a measure of the cloudiness of water. It is used to indicate water quality and filtration effectiveness (e.g., whether disease causing organisms are present). Higher turbidity levels are often associated with higher levels of disease causing microorganisms such as viruses, parasites, and some bacteria. These organisms can cause symptoms such as nausea, cramps, diarrhea, and associated headaches.	soil runoff
viruses (enteric)	0	TT[c]	gastrointestinal illness (e.g., diarrhea, vomiting, cramps)	human and animal fecal waste

disinfection products	MCLG[a] (mg/L)[b]	MCL or TT[a] (mg/L)[b]	potential health effects from ingestion of water	sources of contaminant in drinking water
bromate	0	0.010	increased risk of cancer	by-product of drinking-water disinfection
chlorite	0.8	1.0	anemia in infants and young children; nervous system effects	by-product of drinking-water disinfection
haloacetic acids (HAA5)	n/a[f]	0.060	increased risk of cancer	by-product of drinking-water disinfection
total trihalomethanes (TTHMs)	n/a[f]	0.080	liver, kidney, or central nervous system problems; increased risk of cancer	by-product of drinking-water disinfection

(continued)

APPENDIX 5.B *(continued)*
National Primary Drinking Water Regulations Code of Federal Regulations (CFR), Title 40, Ch. I, Part 141, October 2003

disinfectants	MCLG[a] $(mg/L)^b$	MCL or TT[a] $(mg/L)^b$	potential health effects from ingestion of water	sources of contaminant in drinking water
chloramines (as Cl_2)	4[a]	4.0[a]	eye/nose irritation, stomach discomfort, anemia	water additive used to control microbes
chlorine (as Cl_2)	4[a]	4.0[a]	eye/nose irritation, stomach discomfort	water additive used to control microbes
chlorine dioxide (as ClO_2)	0.8[a]	4.0[a]	anemia in infants and young children, nervous system effects	water additive used to control microbes

inorganic chemicals	MCLG[a] $(mg/L)^b$	MCL or TT[a] $(mg/L)^b$	potential health effects from ingestion of water	sources of contaminant in drinking water
antimony	0.006	0.006	increase in blood cholesterol; decrease in blood sugar	discharge from petroleum refineries; fire retardants; ceramics; electronics; solder
arsenic	0[g]	0.010 as of January 23, 2006	skin damage or problems with cirulatory systems; may increase cancer risk	erosion of natural deposits; runoff from orchards; runoff from glass and electronics production wastes
asbestos (fiber > 10 micrometers)	7 million fibers per liter	7 MFL	increased risk of developing benign intestinal polyps	decay of asbestos cement in water mains; erosion of natural deposits
barium	2	2	increase in blood pressure	discharge of drilling wastes; discharge from metal refineries; erosion of natural deposits
beryllium	0.004	0.004	intestinal lesions	discharge from metal refineries and coal-burning factories; discharge from electrical, aerospace, and defense industries
cadmium	0.005	0.005	kidney damage	corrosion of galvanized pipes; erosion of natural deposits; discharge from metal refineries; runoff from waste batteries and paints
chromium (total)	0.1	0.1	allergic dermatitis	discharge from steel and pulp mills; erosion of natural deposits
copper	1.3	TT[h], action level = 1.3	short-term exposure: gastrointestinal distress long-term exposure: liver or kidney damage People with Wilson's disease should consult their personal doctor if the amount of copper in their water exceeds the action level.	corrosion of household plumbing systems; erosion of natural deposits
cyanide (as free cyanide)	0.2	0.2	nerve damage or thyroid problems	discharge from steel/metal factories; discharge from plastic and fertilizer factories

(continued)

APPENDIX 5.B *(continued)*

National Primary Drinking Water Regulations Code of Federal Regulations (CFR), Title 40, Ch. I, Part 141, October 2003

inorganic chemicals	MCLG[a] (mg/L)[b]	MCL or TT[a] (mg/L)[b]	potential health effects from ingestion of water	sources of contaminant in drinking water
fluoride	4.0	4.0	bone disease (pain and tenderness of the bones); children may get mottled teeth	water additive that promotes strong teeth; erosion of natural deposits; discharge from fertilizer and aluminum factories
lead	0	TT[h], action level = 0.015	infants and children: delays in physical or mental development; children could show slight deficits in attention span and learning disabilities adults: kidney problems, high blood pressure	corrosion of household plumbing systems; erosion of natural deposits
mercury (inorganic)	0.002	0.002	kidney damage	erosion of natural deposits; discharge from refineries and factories; runoff from landfills and croplands
nitrate (measured as nitrogen)	10	10	Infants below the age of six months who drink water containing nitrate in excess of the MCL could become seriously ill and, if untreated, may die. Symptoms include shortness of breath and blue baby syndrome.	runoff from fertilizer use; leaching from septic tanks/sewage; erosion of natural deposits
nitrite (measured as nitrogen)	1	1	Infants below the age of six months who drink water containing nitrite in excess of the MCL could become seriously ill and, if untreated, may die. Symptoms include shortness of breath and blue baby syndrome.	runoff from fertilizer use; leaching from septic tanks/sewage; erosion of natural deposits
selenium	0.05	0.05	hair and fingernail loss; numbness in fingers or toes; circulatory problems	discharge from petroleum refineries; erosion of natural deposits; discharge from mines
thalium	0.0005	0.002	hair loss; changes in blood; kidney, intestine, or liver problems	leaching from ore-processing sites; discharge from electronics, glass, and drug factories

organic chemicals	MCLG[a] (mg/L)[b]	MCL or TT[a] (mg/L)[b]	potential health effects from ingestion of water	sources of contaminant in drinking water
acrylamide	0	TT[i]	nervous system or blood problems; increased risk of cancer	added to water during sewage/ wastewater treatment
alachlor	0	0.002	eye, liver, kidney, or spleen problems; anemia; increased risk of cancer	runoff from herbicide used on row crops
atrazine	0.003	0.003	cardiovascular system or reproductive problems	runoff from herbicide used on row crops
benzene	0	0.005	anemia; decrease in blood platelets; increased risk of cancer	discharge from factories; leaching from gas storage tanks and landfills
benzo(a)pyrene (PAHs)	0	0.0002	reproductive difficulties; increased risk of cancer	leaching from linings of water storage tanks and distribution lines

(continued)

APPENDIX 5.B (continued)
National Primary Drinking Water Regulations Code of Federal Regulations (CFR), Title 40, Ch. I, Part 141, October 2003

organic chemicals	MCLG[a] (mg/L)[b]	MCL or TT[a] (mg/L)[b]	potential health effects from ingestion of water	sources of contaminant in drinking water
carbofuran	0.04	0.04	problems with blood, nervous system, or reproductive system	leaching of soil fumigant used on rice and alfalfa
carbon tetrachloride	0	0.005	liver problems; increased risk of cancer	discharge from chemical plants and other industrial activities
chlordane	0	0.002	liver or nervous system problems; increased risk of cancer	residue of banned termiticide
chlorobenzene	0.1	0.1	liver or kidney problems	discharge from chemical and agricultural chemical factories
2,4-D	0.07	0.07	kidney, liver, or adrenal gland problems	runoff from herbicide used on row crops
dalapon	0.2	0.2	minor kidney changes	runoff from herbicide used on rights of way
1,2-dibromo-3-chloropropane (DBCP)	0	0.0002	reproductive difficulties; increased risk of cancer	runoff/leaching from soil fumigant used on soybeans, cotton, pineapples, and orchards
o-dichloro-benzene	0.6	0.6	liver, kidney, or circulatory system problems	discharge from industrial chemical factories
p-dichloro-benzene	0.007	0.075	anemia; liver, kidney, or spleen damage; changes in blood	discharge from industrial chemical factories
1,2-dichloroethane	0	0.005	increased risk of cancer	discharge from industrial chemical factories
1,1-dichloroethylene	0.007	0.007	liver problems	discharge from industrial chemical factories
cis-1,2-dichloroethylene	0.07	0.07	liver problems	discharge from industrial chemical factories
trans-1,2-dichloroethylene	0.1	0.1	liver problems	discharge from industrial chemical factories
dichloromethane	0	0.005	liver problems; increased risk of cancer	discharge from industrial chemical factories
1,2-dichloropropane	0	0.005	increased risk of cancer	discharge from industrial chemical factories
di(2-ethylhexyl) adipate	0.4	0.04	general toxic effects or reproductive difficulties	discharge from industrial chemical factories
di(2-ethylhexyl) phthalate	0	0.006	reproductive difficulties; liver problems; increased risk of cancer	discharge from industrial chemical factories
dinoseb	0.007	0.007	reproductive difficulties	runoff from herbicide used on soybeans and vegetables
dioxin (2,3,7,8-TCDD)	0	0.00000003	reproductive difficulties; increased risk of cancer	emissions from waste incineration and other combustion; discharge from chemical factories
diquat	0.02	0.02	cataracts	runoff from herbicide use
endothall	0.1	0.1	stomach and intestinal problems	runoff from herbicide use
endrin	0.002	0.002	liver problems	residue of banned insecticide
epichlorohydrin	0	TT[i]	increased cancer risk; over a long period of time, stomach problems	discharge from industrial chemical factories; an impurity of some water treatment chemicals

(continued)

APPENDIX 5.B *(continued)*
National Primary Drinking Water Regulations Code of Federal Regulations (CFR), Title 40, Ch. I, Part 141, October 2003

organic chemicals	MCLG[a] (mg/L)[b]	MCL or TT[a] (mg/L)[b]	potential health effects from ingestion of water	sources of contaminant in drinking water
ethylbenzene	0.7	0.7	liver or kidney problems	discharge from petroleum refineries
ethylene dibromide	0	0.00005	problems with liver, stomach, reproductive system, or kidneys; increased risk of cancer	discharge from petroleum refineries
glyphosphate	0.7	0.7	kidney problems; reproductive difficulties	runoff from herbicide use
heptachlor	0	0.0004	liver damage; increased risk of cancer	residue of banned termiticide
heptachlor epoxide	0	0.0002	liver damage; increased risk of cancer	breakdown of heptachlor
hexachlorobenzene	0	0.001	liver or kidney problems; reproductive difficulties; increased risk of cancer	discharge from metal refineries and agricultural chemical factories
hexachloro-cyclopentadiene	0.05	0.05	kidney or stomach problems	discharge from chemical factories
lindane	0.0002	0.0002	liver or kidney problems	runoff/leaching from insecticide used on cattle, lumber, and gardens
methoxychlor	0.04	0.04	reproductive difficulties	runoff/leaching from insecticide used on fruits, vegetables, alfalfa, and livestock
oxamyl (vydate)	0.2	0.2	slight nervous system effects	runoff/leaching from insecticide used on apples, potatoes, and tomatoes
polychlorinated biphenyls (PCBs)	0	0.0005	skin changes; thymus gland problems; immune deficiencies; reproductive or nervous system difficulties; increased risk of cancer	runoff from landfills; discharge of waste chemicals
pentachlorophenol	0	0.001	liver or kidney problems; increased cancer risk	discharge from wood preserving factories
picloram	0.5	0.5	liver problems	herbicide runoff
simazine	0.004	0.004	problems with blood	herbicide runoff
styrene	0.1	0.1	liver, kidney, or circulatory system problems	discharge from rubber and plastic factories; leaching from landfills
tetrachloroethylene	0	0.005	liver problems; increased risk of cancer	discharge from factories and dry cleaners
toluene	1	1	nervous system, kidney, or liver problems	discharge from petroleum factories
toxaphene	0	0.003	kidney, liver, or thyroid problems; increased risk of cancer	runoff/leaching from insecticide used on cotton and cattle
2,4,5-TP (silvex)	0.05	0.05	liver problems	residue of banned herbicide
1,2,4-trichlorobenzene	0.07	0.07	changes in adrenal glands	discharge from textile finishing factories
1,1,1-trichloroethane	0.20	0.2	liver, nervous system, or circulatory problems	discharge from metal degreasing sites and other factories

(continued)

APPENDIX 5.B (continued)
National Primary Drinking Water Regulations Code of Federal Regulations (CFR), Title 40, Ch. I, Part 141, October 2003

organic chemicals	MCLG[a] (mg/L)[b]	MCL or TT[a] (mg/L)[b]	potential health effects from ingestion of water	sources of contaminant in drinking water
1,1,2-trichloroethane	0.003	0.005	liver, kidney, or immune system problems	discharge from industrial chemical factories
trichloroethylene	0	0.005	liver problems; increased risk of cancer	discharge from metal degreasing sites and other factories
vinyl chloride	0	0.002	increased risk of cancer	leaching from PVC pipes; discharge from plastic factories
xylenes (total)	10	10	nervous system damage	discharge from petroleum factories; discharge from chemical factories

radionuclides	MCLG[a] (mg/L)[b]	MCL or TT[a] (mg/L)[b]	potential health effects from ingestion of water	sources of contaminant in drinking water
alpha particles	none[g]	15 pCi/L	increased risk of cancer	erosion of natural deposits of certain minerals that are radioactive and may emit a form of radiation known as alpha radiation
beta particles and photon emitters	none[g]	4 mrem/yr	increased risk of cancer	decay of natural and artificial deposits of certain minerals that are radioactive and may emit forms of radiation known as photons and beta radiation
radium 226 and radium 228 (combined)	none[g]	5 pCi/L	increased risk of cancer	erosion of natural deposits
uranium	0	30 μg/L as of December 8, 2003	increased risk of cancer; kidney toxicity	erosion of natural deposits

[a]Definitions:
Maximum Contaminant Level (MCL): The highest level of a contaminant that is allowed in drinking water. MCLs are set as close to MCLGs as feasible using the best available treatment technology and taking cost into consideration. MCLs are enforceable standards.
Maximum Contaminant Level Goal (MCLG): The level of a contaminant in drinking water below which there is no known or expected risk to health. MCLGs allow for a margin of safety and are non-enforceable public health goals.
Maximum Residual Disinfectant Level (MRDL): The highest level of a disinfectant allowed in drinking water. There is convincing evidence that addition of a disinfectant is necessary for control of microbial contaminants.
Maximum Residual Disinfectant Level Goal (MRDLG): The level of a drinking water disinfectant below which there is no known or expected risk to health. MRDLGs do not reflect the benefits of the use of disinfectants to control microbial contaminants.
Treatment Technique: A required process intended to reduce the level of a contaminant in drinking water.
[b]Units are in milligrams per liter (mg/L) unless otherwise noted. Milligrams per liter are equivalent to parts per million.

(continued)

APPENDIX 5.B *(continued)*
National Primary Drinking Water Regulations Code of Federal Regulations (CFR), Title 40, Ch. I, Part 141, October 2003

[c]The EPA's surface water treatment rules require systems using surface water or ground water under the direct influence of surface water to (1) disinfect their water, and (2) filter their water or meet criteria for avoiding filtration so that the following contaminants are controlled at the following levels.

- Cryptosporidium (as of January 1, 2002, for systems serving $>10{,}000$ and January 14, 2005, for systems serving $< 10{,}000$): 99% removal
- *Giardia lamblia*: 99.9% removal/inactivation
- *Legionella*: No limit, but the EPA believes that if *Giardia* and viruses are removed/inactivated, *Legionella* will also be controlled.
- Turbidity: At no time can turbidity (cloudiness of water) go above 5 nephelolometric turbidity units (NTU); systems that filter must ensure that the turbidity go no higher than 1 NTU (0.5 NTU for conventional or direct filtration) in at least 95% of the daily samples in any month. As of January 1, 2002, turbidity may never exceed 1 NTU, and must not exceed 0.3 NTU in 95% of daily samples in any month.
- Heterotrophic plate count (HPC): No more than 500 bacterial colonies per milliliter.
- Long Term 1 Enhanced Surface Water Treatment (as of January 14, 2005): Surface water systems or ground water under direct influence (GWUDI) systems serving fewer than 10,000 people must comply with the applicable Long Term 1 Enhanced Surface Water Treatment Rule provisions (e.g., turbidity standards, individual filter monitoring, cryptosporidium removal requirements, updated watershed control requirements for unfiltered systems).
- Filter Backwash Recycling: The Filter Backwash Recycling Rule requires systems that recycle to return specific recycle flows through all processes of the systems' existing conventional or direct filtration system or at an alternate location approved by the state.

[d]More than 5.0% of samples are total coliform-positive in a month. (For water systems that collect fewer than 40 routine samples per month, no more than one sample can be total coliform-positive per month.) Every sample that has total coliform must be analyzed for either fecal coliforms or *E. coli*. If two consecutive samples are TC-positive, and one is also positive for *E. coli* or fecal coliforms, the system has an acute MCL violation.

[e]Fecal coliform and *E. coli* are bacteria whose presence indicates that the water may be contaminated with human or animal wastes. Disease-causing microbes (pathogens) in these wastes can cause diarrhea, cramps, nausea, headaches, or other symptoms. These pathogens may pose a special health risk for infants, young children, and people with severely compromised immune systems.

[f]Although there is no collective MCLG for this contaminant group, there are individual MCLGs for some of the individual contaminants.

- Haloacetic acids: dichloroacetic acid (0); trichloroacetic acid (0.3 mg/L). Monochloroacetic acid, bromoacetic acid, and dibromoacetic acid are regulated with this group but have no MCLGs.
- Trihalomethanes: bromodichloromethane (0); bromoform (0); dibromochloromethane (0.06 mg/L). Chloroform is regulated with this group but has no MCLG.

[g]MCLGs were not established before the 1986 Amendments to the Safe Drinking Water Act. Therefore, there is no MCLG for this contaminant.

[h]Lead and copper are regulated by a treatment technique that requires systems to control the corrosiveness of their water. If more than 10% of tap water samples exceed the action level, water systems must take additional steps. For copper, the action level is 1.3 mg/L, and for lead it is 0.015 mg/L.

[i]Each water system agency must certify, in writing, to the state (using third party or manufacturers' certification) that, when acrylamide and epichlorohydrin are used in drinking water systems, the combination (or product) of dose and monomer level does not exceed the levels specified, as follows.

- acrylamide = 0.05% dosed at 1 mg/L (or equivalent)
- epichlorohydrin = 0.01% dosed at 20 mg/L (or equivalent)

APPENDIX 5.C
National Secondary Drinking Water Regulations
Code of Federal Regulations (CFR),
Title 40, Ch. I, Part 143, July 2010

contaminant	secondary standard
aluminum	0.05–0.2 mg/L
chloride	250 mg/L
color	15 (color units)
copper	1.0 mg/L
corrosivity	noncorrosive
fluoride	2.0 mg/L
foaming agents	0.5 mg/L
iron	0.3 mg/L
manganese	0.05 mg/L
odor	3 threshold odor number
pH	6.5–8.5
silver	0.10 mg/L
sulfate	250 mg/L
total dissolved solids	500 mg/L
zinc	5 mg/L

Appendices

Index

INDEX - L

Y

Z

Exam Cafe
Get Immediate Results with
Online Civil PE Sample Exams and Practice Problems

Visit www.ppi2pass.com/examcafe today.

PPI's Exam Cafe is an online collection of over 1,300 problems similar in format and level of difficulty to the ones found on the Civil PE exam. Easily create realistic timed exams, or work through problems one at a time, going at your own pace. Since Exam Cafe is online, it is available to you 24 hours a day, seven days a week, so you can practice for your exam anytime, anywhere.

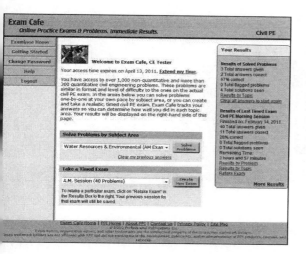

Solve Over 1,300 Civil PE Exam-Like Problems

- Problems are similar in format and level of difficulty to those found on the Civil PE exam.
- Get complete Civil PE exam coverage for the morning session and each of the afternoon sections (construction, geotechnical, structural, transportation, and water resources and environmental).

Create Realistic, Timed Civil PE Exams

Automatically create timed Civil PE exams in seconds.

Take exams that mirror the format, level of difficulty, and time constraints of the actual Civil PE exam.

Prepare yourself for the pressure of working under timed conditions.

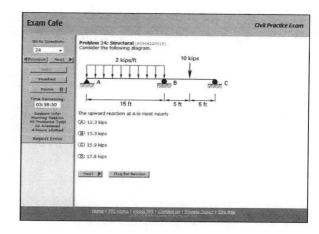

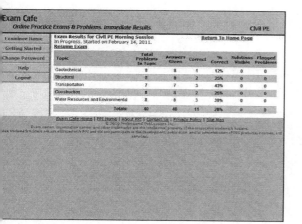

Get Immediate Online Solutions and Results

- Get immediate feedback with step-by-step solutions.
- View results of timed exams instantaneously.
- Analyze results by topic or by problem.
- Assess your strengths and weaknesses.

Visit Exam Cafe today at
www.ppi2pass.com/examcafe.

The Power to Pass®
www.ppi2pass.com